8.25
8.23
8.9
8.17

Surface Analysis – The Principal Techniques

Edited by

John C. Vickerman

Surface Analysis Research Centre, Department of Chemistry UMIST,
Manchester, UK

JOHN WILEY & SONS
Chichester • New York • Weinheim • Brisbane • Singapore • Toronto

Other Wiley Editorial Offices

John Wiley & Sons, Inc., 605 Third Avenue,
New York, NY 10158-0012, USA

VCH Verlagsgesellschaft mbH,
Pappelallee 3, D-69469 Weinheim, Germany

Jacaranda Wiley Ltd, 33 Park Road, Milton,
Queensland 4064, Australia

John Wiley & Sons (Canada) Ltd, 22 Worcester Road,
Rexdale, Ontario M9W 1L1, Canada

John Wiley & Sons (Asia) Pte Ltd, 2 Clementi Loop #02-01,
Jin Xing Distripark, Singapore 129809

British Library Cataloguing in Publication Data

A catalogue record for this book is available from the British Library

ISBN 0–471–95939 1 (cloth)
ISBN 0–471–97292 4 (paper)

Typeset in 10/12pt Times by Laser Words, Madras, India
Printed and bound in Great Britain by Biddles Ltd, Guildford, Surrey
This book is printed on acid-free paper responsibly manufactured from sustainable forestation,
for which at least two trees are planted for each one used for paper production.

Contents

List of Contributors

PROFESSOR JOHN. C. VICKERMAN

Surface Analysis Research Centre, Department of Chemistry, UMIST, PO Box 88, Manchester M60 1QD, UK

DR ROD WILSON

Graseby Ionics, Odhams Trading Estate, St. Albans Road, Watford, Herts, WD2 5JX

PROFESSOR BUDDY D. RATNER and PROFESSOR DAVID CASTNER

National ESCA and Surface Analysis Center for Biomedical Problems (NESAC/BIO), Department of Chemical Engineering and Center for Bioengineering, BF-10 University of Washington, Seattle, WA 98195 USA

PROFESSOR HANS JÖRG MATHIEU

EPFL, Départment des Matériaux, CH-1015 Lausanne, SWITZERLAND

DR ANDREW SWIFT

CSMA Limited, Armstrong House, Oxford Road Manchester, UK

PROFESSOR EDMUND TAGLAUER

Max-Planck-Institut für Plasmaphysik, EURATOM Association, D-8046 Garching bei München, GERMANY

PROFESSOR MARTYN E PEMBLE

Department of Chemistry University of Salford, The Crescent, Salford M5 4WT, UK

DR WENDY R. FLAVELL

Department of Physics, UMIST, PO Box 88, Manchester M60 1QD, UK

DR GRAHAM J. LEGGETT

Department of Materials Engineering and Materials Design, The University of Nottingham, University Park, Nottingham NG7 2RD, UK

Preface

In today's world there are vast areas of materials technology which would benefit from the application of surface analysis techniques in both research and quality control. Over the years an enormous number of techniques have been developed to probe different aspects of the physics and chemistry of surfaces, however, only a few have found wide application in basic surface science and applied surface analysis. This book seeks to introduce the reader to the principal techniques used in these fields. Each technique is introduced by an expert practitioner. The coverage includes the basic theory and practice of each technique together with practical examples of its use and application and most chapters are followed by some review questions to enable the reader to develop and test their understanding. The aim has been to give a thorough grounding without being too detailed.

Following an introduction to surface analysis in *Chapter 1*, since most though not all, surface analysis techniques are carried out in vacuum based equipment the book starts with a chapter by Dr Rod Wilson on Vacuum Technology. Electron Spectroscopy for Surface Analysis (ESCA) or X-ray Photoelectron Spectroscopy (XPS) is probably the most widely used surface analysis technique. It has been extremely effective for the solution of an enormous number of problems in both basic surface science and in applied analysis. Professors Buddy Ratner and Dave Castner from Washington State University have exploited the technique very successfully for polymer and biomaterials analysis and they introduce this technique in *Chapter 3*.

In *Chapter 4* Professor Hans Jörg Mathieu from Ecole Polytechnique, Lausanne introduces perhaps the oldest widely used technique of surface analysis–Auger Electron Spectroscopy (AES). This technique is exploited extensively and extremely effectively in Lausanne for metal and alloy analysis.

Secondary ion mass spectrometry (SIMS), introduced in *Chapter 5* by Professor John Vickerman and Dr Andrew Swift is a very powerful technique because of the mass spectral nature of the data. The group in Manchester have contributed particularly to the development of static SIMS for surface analysis and they have shown that together with ESCA it can be exploited very effectively to solve industrial problems.

Low energy ion scattering (LEIS) and Rutherford backscattering (RBS) are powerful for probing the elemental composition and structure of surfaces. Professor Edmund Taglauer from the Max Planck Institute in Garching is a widely recognised authority on these elegant techniques which are introduced in *Chapter 6*.

Vibrational spectroscopy is very widely used in Chemistry for compound identification and analysis. There are now many variants which can be applied to the study of surfaces and particularly of molecules on surfaces. Professor Martyn Pemble of Salford University has been involved in the development of several of the techniques and uses them in research associated with the growth of electronic materials and he discusses a number of these variants in *Chapter 7*.

In *Chapter 8* Dr Wendy Flavell from UMIST, Manchester introduces techniques which use diffraction and other Interference based methods for the analysis of surface structure. Low energy electron diffraction (LEED) has been an important technique in basic surface science for many years, however more recently extended x-ray absorption fine structure (EXAFS) and the related techniques which probe local short range surface structure have become extremely valuable and are used extensively in Dr Flavell's research into oxide surface structure.

Finally in the last few years surface studies have been significantly advanced by the new scanning probe techniques - scanning tunnelling microscopy (STM) and atomic force microscopy (AFM). The impressive images with atomic resolution of metal surfaces have excited many surface analysts. The extension of the capabilities to bio-organic materials promises considerable insights into the surface behaviour of these materials. Dr Graham Leggett, who is exploring the use of these techniques to study bio-organic surfaces at the University of Nottingham, describes the theory and practice of these techniques in *Chapter 9*.

Most surface problems, be they in basic surface science or applied surface analysis, require careful selection of the most appropriate technique to answer the questions posed. Frequently more than one technique will be required. It is anticipated that the reader of this book will be equipped to make the judgements required. Thus the book should be of value to those who need to have a wide overview of the techniques in education or in industrial quality control or R&D laoratories. For those who wish to further develop their knowledge and practice of particular techniques, it should also give a good basic understanding from which to build.

John C Vickerman
Manchester, UK

CHAPTER 1

Introduction

JOHN C. VICKERMAN

UMIST, Manchester, UK

The surface behaviour of materials is crucial to our lives. The obvious problems of *corrosion* are overcome by special surface treatments. The *optical behaviour* of glass can be modified by *surface coatings* or by changing the surface composition. The surface chemistry of polymers can be tuned so that they cling for packaging, are non-stick for cooking or can be implanted into our bodies to feed in drugs or replace body components. The *auto-exhaust catalyst* which removes some of the worst output of the combustion engine is a masterpiece of surface chemistry as are the industrial catalysts which are vital for about 90% of the output of the chemical industry. Thus whether one considers a car body shell, a biological implant, a catalyst, a solid state electronic device or a moving component in an engine, it is the surface which interfaces with its environment. The surface reactivity will determine how well the material behaves in its intended function. It is therefore vital that the surface properties and behaviour of materials used in our modern world are thoroughly understood. Techniques are required which enable us to analyse the surface chemical and physical state and clearly distinguish it from that of the underlying solid.

1 HOW DO WE DEFINE THE SURFACE?

It is obvious that the surface properties of solids are influenced to a large extent by the solid state properties of the material. The question arises as to how we define the surface. Since the top layer of surface atoms are those which are the immediate interface with the other phases (gas, liquid or solid) impinging on it, this could be regarded as the surface. However, the structure and chemistry of that top layer of atoms will be significantly determined by the atoms immediately below. In a very real sense therefore, the surface could be said to be the top 2–10 atomic layers (say, 0.5–3 nm). However, many technologies apply surface films to devices and components, to protect, lubricate, change the surface optical properties, etc. These films are in the range 10–100 nm or sometimes even thicker, but the surface may be thought of in this depth range. However, beyond 100 nm it is

Surface Analysis — The Principal Techniques. Edited by John C. Vickerman
© 1997 John Wiley & Sons Ltd

more appropriate to begin to describe such a layer in terms of its bulk solid state properties. Thus we can consider the surface in terms of three regimes: the top surface monolayer; the first ten or so layers; and the surface film, no greater than 100 nm. To understand fully the surface of a solid material, we need techniques which not only distinguish the surface from the bulk of the solid, but also ones that distinguish the properties of these three regimes.

2   HOW MANY ATOMS IN A SURFACE?

It will be appreciated that it is not straightforward to probe a surface layer of atoms analytically and distinguish their structure and properties from that of the rest of the solid. One has only to consider the relatively small number of atoms involved in the surface layer(s) to see that high sensitivity is required. How many atoms are we dealing with at the surface and in the bulk of a solid? We can consider a 1 cm cube of metal. One of the 1 cm^2 surfaces has roughly 10^{15} atoms in the surface layer. Thus the total number of atoms in the cube will be $\approx 10^{23}$. Therefore the percentage of surface to bulk atoms will be

$$s/b \approx 10^{-8} \times 100 \approx 10^{-6}\%$$

Typically, a surface analysis technique may be able to probe in the region of 1 mm^2. Thus in the top monolayer there will be about 10^{13} atoms. In the top ten layers there will be 10^{14} atoms or 10^{-10} mol. Clearly in comparison with conventional chemical analysis we are considering very low concentrations. Things become more demanding when we remember that frequently the chemical species which play an important role in influencing surface reactivity may be present in very low concentration, so the requirement will be to analyse an additive or contaminant at the 10^{-3} or even 10^{-6} (ppm) atomic level, ie 10^{10} or 10^{7} atoms or 10^{-14} or 10^{-17} mole levels respectively, perhaps even less.

Similar demands arise if the analysis has to be carried out with high spatial resolution. The requirement to map variations in chemistry across a surface can arise in a wide variety of technologies. There may be a need to monitor the homogeneity of an optical or a protective coating or the distribution of catalyst components across a support or a contaminant on an electronic device etc. It is not unusual for 1 μm spatial resolution to be demanded, frequently even less would be beneficial. In an area of 1 μm^2 (10^{-12} m^2 or 10^{8} cm^2) there are only $\approx 10^{7}$ atoms, so if we want to analyse to the 10^{-3} atom fraction level, there are only 10^{4} atoms.

Thus surface analysis is demanding in terms of its surface resolution and sensitivity requirements. However, there are in fact many surface analysis techniques, all characterised by distinguishing acronyms — LEED, XPS, AES, SIMS etc. Most were developed in the course of fundamental studies of surface phenomena on single crystal surface planes. Such studies which comprise the research field known as *surface science* seek to provide an understanding of surface processes at the atomic and molecular level. Thus for example, in the area of catalysis, there has been an enormous research effort directed towards understanding the role of surface atomic structure, composition, electronic state etc. on the adsorption and surface reactivity of reactant molecules at the surface of the catalyst. To simplify and systematically control the variables involved, much of the research to date has focused on single crystal surfaces of catalytically important metals. The surface analysis techniques developed in the course of these and related research are, in the main, based on bombarding the surface to be studied with electrons, photons or ions and detecting the emitted electrons, photons or ions.

3 INFORMATION REQUIRED

To understand the properties and reactivity of a surface, the following information is required: the physical topography; the chemical composition; the chemical structure; the atomic structure; the electronic state; a detailed description of bonding of molecules at the surface. No one technique can provide all these different pieces of information. A full investigation of a surface phenomenon will always require several techniques. To solve particular problems it is seldom necessary to have all these different aspects covered, however, it is almost always true that understanding is greatly advanced by applying more than one technique to a surface study. This book does not attempt to cover all the techniques in existence. A recent count identified over 50! The techniques introduced here are those (excluding electron microscopy which is not covered but for which there are numerous introductions) which have made the most significant impact in both fundamental *and* applied surface analysis. They are tabulated (via their acronyms) in Table 1.1 according to the principal information they provide and the probe/detection system they use. The number after each technique indicates the chapter in which it is described.

It is a characteristic of most techniques of surface analysis that they are carried out in vacuum. This is because electrons and ions are scattered by molecules in the gas phase. While photon based techniques can in principle operate in the ambient, sometimes gas phase absorption of photons can occur and as a consequence these may also require vacuum operation. This imposes a restriction on some of the surface processes which can be studied. For example, to study the surface gas or liquid interface it will usually be necessary to use a photon based technique, or one of the scanning probe techniques.

However, the vacuum based methods allow one to control the influence of the ambient on the surface under study. To analyse a surface uncontaminated by any adsorbate it is necessary to operate in ultra-high vacuum ($<10^{-9}$ mm Hg since at 10^{-6} mm Hg a surface can be covered by one monolayer of adsorb

Table 1.1. Surface analysis techniques and the information they can provide

Radiation IN	photon	photon	electron	ion	neutron
Radiation DETECTED	electron	photon	electron	ion	neutron
SURFACE INFORMATION					
Physical topography			SEM STM (9)		
Chemical composition	ESCA/XPS (3)		AES (4)	SIMS (5) ISS (6)	
Chemical structure	ESCA/XPS (3)	EXAFS (8) IR & SFG (7)	EELS (7)	SIMS (5)	INS (7)
Atomic structure		EXAFS (8)	LEED RHEED (8)	ISS (6)	
Adsorbate bonding		EXAFS (8) IR (7)	EELS (7)	SIMS (5)	INS (7)

ESCA/XPS — Electron analysis for chemical analysis/X-ray photoelectron spectroscopy: X-ray photons of precisely defined energy bombard the surface, electrons are emitted from the orbitals of the component atoms, electron kinetic energies are measured, their electron binding energies can be determined enabling the component atoms to be determined.

AES — Auger electron spectroscopy: Basically very similar to above except that a keV electron beam may be used to bombard the surface.

SIMS — Secondary ion mass spectrometry: A beam of high energy (keV) *primary* ions bombards the surface, secondary atomic and cluster ions are emitted and analysed with a mass spectrometer.

ISS — Ion scattering spectrometry: An ion beam bombards the surface and is scattered from the atoms in the surface. The scattering angles and energies are measured and used to compute the composition and surface structure of the sample target.

IR — Infra-red spectroscopy: Various variants on the classical methods — irradiate with infrared photons which excite vibrational frequencies in the surface layers, photon energy losses are detected to generate spectra.

EELS — Electron energy loss spectroscopy: Low energy (few eV) electrons bombard the surface and excite vibrations the resultant energy loss is detected and related to the vibrations excited.

INS — Inelastic neutron scattering: Bombard a surface with neutrons, energy loss occurs due to the excitation of vibrations. It is most efficient in bonds containing hydrogen.

SFG — Sum frequency generation: Two photons irradiate and interact with an interface (solid/gas or solid liquid) such that a single photon merges resulting in electronic or vibrational information about the interface region.

LEED — Low energy electron diffraction: A beam of low energy (tens of eV) electrons bombard a surface, the electrons are diffracted by the surface structure enabling the structure to be deduced.

RHEED — Reflection high energy electron diffraction: A high energy beam (keV) of electrons is directed at a surface at glancing incidence. The angles of electron scattering can be related to the surface atomic structure.

EXAFS — Extended X-ray absorption fine structure: The fine structure of the absorption spectrum resulting from X-ray irradiation of the sample is analysed to obtain information on local chemical and electronic structure.

STM — Scanning tunnelling microscopy: A sharp tip is scanned over a conducting surface at a very small distance above the surface. The electron current flowing between the surface and the tip is monitored, physical and electron density maps of the surface can be generated with high spatial resolution.

AFM — Atomic force microscopy (not on table): Similar to STM but applicable to non-conducting surfaces. The forces developed between the surface and the tip are monitored. A topographical map of the surface is generated.

species within 1 s *if* the sticking coefficient (probability for adsorption) is 1. Controlled exposure of the surface to adsorbates or other surface treatments can then be carried out to monitor effects in a controlled manner. Chapter 1 on Vacuum Technology has been included to introduce the reader to the concepts and equipment requirements in the generation of vacua.

4 SURFACE SENSITIVITY

To generate the information, we require that a surface analysis technique should derive its data as near exclusively as possible from within the depth range discussed in Section 1.2. The extent to which a technique does this is a measure of its surface sensitivity. Ion scattering spectrometry (ISS) derives almost all its information from the top monolayer. It is very surface sensitive. Electron Spectroscopy for Chemical Analysis (ESCA) or X-ray Photoelectron Spectroscopy (XPS) samples in the top 10 or so layers of the surface. Whilst, infra-red (IR) spectroscopy is not very surface sensitive and will sample deep into the solid, unless it is used as a reflection mode.

In general the surface sensitivity of an analytical method is dependent on the radiation *detected*. As already indicated, most of the methods of surface analysis involve bombarding the surface with a form of radiation — electrons, photons, ions, neutrons — and then collecting the resulting emitted radiation — electrons, photons, ions, neutrons. The scanning probe methods are a little different, although one could say that scanning tunnelling microscopy (STM) detects electrons. (Atomic force microscopy monitors the forces between the surface and a sharp tip, Chapter 9.) The surface sensitivity depends on the depth of origin of the detected species. Thus in XPS whilst the X-ray photons which bombard the surface can penetrate deep into the solid, the resultant emitted electrons which can be detected without loss of energy can only arise from within 1–4 or 8 nm of the surface. Electrons generated deeper in the solid may escape, but on the way out they will have collided with other atoms and lost energy. They are no use for analysis. Thus the surface sensitivity of ESCA is a consequence of the short distance electrons can travel in solids without being scattered (known as the *inelastic mean free path*). Similarly, in secondary ion mass spectrometry (SIMS) the surface is bombarded by high energy ions. They deposit their energy down to 30 or 40 nm. However, 95% of the secondary ions which are knocked out (sputtered) of the solid, arise from the top two layers.

There are techniques like infra-red spectroscopy which, although they are not intrinsically very surface sensitive, can be made so by the methods used to apply them. Thus with IR a reflection approach can be used in which the incoming radiation is brought in at a glancing incidence. This enables vibrational spectra to be generated from adsorbates on single crystal surfaces. The technique is very surface sensitive. Surface sensitivity can be significantly increased even in surface sensitive methods like ESCA by irradiating the surface at glancing incidence, see Chapter 2.

Various terms are used to define surface sensitivity. With all the techniques described in this book the total signal detected will originate over a range of depths from the surface. An *information depth* may be specified which is usually defined as the average distance (in nm) normal to the surface from which a specified percentage (frequently 90, 95 or 99%) of the detected signal originates. Sometimes, as in ESCA, a *sampling depth*, is defined. This is three times the *inelastic mean free path*, and turns out to be the information depth where the percentage is 95%. Obviously a very small proportion of the detected signal does arise from deeper in the solid, but the vast majority of the useful analytical information arises from within the sampling depth region.

In static SIMS the information depth is the depth from which 95% of the secondary ions originate. For most materials this is believed to be about two atomic layers, about 0.6 nm. However, it is sometimes difficult to be sure what a *layer* is. For example, there are surface layers used to generate new optical properties which are composed of long organic chains bonded to metal or oxide surfaces. The organic layer is much less dense than the substrate underneath. SSIMS studies of these materials suggests that the analytical process may remove the whole molecular chain which can easily be > 20 nm long. Surface sensitivity in this case is a very different concept from that which would apply to the surface of a metal or inorganic compound.

5 RADIATION EFFECTS — SURFACE DAMAGE

To obtain the surface information required entails 'interfering' with the surface state in some way! Most of the techniques require the surface to be bombarded with photons, electrons or ions. They will affect the chemical and physical state of the surface being analysed. Thus in the course of analysing the surface, the surface may be changed. It is important to understand the extent to which this may happen, otherwise the information being generated from the surface may not be characteristic of the surface before analysis, rather it may reflect a surface damaged by the incident radiation.

Table 1.2 shows the penetration depth and influence of the 1000 eV particles. It can be seen that most of the energy is deposited in the near surface under ion and electron bombardment, so in general terms it would be expected that the extent of surface damage would vary as photons < electrons < ions. Consequently, it is sometimes carelessly suggested that ESCA/XPS is a low damage technique. However, the power input to the surface in the course of an experiment is considerably less in the ion bombardment method of SIMS compared to photon bombardment in ESCA, Table 1.3. SIMS is very obviously a phenomenon which depends on damage — ions bombard to knock out other ions! Without damage there is no information, but as will be seen in Chapter 5 it can be operated in a

Table 1.2. Penetration depths of particles

Particle	Energy (eV)	Depths (Å)
Photon	1000	10 000
Electron	1000	20
Ions	1000	10

Table 1.3. Comparison of typical primary particle flux densities and energies and the resulting power dissipated in SSIMS, LEED, and X-ray photoelectron experiments

	Primary flux (cm^{-2})	Primary energy	Power (W cm^{-2})
SIMS	10^{10} ions s^{-1}	3 keV	3×10^{-6}
LEED	10^{15} electrons s^{-1}	50 eV	5×10^{-3}
XPS	10^{14} photons s^{-1}	1.4 keV	2×10^{-2}

low damage mode to generate significant surface information. The X-ray photons which bombard the surface in XPS, penetrate deep into the solid. However if the material is delicate, eg a polymer, and if the power input is too high or the time under the beam too long, the sample can be literally fried. The same effect is even more obvious for the methods involving electron irradiation. It is consequently very difficult to analyse the surface of organic materials using any technique which relies on electron bombardment.

Surface analysis techniques have been enormously successful in developing our understanding of surface phenomena. There are vast numbers of areas of technology which would benefit from the application of surface analysis techniques in both research and development and in quality control. These techniques are not being applied because of a lack knowledge and understanding of how they can help. Hopefully, this book will help to develop increased awareness such that surface analysis will be increasingly applied to further our understanding of the surface state at both the fundamental and applied levels.

None of the techniques are analytical 'black boxes' delivering answers to problems at the push of a button. Two general rules should be remembered in surface analysis. (a) In every case it is important to understand the capabilities and limitations of the technique being used with regard to the material being studied and the information required. (b) No one technique gives the whole story.

CHAPTER 2

Vacuum Technology for Applied Surface Science

ROD WILSON
Grasby Ionics, UK

1 INTRODUCTION: GASES AND VAPOURS

A prerequisite of the vast majority of surface analysis techniques is a 'vacuum' environment in which the particular technique can be applied. The reasons for needing a high vacuum environment are manifold, the most fundamental being the requirement for long mean free paths for particles used in studying surfaces. High vacuum conditions will mean that the trajectories of particles such as ions and electrons used in surface analysis will remain undisturbed when in a surface analysis system. It is often necessary also to keep a surface free from absorbed gases during the course of a surface analysis experiment, this requires tighter vacuum constraints to keep the so-called monolayer time long enough to gather data from a surface. The need to be able to sustain high voltages in a surface analysis system without breakdown or glow discharges being created also imposes vacuum constraints. An important part therefore of understanding surface analysis and its applications is understanding vacuum technology and what 'vacuum' physically means. Since vacuum technology deals with gases and vapours it is important to develop a physical picture of a gas or vapour.

Gases are a low density collection of atoms or molecules which can often be pictured as simple hard spheres and are generally treated as having no forces acting between them, except at the instant of collision. A picture of the instantaneous structure of a small volume of gas is shown in Figure 2.1. The molecules are usually a large distance apart compared with their diameter and there is no regularity in their arrangement in space. The molecules are distributed at random throughout the whole volume they occupy and are moving randomly and, at room temperature, will have a mean velocity typically of the order of 10^2 ms^{-1}. It is worth noting that the noble gases present a close physical realisation of this ideal behaviour.

This postulate that a gas consists of a number of discrete particles between which no forces are acting led to a series of theoretical considerations which

Surface Analysis — The Principal Techniques. Edited by John C. Vickerman
© 1997 John Wiley & Sons Ltd

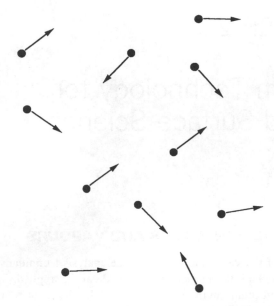

Figure 2.1. An instantaneous picture of a small volume of gas, the arrows showing the random nature of the motion of the particles. (N.B. If the relative sizes of the particles and their seperation was to scale then this picture would correspond to a cube of ≈30Å at a pressure of ≈20 atmospheres.)

are referred to as 'the kinetic theory of gases'. This theory tries to explain the macroscopic properties of gases, such as pressure and temperature by considering the microscopic behaviour of the molecules of which they consist. One of the first and most important results from this type of treatment was to relate the pressure, p of a gas to the gas density, ρ and the mean square velocity, $\langle c^2 \rangle$ of the gas molecules each of mass, m. This relation is given by,

$$p = \tfrac{1}{3}\rho\langle c^2 \rangle \tag{1}$$

or

$$p = \tfrac{1}{3}nm\langle c^2 \rangle \tag{1a}$$

where

$$\langle c^2 \rangle = 3\frac{kT}{m} \tag{2}$$

and,

   n = number density of particles, (m^{-3}),

   k = Boltzmann constant, (JK^{-1}),

   T = Absolute temperature, (K),

The gas molecules will travel with a distribution of velocities in straight lines and collide with the walls of the container they are in and also collide elastically with each other. The average numbers of collisions per second between particles is called the collision rate, z and the path which each particle makes on average between collisions is called the mean free path, λ. Both these parameters are functions of the mean particle velocity, $\langle c \rangle$, the particle diameter, $2r$ and the number density of particles. To a good approximation they are given by,

$$z = \frac{\langle c \rangle}{\lambda} \tag{3}$$

where,

$$\langle c \rangle = \sqrt{\frac{8kT}{pm}} \tag{4}$$

and using simple geometry to show that if particles cross such that their diameters are less than $2r$ apart then a collision will occur it can be shown that,

$$\lambda = \frac{1}{\sqrt{2} n \pi (2r)^2} \tag{5}$$

The quantity $\pi(2r)^2$ is called the collision cross-section of the molecule and is often denoted by the symbol σ. It follows from this that the mean free path is inversely proportional to the number density of molecules and therefore the gas pressure p. At constant temperature, for every gas, the product $p\lambda$ is a constant.

Understanding the concept of mean free path is an important stage in understanding what is happening on a molecular level inside your vacuum system and often the need for a vacuum is governed by a need to increase the mean free path of the molecules in the system.

The main aim of vacuum technology in itself is to simply reduce the number density of atoms or molecules in a defined volume. Before the methodology for achieving this reduction in number density can be explained it is important to define some of the terminology and fundamental quantities used in vacuum technology and the principles involved.

The most common term used in vacuum technology is pressure, and is most often denoted by the symbol p. This is defined as 'the quotient of the perpendicular force on a surface and the area of this surface'. More simply it can be considered as the force per unit area applied to a surface by a fluid and is given the units of force per unit area. In the SI system of units this is given as Newtons per metre^2 (Nm^{-2}) or Pascals (P). In the field of vacuum technology, however, it is often more convenient to refer to pressure in the units of millibar (mbar). Previously in vacuum technology it was common to refer to pressure in the units 'torr'. This is no longer a recommended unit but should be mentioned as it is descriptive of the physical picture. 1 torr is understood as the fluid pressure which

Table 2.1. Table showing conversion of pressure units

mbar	Pa	torr[1]	bar	atm[2]	at[3]	% vac[4]	mm water[5]
1,013	101 300	760.00	1.01	1.0	1.03	0	1.03×10^4
1,000	100 000	750.00	1.00	0.987	1.02	1.3	1.02×10^4
981	98 100	735.75	0.981	0.968	1	3.2	10^4
900	90 000	675.00	0.90	0.888	0.918	11.1	9 177
600	80 000	600.00	0.80	0.789	0.816	21.0	8 157
700	70 000	525.00	0.70	0.691	0.714	30.9	7 137
600	60 000	450.00	0.60	0.592	0.612	40.8	6 118
500	50 000	375.00	0.50	0.494	0.510	50.6	5 098
400	40 000	300.00	0.40	0.395	0.408	60.5	4 078
300	30 000	225.00	0.30	0.296	0.306	70.4	3 059
200	20 000	150.00	0.20	0.197	0.204	80.2	2 039
100	10 000	75.00	0.10	0.099	0.102	90.1	1 019
90	9 000	67.50	0.09	0.089	0.092	91.1	918
80	8 000	60.00	0.08	0.079	0.082	92.1	816
70	7 000	52.50	0.07	0.069	0.071	93.1	714
60	6 000	45.00	0.06	0.059	0.061	94.1	612
50	5 000	37.50	0.05	0.049	0.051	95.1	510
40	4 000	30.00	0.04	0.040	0.041	96.1	408
30	3 000	22.50	0.03	0.030	0.031	97.0	306
20	2 000	15.00	0.02	0.020	0.020	98.0	204
10	1 000	7.50	0.01	0.010	0.010	99.0	102
5	500	3.75	0.005	0.005	0.005	99.5	51
1	100	0.75	0.001	0.001	0.001	99.9	10
0.5	50	0.375	5×10^{-4}	5×10^{-4}	5×10^{-4}	99.9	5
0.1	10	0.075	1×10^{-4}	1×10^{-4}	1×10^{-4}	99.99	1
1×10^{-2}	1	7.5×10^{-3}	1×10^{-5}	1×10^{-5}	1×10^{-5}	99.99	0.1
1×10^{-n}	$1 \times 10^{-(n+2)}$	$7.5 \times 10^{-(n+1)}$	$1 \times 10^{-(n+3)}$	$1 \times 10^{-(n+3)}$	$1 \times 10^{-(n+3)}$		$1 \times 10^{-(n-1)}$
0	0	0	0	0	0	100	0

[1] 1 torr = 1 mm Hg
[2] 1 atm (= standard atmosphere) = 101 325 Pa (standard reference pressure)
[3] 1 at = 1 techn. atmosphere
[4] α_0 vacuum: a pressure increase (or decrease) of approx.
1 mbar corresponds to a change of vacuum of 0.1%
[5] 1 mm water (column) = 10^{-4} at = 9.8 mbar

is able to balance, at 0°C, a column of mercury 1 mm in height (for conversion factors see Table 2.1).

Although pressure is by far the most common term used in vacuum technology it is in itself of little relevance when applied to the field of surface science. The chief objective of vacuum technology here is to reduce the number density of particles, n in a given volume, V. The number density of particles is however related to the gas pressure p and the thermodynamic temperature, T by the laws of the kinetic theory of gases. This relationship is most often expressed as:-

$$p = nkT \qquad (6)$$

where k is the Boltzmann constant.

It can be seen that, at constant temperature, a reduction in the number density of particles is always equivalent to a reduction in pressure. It must be emphasised, however, that a pressure decrease (keeping the volume constant) may also be attained by a reduction in the temperature of the gas. This must always be taken into account if the same temperature does not prevail throughout the volume of interest.

It can also be seen from equation (6) that, at a given temperature, the pressure depends only on the number density of molecules and not on the nature of the gas (i.e. the pressure of a gas is independent of chemical species for a given temperature and number density of particles).

When refering to pressure in vacuum technology it is invariably the absolute pressure, not referenced to any other pressure (analogous to absolute temperature) however, in certain cases it is necessary to be more specific and add an index to the symbol p. Several examples of this follow.

i) The total pressure, p_{tot} in a container is equal to the sum of the partial pressures of all the gases and vapours within it.

ii) The partial pressure, p_{part} of a given gas or vapour in a container is that pressure that the gas or vapour would have if it were present alone in the container.

iii) The vapour pressure, p_d is the pressure due to vapours in a system as opposed to gases.

iv) The pressure of a saturated vapour is called the saturation vapour pressure, p_s and for a given material it is only a function of temperature.

v) The ultimate pressure, p_{ult} in a vacuum container is the lowest attainable pressure in that container for a given pumping speed.

At this stage it is also necessary to explain what distinguishes a gas from a vapour. 'gas' generally refers to material in the gaseous state which is not condensable at the operating temperature. 'Vapour' likewise refers to material in the gaseous state which is, however, condensable at the ambient temperature. In the following text, the distinction between gases and vapours will only be made when it is required for understanding.

The volume, V, expressed in litres or metre^3 (ltr, m^3), is another frequently referred to term in vacuum technology. Physically, this is the purely geometric volume of the vacuum container or whole vacuum system. Volume, however, can also be used to indicate the pressure dependent volume of a gas which is, for example, transferred by a vacuum pump or cleaned up by sorption materials. The volume of gas which flows through a conducting element per unit time, at a specified temperature and pressure, defines the volume flow rate, q_v of that conducting element. It must be made clear that the number of particles transferred at a given volume flow rate is different for different temperatures and pressures.

The idea of volume flow rate leads to a parameter which is very important in defining the performance of a pump or pumping system, the pumping speed, S. The pumping speed of a pump (usually expressed in ltr s^{-1} or m^3 s^{-1} or similar) is the volume flow rate though the inlet aperture of the pump.

Expressed as:-

$$S = dV/dt \tag{7}$$

or, if S is constant during the pumping process, as is the case over the operating ranges of most high vacuum pumps, the differential quotient

$$S = \partial V / \partial t \tag{7a}$$

As stated previously, the volume flow rate does not indicate directly the number of particles flowing per unit time. It will also be a function of both the temperature and the pressure. Therefore it can often be more informative to define the flow rate of a quantity of gas through an element. A quantity of gas can be specified in terms of its mass G in the unit of grams or kilograms, in vacuum technology however the product pV is often of far more relevance. It can be seen from the following equation,:-

$$G = pV \frac{M}{RT} \tag{8}$$

a simple re-arrangement of the ideal gas equation

where G = mass, (kg),

$\quad M$ = Molar mass, (kg mol^{-1})

$\quad R$ = Molar gas constant, (R = 8.314 J mol^{-1} K^{-1})

that the mass can be readily calculated from a knowledge of the nature of the gas and the temperature.

The quantity of gas flowing through an element can therefore be expressed as the mass flow rate, q_m where,

$$q_m = G/\Delta t \tag{9}$$

or as the throughput, q_{pV} where,

$$q_{pV} = p(\Delta V/\Delta t) \tag{10}$$

(units = Pa m^3 s^{-1} or more commonly, mbar I s^{-1})

The throughput of a pump or pumping system is often used to define its performance and is the throughput through the intake aperture of the pump where p is the pressure at the intake side of the pump. If p and ΔV are constant, which they approximately are after pumpdown, then the throuput of the pump is given by the simple relation,

$$q_{pV} = pS \tag{11}$$

where S is the pumping speed of the pump at the intake pressure p.

The concept of pump throughput is of great importance in understanding how pumping systems work and how to design one. It is important that the concepts of pump throughput and pumping speed be fully understood and not confused with each other.

The ability of a pump to remove gas from a system will not only be determined by the throughput of the pump but also and often more importantly by the ability of the elements in the pumping system to transmit the gas. The throughput of gas through any conducting element, for example a hose or a valve or an aperture etc., is given by

$$q_{pV} = C(p_1 - p_2) \qquad (12)$$

Here $(p_1 - p_2)$ is the difference between the pressures at the intake and exit of the conducting element. The constant of proportionality C is called the conductance of the element and is determine by the geometric nature of the element.

This conductance is analogous to electrical conductance and can be treated in a similar way using an analogy of Ohms law in vacuum technology. For example if several elements, A, B, C etc. are connected in series then their total conductance is given by,

$$\frac{1}{C_{tot}} = \frac{1}{C_A} + \frac{1}{C_B} + \frac{1}{C_C} \qquad (13)$$

and if connected in parallel, it is given by,

$$C_{tot} = C_A + C_B + C_C \qquad (14)$$

2   THE PRESSURE REGIONS OF VACUUM TECHNOLOGY AND THEIR CHARACTERISTICS

As the pressure of a gas changes, then some of the characteristics of its behaviour also change. These changes can often determine the pressure conditions required for a certain experiment. It has therefore become customary in vacuum technology to subdivide the pressure region we call 'vacuum' into smaller regions in which the behaviour of the gas would have similar characteristics. In general these subdivisions are,

Rough vacuum	1000–1 mbar
Medium vacuum	$1-10^{-3}$ mbar
High vacuum	$10^{-3}-10^{-7}$ mbar
Ultra-high vacuum	10^{-7} and below

These divisions are somewhat arbitrary, and boundaries between the regions cannot be considered sharp. The characteristic which changes most strikingly between the regions is the nature of gas flow.

In the rough vacuum region, viscous flow prevails almost exclusively. Here the mutual interaction of the particles with one another determines the character of the flow, i.e. the viscosity or inner friction of the streaming material plays the dominant role. If vortex motion appears in the streaming fluid, then it is referred to as turbulent flow. If however, the various layers of the streaming medium slide over each other it is referred to as laminar flow. The criterion for viscous flow is that the mean free path of the particles is smaller than the diameter of the conducting tube, i.e. $\lambda < d$.

In the high and ultra-high vacuum regions, the particles can move freely, virtually without any mutual hindrance and the type of flow dominant here is called molecular flow. The criteria here is that the mean free path of the particles is greater than the diameter of the conducting tube, i.e. $\lambda > d$.

In the medium vacuum region there is a transition from the viscous type flow to the molecular type flow, this type of flow is called Knudsen flow. For this type of flow the mean free path of the particles must be of the same order as the diameter of the conducting tube, i.e. $\lambda \approx d$.

It is often important to have a mental picture of what the behaviour of the gas on a molecular level is like in the different pressure regions with the different types of flow. In the viscous flow region the preferred direction of travel for all the gas molecules is the same as the macroscopic direction of flow of the steaming gas. The particles forming the gas are densely packed and they will collide much more frequently with each other than with the boundary walls of the containing vessel. In the region of molecular flow, however the collisions of the particles with the vessel walls predominate. As a result of elastic reflections from the walls and desorption of gas particles off the vessel walls gas particles in the high vacuum region can have any random direction, it is incorrect to think of streaming of the gas in the macroscopic sense. This is why in the molecular flow region the conductance of a pumping system is controlled totally by the geometry of the system, since the gas particles only arrive at the apertures or openings by chance. It is important to understand this characteristic of gases in a high vacuum environment as it is frequently misunderstood.

These different characteristics of flow allow one to distinguish easily between the rough, medium, and high vacuum regions. To distinguish between the high and ultra-high vacuum regions, however, requires the introduction of another parameter called the monolayer time, τ. In the high and ultra-high vacuum regions it is the nature of the container walls which is of most significance, since at pressures below 10^{-3} mbar there are more gas molecules on the surfaces of the vacuum vessel or chamber than there are in the gas space itself. It is therefore an important parameter in characterising this pressure region, to consider the time it takes to form a single molecular or atomic layer, a so-called monolayer, on a gas free surface in the vacuum. The value of this monolayer time is calculated with the assumption that every particle that impinges on the surface remains bonded to it. This monolayer time is obviously related to the number of particles which are incident upon unit surface area per unit time, the so-called impingement rate, Z_A. In a gas in the static state the impingement rate is related to the number density of particles and the mean velocity of particle as given by,

$$Z_A = \frac{n\langle c \rangle}{4} \tag{15}$$

If a surface has a number of free places per unit surface area, a then the monolayer lifetime is given by,

$$\tau = \frac{a}{Z_A} = \frac{4a}{n\langle c \rangle} \tag{16}$$

Assuming therefore that a monolayer is absorbed on the inner wall of an evacuated sphere of 1 litre volume, then the ratio of the number of absorbed particles to the number of free particles in the space is,

at 1 mbar	10^{-2}
at 10^{-6} mbar	10^4
at 10^{-11} mbar	10^9

Using this information and the concept of monolayer time, we can understand the need for the boundary between the high vacuum and ultra-high vacuum regions. In the high vacuum region the monolayer time amounts to only a fraction of a second, in the ultra-high vacuum region however, it is of the order of minutes or hours.

To achieve and maintain a surface which is free of absorbed gas for any practical length of time it is obviously necessary to keep the surface in an ultra-high vacuum environment. In many areas of surface analysis, for example when studying adsorption processes, it is obvious that such conditions are required.

Flow and monolayer time are not the only properties which change significantly as the pressure changes: other physical properties, for example thermal conductivity and viscosity of gases, are also strongly dependent on pressure. It is understandable therefore that the pumps needed for the production of vacuum in the different regions employ various physical methods, as do the vacuum gauges which are applicable to the measurement of these pressures.

3 PRODUCTION OF A VACUUM

3.1 TYPES OF PUMP

In order to reduce the number density of gas particles, and thereby the pressure, in a vessel, gas particles must be removed from it. This is the purpose of the vacuum pump. Vacuum pumps come in many shapes and sizes and with many different mechanisms of operation, however, fundamentally a distinction can be made between two classes of pump.

i) Those which remove gas particles from the pumped volume and convey them to the atmosphere in one or more stages of compression. These are called compression or gas transfer pumps,

ii)   Those which condense or chemically bind the particles to be removed to a
      solid wall, which is often part of the vessel being pumped. These are called
      entrapment pumps.

Within these two classes there are subsections which further distinguish the
method of operation of a pump. In the class of compression pumps there are:

a)   pumps which operate by creating periodically increasing and decreasing
     chamber volumes,

b)   pumps which transfer gas from a low pressure to a high pressure side in
     which the pump chamber volume is constant,

c)   pumps in which the pumping action is due to diffusion of gases in a stream
     of high velocity particles.

In the class of entrapment pumps there are:

a)   pumps which pump vapours by condensation or remove gases by condensa-
     tion at very low temperatures,

b)   pumps which bind or embed gases at extensive gas-free surfaces by adsorp-
     tion or absorption.

To give detailed descriptions of all types of pump is beyond the scope of
this book (for more detail see reference [1]), however, an understanding of the
operation of those most commonly found in surface science is essential to anyone
wishing to work in this field and therefore an outline of the mechanisms of
operation of these will follow. Examples of specific applications in the field of
surface analysis will follow later chapters.

One of the most common types of pump found in vacuum technology is
the rotary pump. These are mechanical pumps which belong to the group of
compression pumps operating by creating a periodically increasing and decreasing
chamber volume. These again come in a few different designs the most common
of these found in surface science being the rotary vane pump (see Figure 2.2).
These consist of a cylindrical housing or stator in which rotates an eccentrically
mounted, slotted rotor. The rotor contains vanes which are forced apart either by
centrifugal force or, in some models, by springs. These vanes slide along the walls
of the stator and thereby push forward the low pressure gas drawn in at the inlet
and eject it finally at increased pressure through the outlet or discharge valve. The
oil charge of the rotary vane pump not only serves for lubrication and cooling
but also as the sealing medium, filling up dead space and any gaps in the pump.
Rotary vane models come in single (as in Figure 2.2) and two-stage models (as
is represented in Figure 2.3). Lower ultimate pressures can be produced by the
two stage model, at the expense of cost and to a certain extent reliability.

There are several other types of rotary pump used in vacuum technology for
example, rotary piston pumps, trochoid pumps and the high pumping speed and

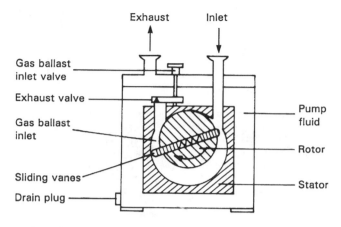

Figure 2.2. Cross-section of a single-stage sliding vane rotary pump

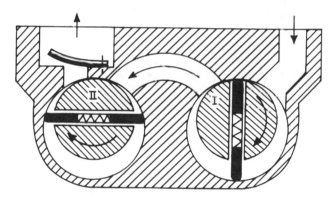

Figure 2.3. Cross-section of a two-stage sliding vane rotary pump. I is the high vacuum stage and II is the rough vacuum stage

lower ultimate pressure roots pumps. However, these are less common in surface analysis and for descriptions of these the reader should see [1].

An important part of the modern rotary pump is the **gas ballast**. If vapours are pumped by a rotary pump they can only be compressed to their saturation vapour pressure at the temperature of the pump. For example, if water vapour is pumped at a pump temperature of 70°C it can only be compressed to 312 mbar. On further compression the water vapour condenses without increase in pressure. This is insufficient pressure to open the discharge valve of the pump and the water (in liquid form) remains in the pump and emulsifies with the pump oil. As a result of this the lubricating properties of the pump decrease rapidly and the overall performance of the pump is affected. To overcome this problem the gas ballast device, as developed by Gaede in 1935, can be used, preventing possible

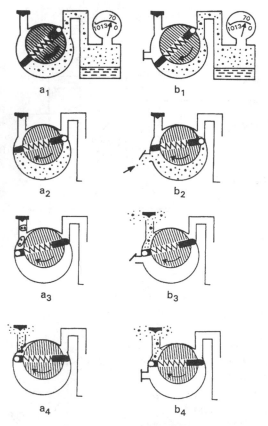

a_1 b_1

a_2 b_2

a_3 b_3

a_4 b_4

Without gas ballast

a_1) The pump is connected to the vessel which is already almost empty of air (approx. 70 mbar). It must therefore transport mostly vapour particles. It works without gas ballast.

a_2) The pump chamber is separated from the vessel. Compression begins.

a_3) The content of the pump chamber is already so far compressed that the vapour condenses to form droplets. Overpressure is not yet reached.

a_4) The residual air only now produces the required overpressure and opens the discharge outlet valve. But the vapour has already condensed and the droplets are precipitated in the pump.

With gas ballast

b_1) The pump is connected to the vessel which is already almost empty of air (approx. 70 mbar). It must therefore transport mostly vapour particles.

b_2) The pump chamber is separated from the vessel. Now the gas-ballast valve opens, through which the pump chamber is filled with additional air from outside. This additional air is called "gas ballast".

b_3) The discharge outlet valve is pressed open; particles of vapour and gas are pushed out. The overpressure required for this to occur is reached very early because of the supplementary gas-ballast air, as at the beginning of the whole pumping process. Condensation cannot occur.

b_4) The pump discharges further air and vapour.

condensation of water vapours in the pump. This device works as is shown in Figure 2.4. Before the compression action can begin, an exactly regulated quantity of air, namely the 'gas ballast', is let into the pump's compression stage. Now the vapours can be compressed with the gas ballast before their condensation point is reached and they can be ejected out of the pump.

The main application of such rotary pumps is to achieve pressures in the rough and medium vacuum regions or to act as backing pumps, removing the gas compressed by high vacuum pumps which will then achieve pressures in the high and ultra-high vacuum regions.

Probably the dominant pump used for achieving pressures in the high and ultra-high pressure regions in surface analysis instrumentation is the **turbomolecular pump**. This kind of pump falls into the classification of compression pumps which transport gas from a low pressure to a high pressure region where the chamber volume is constant. A sectional drawing of a typical turbomolecular pump is shown in Figure 2.5 The principle of operation of molecular pumps has been well known since 1913 and depends on the fact that the gas particles to be pumped will receive, through impact with the rapidly moving surface of a rotor, an impulse in the required flow direction. Early molecular pumps which used rotor blades simply in the form of discs, suffered from constructional difficulties and a high susceptibility to mechanical failure. More recently the blading of the rotors in molecular pumps was constructed in the form of a turbine, which allowed easier construction and greater reliability of operation, and in this form has developed into the turbomolecular pump of today (as shown in Figure 2.5).

The gas to be pumped arrives directly through the aperture of the pump inlet flange, giving maximum possible conductance. The top blades on the rotor, the so-called 'vacuum stage', are of large radial span, allowing an large annular inlet area. The gas captured in the high vacuum stages is transferred to the lower 'compression stages' which have blades of shorter radial span, here the gas is compressed to the backing pressure. From here it is removed by a backing pump, which is most commonly a rotary vane pump.

The pumping speed characteristics of turbomolecular pumps are shown in Figure 2.6. The pumping speed is constant over the whole working pressure range. It decreases however, at intake pressures greater than 10^{-2} mbar, where the transition between molecular flow and laminar viscous flow takes place. Although a turbomolecular pump backed by a rotary pump can pump a chamber directly from atmosphere, at pressures above 10^{-2} mbar, due to the viscous nature of the gas, the pump will be operating under strain and should therefore be only exposed to these pressures for a short period of time.

Figure 2.4. Illustration of the pumping process in a rotary vane pump, without (left) and with (right) gas ballast device when pumping condensable substances

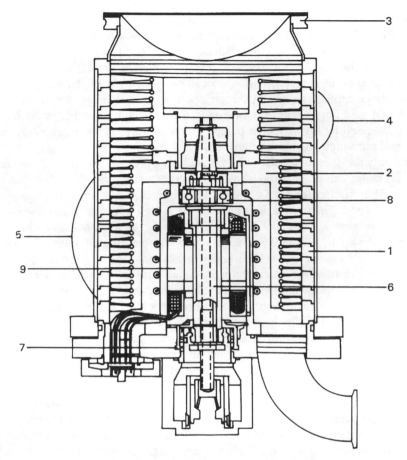

Figure 2.5. Cross-sectional picture of a turbomolecular pump of a single-ended axial flow design. 1, Strator blades; 2, Rotor body; 3, Intake flang; 4, Blades of the suction stage; 5, Blades of the compression stage; 6, Drive shaft; 7 & 8, Ball bearings; 9, High-frequency motor

Figure 2.6 also shows that the pumping speed very much depends on the type of gas. Due to its high pumping speed for high mass hydrocarbon molecules, a turbomolecular pump can be fitted directly to a vacuum chamber without any need for cooled baffles or traps. When these pumps are switched off, however, they must be vented to atmosphere, or oil from the pump and backing pump will be sucked back into the vacuum system. If venting fails during shut-down or does not operated correctly, then oil vapours can get through into the vacuum chamber. To impede this isolation, valves are often fitted between these pumps and the vacuum chamber so that the pump can be vented independently of the chamber.

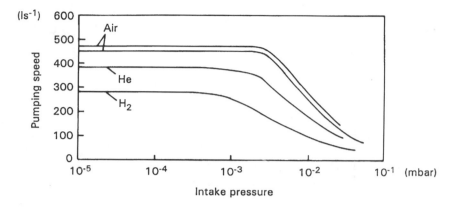

Figure 2.6. The pumping speed characteristic curves for a nominally 450 ls^{-1} turbo-molecular pump for different gases

In recent times, some sub-species of the turbomolecular pump have been developed for specific applications. Of these the most used is a turbomolecular pump fitted with magnetically suspended blades, the so-called 'Maglev' pump, these have the advantages of very low vibration level for imaging applications and lower oil vapours than conventional turbomolecular pumps.

Another common pump used often in surface science for achieving high and ultra-high vacuum is the **diffusion pump** although these have been superseded in recent years by the turbomolecular pump. These come under the classification fluid entrainment pumps where the pumping action is due to the diffusion of gases in a stream of high velocity particles. In these pumps the pumped gas molecules are removed from the vessel into a fast moving stream of pump fluid, most often in vapour form (most typically oil or mercury). This conveys the gas in the pumping direction by impact, thereby the pumped gas is transported into a space at higher pressure. The pump fluid itself after leaving the nozzle in the form of a vapour, condenses on the cooled outer walls of the pump.

The are several types of fluid entrainment pump used depending on the pressure conditions required, the most common in surface science, however, are low vapour density oil diffusion pumps, with a working pressure region below 10^{-3} mbar. A schematic diagram showing the mode of operation of such a pump is shown in Figure 2.7 These pumps consist basically of a pump body with a cooled wall and a three- or four-stage nozzle system. The low vapour density oil is in the boiler and is vapourised here by electrical heating. The oil vapour streams through the chimneys and emerges with supersonic speed through the different nozzles. The so-formed jet of oil vapour widens like an umbrella until it reaches the walls of the pump where it condenses and flows back in liquid form to the boiler. Diffusion pumps have high pumping speed over a wide pressure

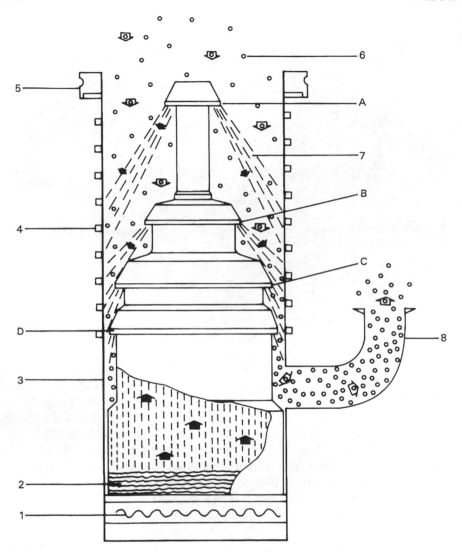

Figure 2.7. Schematic diagram showing the mode of operation of a diffusion pump. 1, Heater; 2, Boiler 3, Pumping body; 4, Cooling coil 5, High vacuum flange connection; 6, Gas particles; 7, Vapour jet; 8, Backing vacuum connection port; A–D, Nozzles

range, it is also practically constant over the whole region lower than 10^{-3} mbar (data for this is presented in Figure 2.8).

The cooling of the walls of diffusion pumps is critical to their operation and almost all large diffusion pumps are water cooled, but some smaller ones are air cooled. Thermally operated protection switches or water flow switches are

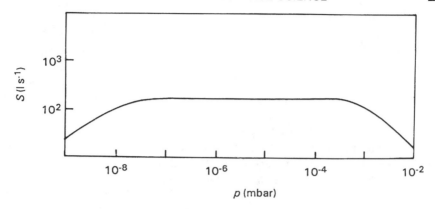

Figure 2.8. Characteristic pumping speed curve for a nominally 150 ls^{-1} diffusion pump (without cryo-trap)

often fitted to diffusion pumps which will switch off the pump heater if there is a cooling failure.

Since a certain pressure is needed in the pump before the vapour will form, the pump providing the backing (most typically a rotary pump) must have attained a certain pressure before the heater can be switched on, a backing pressure of typically 10^{-1} or 10^{-2} mbar. Because of this it is important that these pumps are protected against running with a pressure in the backing stage which is too high, and they should be fitted with a vacuum trip from the backing pressure gauge which operates such that if the backing pressure exceeds 10^{-1} mbar the pump heater will switch off. The failure to fit such a trip can have serious consequences to the pump, because if the pump runs hot with too high a backing pressure, the pump will stall and backstreaming of oil vapours into vacuum system will occur. If this condition continues for a period of a few tens of minutes, the oil in the pump can pyrolise and is then useless, this can only be remedied by completely stripping down the pump and putting in a new fill of oil.

For the lowest ultimate pressure from such pumps, backstreaming of the pump fluid into the vacuum chamber should be reduced as far as possible. For this purpose therefore cold traps are used which are cooled with liquid nitrogen so that the temperature of a baffle between the pump and the vacuum chamber is maintained at $-196°C$. The vapours will condense on the baffle at these temperatures and therefore will be effectively removed from the vacuum system until the temperature of the baffle rises.

A large classification of pumps which are less generally applicable than those mentioned previously but none the less have areas of application in surface science are sorption pumps. Within the concept of sorption pumps we include all arrangements where the pumped gas particles are removed from the space by being bound to a surface or embedded in the surface. This process can either be on the basis of temperature dependent adsorption by the so-called 'Van de Waals forces', chemisorption, absorption or by the particles becoming embedded in the course of the continuous formation of a new surface. By comparing their operation principles we can therefore further distinguish between adsorption pumps in which the sorption of gas takes place by simple temperature dependent adsorption processes and getter pumps where the sorption and retention of gases is due to the formation of chemical compounds.

Adsorption pumps work by the adsorption of gases on the surface of molecular sieves or other materials. Materials are chosen which possess an extraordinarily high surface area for a given mass of the material. This can be of order of 10^3 m^2 per gramme of the solid substance. A typical such material is Zeolite 13X. In practice the pump is connected to the vacuum chamber through a valve and it is only on immersing the pump in liquid nitrogen that the sorption effect becomes useful and the valve can be opened. These pumps are most often used for the clean roughing of a system prior to the use of a high vacuum pump such as a cryo or ion pump.

The pumping speeds of such pumps are very dependent on the gases being pumped, typically the best values being achieved for nitrogen, carbon dioxide, water vapour and hydrocarbon vapours, where light noble gases are hardly pumped at all. For pumping down a vessel which contains atmospheric air, where light noble gases are only present at a few p.p.m., the pressure of $< 10^{-2}$ mbar can be obtained without difficulty. After the pumping process the pump has to be warmed to room temperature for the adsorbed gases to be given off and the zeolite be regenerated for use.

One of the more common sorption pump used in surface science are sublimation pumps. These are a form of getter pump in which the getter material is evaporated and deposited on the inner wall of the vacuum chamber as a getter film. Titanium is almost exclusively the getter material in sublimation pumps and is evaporated from a wire of high titanium content alloy by resistance heating. Particles from the gas which impinge on the getter film become bound to it by chemisorption and form stable compounds with the titanium, which have immeasurably low vapour pressures.

These pumps are not used continuously but are switched on in short bursts of a few minutes and are often controlled by a timer such that they will come on at regular intervals, say typically every few hours. As such, these pumps are used to supplement the pumping from other pumps on the system, their incredibly high pumping speeds for active gases means that they can be used periodically to keep a system clean from such gases or where a sudden evolution of such gases needs to be pumped away quickly.

A similar pump to the sublimation pump but one which can be used continuously is the sputter ion pump. The pumping action of these pumps is due to two processes,

i)   Ions impinge on the cathode surface of a cold cathode discharge system and sputter material off it (the material of the cathode is again titanium). The titanium deposits on surfaces in the pump and acts as a getter film as before.

ii) The energy of the ions is high enough on sputtering incidence for them to become deeply embedded in the cathode where they are in essence adsorbed by ion implantation. The process can pump all types of ions including rare gases.

The ions are created in a Penning gas discharge between the cathode and the anode in the pump. In the pump there are two parallel cathodes made of titanium between which are arranged a system a cylindrical anodes made of stainless steel, see Figure 2.9. The cathode is maintained at high negative electrical potential, of the order of a few kV, with respect to the anode. The whole electrode assembly is maintained in a strong homogeneous magnetic field (flux density ≈ 0.1 T). Electrons near the anode are trapped in the high magnetic field and set up a region of high electron density in the anode cylinders ($n_e \approx 10^{13}$ cm^{-3}). Here the electrons will collide with gas particles in the system and ionise them. Due to their greater mass, these ions will remain relatively unaffected by the magnetic field and will be accelerated toward and impinge on the cathode. At pressures of 10^{-3} mbar or below such a discharge is self-sustaining and does not require a

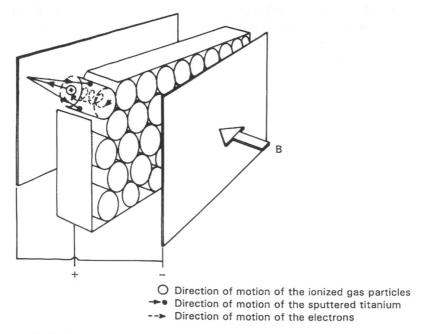

B

○   Direction of motion of the ionized gas particles
→● Direction of motion of the sputtered titanium
--→ Direction of motion of the electrons

Figure 2.9. Schematic showing the mode of operation of a sputter ion pump. (diode type)

hot cathode. The discharge current will obviously be proportional to the number density of gas particles in the system and therefore such a pump can also be used to measure the gas pressure in a system.

The pumping speed of sputter ion pumps will depend on the pressure and on the type of gas being pumped. For air, N_2, CO_2 and H_2O the pumping speeds are practically the same, compared with these gases the pumping speeds for other gases are about,

Hydrogen	150–200%
Methane	100%
Other light hydrocarbons	80–120%
Oxygen	80%
Argon	30%
Helium	28%

These pumps will only operate at pressures below 10^{-2}–10^{-3} mbar, but for them to work all that is required is that the high electric potential is supplied to the pump. One problem with these pumps can be encountered when pumping large volumes of inert gases because as the pump operates, the buried layers of these gases can be re-exposed by the sputtering of the cathode, and therefore the pump can re-emit these gases back into the system. More complicated designs of ion pump reduce this problem.

Care must be taken, when using these pumps on your vacuum system, that stray magnetic fields from the pump do not interfere with the rest of the operation of your system. This may be a particular problem in electron spectroscopy type apparatus. Also stray ions can escape from the pump and care must be taken to put shielding in your system to prevent these ions from interfering with any experiments being undertaken.

Cryopumps are a different type of sorption pump used frequently in MBE systems or ion implanter systems and to some extent in surface science. They consist inherently of a surface cooled down to a temperature of <120 K so that gases and vapours condense or bond to this surface, the cooled surface is situated in the vacuum itself.

The liquid nitrogen cooled vapour traps used frequently with oil diffusion pumps (mentioned previously) are in themselves a type of cryopump. If cooling media of even lower boiling points are used, for example liquid helium with a boiling point of 4.2 K, then gases like O_2, N_2 and H_2 can also be pumped. In order to use such a pump to attain high and ultra-high vacuum conditions the cryosurfaces have to be cooled down to below 20 K.

Various mechanisms are effective when bonding gases to cold surfaces. As well as condensation, cryotrapping and cryosorption also play a part.

In order for a cryopump to pump by condensation, the vapour pressure of the solid condensate has to be significantly lower than the working pressure one wishes to achieve. If a working pressure of 10^{-9} mbar is required, then for gases like air, O_2 and N_2 this requires temperatures below 20 K; for gases like Ne and He, however, this requires the lower temperature only attainable with liquid helium. Hydrogen is of particular interest since it constitutes the major part of the residual gas composition in high and ultra-high vacuum environments. This is very difficult to condense out of a system requiring temperatures of 3.5 K. Therefore to remove hydrogen from a system with such a pump, another mechanism of pumping must be utilised.

One such mechanism which is used for the 'condensation' of difficult gases is 'cryotrapping'. Here a gas which is easily condensable is let into the system such that a mixture of gases is produced. Typical gases which are used are Ar, CO_2, CH_4, NH_3 and other heavier hydrocarbons. The condensate mixtures produced have vapour pressures several orders of magnitude lower than that of pure hydrogen.

This mechanism is automatically present when cryopumping any gas mixture and is of course not confined to the pumping of hydrogen.

Another such mechanism used in the pumping of difficult gases is cryosorption. Here a pre-introduced sorption material, for example, activated charcoal, is cooled and the gases adsorb to this. This has the advantage that no continuous admittance of trapping gas into your system is required.

In principle, a cryopump could be switched on at atmospheric pressure; however, in practice this would lead to the creation of a very thick layer of condensate on the pump at the beginning of the pump-down process, considerably reducing the pumping capacity of the pump. General practice when high and ultra-high vacuums are required, is to rough pump out the chamber to a pressure of $\approx 10^{-3}$ mbar before switching on the cryopump.

In practice the mechanism for achieving the cold surface can be different. For example, there are liquid pool cryopumps, continuous flow cryopumps and refrigerator cryopumps (these are described in [1]) but for all, the principle of pumping is the same.

Cryo-pumps suffer from problems with vibration and are therefore generally not used on systems where imaging applications are being used.

3.2 EVACUATION OF A CHAMBER

One of the first considerations when designing a pumping system is what size of pump is required. If you pick too large a pump you will waste money, too small a pump and you will not achieve the required conditions for your experiment. Basically, two questions arise when choosing the size of pumping system,

i)   How large must the effective pumping speed of the system be so that the pressure will be reduced to the desired value in a given time?

ii) How large must the pumping speed be so that gases released into the vacuum system during operation can be pumped away quickly enough so that the required pressure is not exceeded?

This brings us to the idea of the effective pumping speed, S_{eff}, of the system. This is understood to be the pumping speed of the whole pump arrangement that actually prevails at the vessel, taking into account the conductances of the components between the pump and vessel; for example, valves, apertures baffles, coldtraps, etc. If the conductances of these components are known, and the actual pumping speed of the pump itself is known (this is referred to as the nominal pumping speed, S, of the pump) then the effective pumping speed can be determined. The relationship between the effective pumping speed and the nominal pumping speed of a system is given by the formula,

$$\frac{1}{S_{eff}} = \frac{1}{S} + \frac{1}{C}$$

where, C = the total flow conductance of the tubulation between the pump and the vacuum chamber (baffles, pipes, etc.).
This can be broken down into the conductances of the individual components as,

$$\frac{1}{C} = \frac{1}{C_1} + \frac{1}{C_2} + \cdots \frac{1}{C_N}$$

The conductances of simple tubes can be calculated in the different pressure regimes, (see [1]) but the conductances of geometrically more complicated components such as cold traps, valves, and baffles must usually be determined experimentally.

3.3   CHOICE OF PUMPING SYSTEM

The choice of pumping system will be dependent on the processes which are to be undertaken in the vacuum system and also on the available budget. In general, in vacuum technology the choice of pumping is initially governed by whether these processes fall into the categories of wet or dry processes. Dry processes are those in which there are no significant amounts of vapour to be pumped, whereas wet processes will evolve a significant amount of water vapour which must be pumped away.

In surface science we are almost solely concerned with dry processes and therefore, in this book, limit ourselves to discussing pumping systems relating only to such processes. In most surface science applications, the required vacuum is produced prior to the experimental measurement. Furthermore, in such systems the degassing of the vacuum system itself and the components in it is a critical stage.

When working pressures in the rough and medium vacuum regimes are needed, rotary vane pumps are often ideal. They are especially suitable for pumping down vessels from atmospheric pressure to pressures below 0.1 mbar, in order to work continuously in this lower pressure region. A very common need for medium vacuum is when evacuating a vessel to a pressure such that other high and ultra-high vacuum pumps can be used and subsequently as a backing pump for such pumps. Here two-stage (ultimate pressure = 10^{-3} mbar) rotary vane pumps are ideal.

Pressures in the high vacuum and ultra-high vacuum (UHV) regions can be achieved using diffusion pumps, sputter ion pumps, turbomolecular pumps and cryopumps all fitted with a suitable roughing or backing pump and often used in conjunction with sublimation pumps. Pumping alone, however, will not allow true ultra-high vacuum conditions to be fulfilled, since in this pressure region the major contribution to the pressure comes from gas evolved off the container walls. To achieve such pressures, therefore, the whole vacuum chamber must be baked, whilst pumping it, to temperatures of about 250–350°C in order to desorb the gases off the walls of the chamber and allow them to be pumped away. UHV chambers, therefore, are almost invariably made of stainless steel and are fitted with all-metal seals (described in Section 3.5). When a system has been made leak-tight and leaked test with a helium leak detector, then baking is undertaken. This can extend over several hours or even days. Before the system is allowed to cool fully all components in the system which may desorb gases during their operation must be degassed. This will include hot cathodes or filament assemblies in the system, for example sublimation pump filaments, or filaments in ion or electron guns.

3.4 DETERMINATION OF THE SIZE OF BACKING PUMPS

The size of a backing pump is determined by the fact that the quantity of gas or vapour transported by the high vacuum pump must also be handled by the backing pump such that the maximum permissible backing pressure is not exceeded. If Q is the quantity of gas pumped by the high vacuum pump, with an effective pumping speed S_{eff} at an inlet pressure p_A this quantity of gas must also be transported by the backing pump of speed S_V and backing pressure p_V therefore,

$$Q = p_A S_{eff} = p_V S_V$$

thus the minimum pumping speed of the backing pump can be calculated to be,

$$S_V = \frac{p_A}{p_V} S_{eff}$$

3.5 FLANGES AND THEIR SEALS

Demountable joints in metallic vacuum components are invariably provided with flanges which are sealed by means of a gasket which is compressed or deformed in some way as the flange is tightened and the seal made. Commonly used flange sizes up to 200 mm outer diameter are built to internationally standardised dimensions and come in several different sizes.

Flanges which are suitable for use in the rough, medium and high pressure regions are usually made from a black rubber type material called Viton. Such seals can be used for pressures down to 10^{-8} mbar and are bakeable to $\approx 200°C$. For true UHV pressures, the seals are made from metal and all components should be bakeable to 350°C. The most common type of flange is the so-called

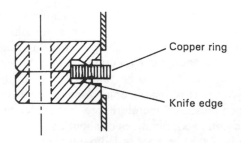

Figure 2.10. A cross-section of a Conflat (or knife-edge seal) flange

'knife edge' or 'ConFlat®' (Varian) flanges which is sealed with copper gaskets. A cross-section through such a flange and gasket is shown in Figure 2.10. On larger or non-circular flanges, other metal seals which use a loop of soft metal, typically gold, indium or aluminium are sometimes used. The seal is made by simply compressing the soft metal loop between the two flat metal surfaces of the flange. Great care must be taken with all sealing surfaces in a UHV system to prevent scratching or damage which will impair the leak-tightness of the seal.

4 MEASUREMENT OF LOW PRESSURES

In modern vacuum technology, pressure measurements have to be made over a range of 16 orders of magnitude from 10^3 mbar to 10^{-13} mbar. It is impossible on fundamental physical grounds to build a gauge which will measure quantitatively over this whole pressure region. Therefore a series of gauges have been developed which have characteristic pressure range, typically extending over a few orders of magnitude.

The types of gauge fall into two categories, those which measure the pressure directly, that is those which measure the pressure in accordance with its definition as the force which acts on a unit area, and those which measure it indirectly where the pressure is measured as a function of a pressure-dependent property of the gas (for example, thermal conductivity, ionisation probability, electrical conductivity). Only in the case of the direct or absolute pressure measurement is the reading independent of the nature of gas (in accordance with equation (6)) it will, however, be dependent on the temperature. In the case of indirect pressure measurement the properties measured are almost invariably dependent on the molar mass of the gas studied as well as the pressure. Consequently such a pressure reading will be dependent on the nature of the gas. On such gauges the scale is always calibrated with air or nitrogen as the test gas. For other gases, correction factors must be used.

The measurement of pressure in the rough vacuum region can be undertaken by gauges with direct pressure measurement, and the pressure can be determined to quite a high level of accuracy. Measurement in the lower pressure regimes, however, is invariably undertaken by indirect methods. This usually means that the accuracy of the measurement is limited by certain fundamental errors. These inaccuracies affect the measurement to such a degree that to make a pressure measurement in the medium and high vacuum regions with an error of less than 50% would take special care from the experimenter. The inaccuracies are even more severe in the ultra-high vacuum region, such that to achieve measurements in these low pressure regimes which are accurate to within a few percent requires special measuring equipment and great care. For this reason the reliability of pressure readings in an experiment has to be treated with careful consideration.

Furthermore, if one wants to make a statement about the pressure in a vacuum chamber recorded by a gauge, firstly, the location of the gauge has to be taken into account. For example in the pressure regions where laminar flow prevails in the system, a pressure gradient due to the pumping process will be present in the system such that a gauge which is situated near the inlet to a pump will record a lower pressure than that which is actually present in the chamber. Also, in this pressure region the conductances of the tube in the chamber can introduce pressure gradients and lead to false readings. In the high and ultra-high vacuum regions the situation is even more complicated, where outgassing of the walls of the chamber and of the gauge itself can have a significant effect on the accuracy of the reading.

4.1   GAUGES FOR DIRECT PRESSURE MEASUREMENT

Gauges for direct pressure measurement work on mechanical principles, measuring the force which the gas particles exert on a surface by virtue of the thermal velocities. This is usually achieved by measuring the displacement of an interface (solid or liquid) between a region at the pressure to be measured and a region at a certain known reference pressure, sometimes the reference is atmospheric pressure.

A large category of mechanical vacuum gauges are diaphragm gauges, the best known of these being the barometer. This contains a hermetically sealed evacuated thin-walled capsule made from a copper–beryllium alloy. This capsule is evacuated to the reference pressure. The external side of the capsule is connected to the vacuum vessel, and as the pressure is reduced, the diaphragm moves outwards. This movement is transfered by a lever system to a pointer which indicates the pressure on a linear scale. Such a reading, due to the sealed reference pressure is independent of atmospheric pressure. A typical type of modern diaphragm gauge is shown in Figure 2.11. Such a pressure gauge will read from atmospheric pressure down to a few mbar with an accuracy of about ± 10 mbar. If accurate readings are needed in the range below 50 mbar, then a gauge with the reference capsule evacuated to below 10^{-3} mbar is superior. Such a gauge can measure pressures in the range 100–1 mbar with an accuracy of 0.3 mbar. Such gauges are quite sensitive to vibrational interference. In a modern gauge

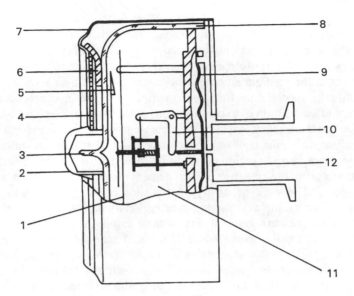

Figure 2.11. A section through a modern diaphragm vacuum gauge. 1, Reflecting cover plate; 2, Protective cap; 3, Seal-off point; 4, Plexiglass disc; 5, Pointer; 6, Scale 7, Metal cover; 8, Glass chamber; 9, Diaphragm; 10, Lever system; 11, Reference vacuum; 12, Connecting flange

the movement of the diaphragm will often be linked to an electrical transducer and then the pressure can be displayed digitally on a panel meter.

One of the simplest but most exact ways of measuring the pressure in the rough vacuum region is the mercury manometer. Here the evacuated limb of the U-tube is maintain at a constant pressure equal to the vapour pressure of mercury at room temperature (10^{-3} mbar). The other limb of the tube is connected to the vacuum vessel. From the difference in levels of the two columns of mercury, the pressure can be directly determined in mbar. The major drawback of such gauges for frequent application is their size and vulnerability to breakage.

A compression type of manometer developed by McLeod in 1874 still has important applications today. The measurement of pressure in such gauges results from the fact that a quantity of gas which occupies a large volume at the pressure which is to be determined is then compressed by trapping it behind a column of mercury. This increased pressure can then be determined in the same way as with the conventional mercury manometer. The absolute pressure in the vessel can then be determined knowing the volume of the enclosed gas and the total volume of the gauge (for more detail of such gauges see [1]). When operating these gauges it must be noted that the pressure reading is not continuous, but for every pressure measurement the mercury must be raised into the gauge. These gauges can provide an accurate determination of the absolute pressure to within

$\pm2\%$ in the rough and medium vacuum regime and can even provide readings extending into the high vacuum region, down to $\approx10^{-5}$ mbar.

With these gauges however, as with any vacuum gauge involving a compression stage, the presence of condensable vapours can influence the pressure reading obtained. If vapours condense out in the gauge, further compression will not increase the pressure in the entrapped volume and a false reading will be given.

Direct pressure measurement is rarely used in surface analysis instrumentation; however, these types of measurement are used to calibrate other gauges or where higher accuracy is needed in the low vacuum region — for example when pre-mixing gases prior to leaking them into the vacuum chamber.

4.2 GAUGES USING INDIRECT MEANS OF PRESSURE MEASUREMENT

Gauges which measure the pressure indirectly, invariably make measurements on the gas of an electrical nature and convert this to a pressure reading. The apparatus consists of a gauge head, which is connected to the vacuum system and a control unit which is normally remote from the head.

One of the most common types of gauge used in vacuum science and indeed surface science is the thermal conductivity or Pirani vacuum gauge. These gauges utilise the variation of mean free path (a function of the number density of particles) and the corresponding variation in thermal conductivity of the gas to monitor the pressure. Such gauges are exploited extensively for measurement in the medium pressure region from 1 to 10^{-3} mbar.

The gauge head of a Pirani gauge has a sensing filament which is open to the vacuum chamber. A current is passed through the filament which produces heat. This heat can be transferred away from the filament by radiation or by thermal transfer to the surrounding gas. In the rough vacuum region, the rate of heat transferred away due to convection is almost independent of pressure. However, as the mean free path of the gas molecules becomes of the order of the diameter of the filament, the convection of heat away becomes strongly dependent on pressure. This continues to be so until the pressure reaches $\approx10^{-3}$ mbar, where the dominant heat transfer process is radiation, which is independent of pressure.

In practice there are two methods used to measure the pressure in this way; those in which the sensing filament is of varying resistance and those in which the resistance of the filament is kept constant. In the first case the sensing filament in the gauge head forms one branch of a Wheatstone's bridge circuit. As the rate of heat transfer changes, so the temperature of the sensing filament changes, for example; if the pressure increases, the rate of heat transfer will increase, the temperature of the filament will decrease and its resistance will decrease, so the bridge becomes out of balance. The bridge current serves as a measure of the pressure which is indicated on a meter. In the second approach to the measurement, the sensing filament is also part of a Wheatstone's bridge, however, in this case the voltage applied to the filament is regulated so as to keep the resistance (and temperature) of the filament constant and the bridge always balanced. As the pressure changes, therefore, the applied voltage must change to compensate for the variation in heat transfer. In this case it is the applied voltage which is a measure of the pressure.

The Pirani gauge of variable resistance can only cope with pressures in the range $10-10^{-3}$ mbar, whereas the constant resistance type can operate in the range 10^3-10^{-3} mbar. Such gauges have an accuracy of typically $\pm10\%$. Due to their pressure range characteristics and their relatively robust nature, these types of gauge are found extremely frequently in surface science, and are most frequently used for monitoring the pressure above a rotary pump. Common examples of this are the monitoring of the backing pressure for a high vacuum pump or indication of the pressure in a gas inlet system. These gauges are always calibrated for nitrogen or for air, but for other small molecular mass species the reading is within the inherent error of the gauge. For large organic molecules, however, the error is increased and can become significant, especially in the low pressure region below 10^{-2} mbar.

Pirani gauges are most frequently used to monitor the low–high vacuum in the backing stages of high vacuum pumps, where they will also activate safety procedures if the pressure in this stage gets too high.

The most common gauges for measuring the pressure in the high and ultra-high vacuum regions are ionisation gauges. These gauges measure the pressure in terms of the number density of molecules in the gas. A portion of the atoms or molecules in the gas are ionised by electron impact and the positive ions thus produced are collected by an electrode in the system and the resulting current is measured. There are two basic types of these gauges distinguishable by the method of generation of the ionising electrons.

In the gauge head of the so-called cold cathode or Penning or inverted Magnetron gauge there are two unheated electrodes the cathode and the anode, between which a self-sustaining discharge is excited by applying a d.c. voltage of about 2 kV across the electrodes. The discharge is maintained by the application of a strong magnetic field perpendicular to the lines of electric field such that the electrons in the discharge have long spiralling paths and subsequently a high probability for collision with gas particles. The positive ions generated in the discharge will travel to the cathode and the pressure is measured by monitoring the generated discharge current which is indicated on a meter. Since the ionisation cross-section is a function of the gas species being ionised, the pressure reading will be gas dependent. There is an upper limit on the measuring region of $\approx10^{-2}$ mbar, this is due to the fact that above this pressure a glow discharge will be generated in the gauge head and in this region the discharge current will be far less dependent on pressure. Although these gauges have a reading limit at 10^{-2} mbar, it is generally safe to operate them at pressures of up to atmospheric pressure, this being one of the great advantages of such gauges, especially where the application requires that the system is frequently let up to atmospheric pressure and then pumped to the high vacuum region again. Penning ionisation gauges have a lower pressure measurement limit of the order of 10^{-8} mbar. Problems can arise from stray magnetic fields from such gauges in techniques which are very sensitive to magnetic fields such as low energy electron spectroscopies.

It can be seen that such gauges work on very similar principles to sputter ion pumps and they therefore have a self-pumping effect with a pumping speed of order of 10^{-2} ltr s^{-1}. This effect leads to quite high inaccuracies in the readings from such gauges, up to $\approx\pm50\%$. Despite this, Penning gauges are frequently found in surface science.

Hot cathode ionisation gauges are one of the most common gauges found in surface science, since they are the only commercially available gauge for measuring pressures in the ultra-high vacuum region. Such gauges use a hot cathode (or filament) as the source of the ionising electrons. In the gauge head of these gauges there are three electrodes, the cathode or filament, the anode and an ion collector (a schematic representation of such an assembly is shown in Figure 2.12). When the filament is heated by passing an electric current through it, it will emit electrons by thermal emission, such a cathode is a very abundant source of electrons. These electrons are then accelerated in an electric field between the cathode and the anode. The anode is in the form of a grid such that

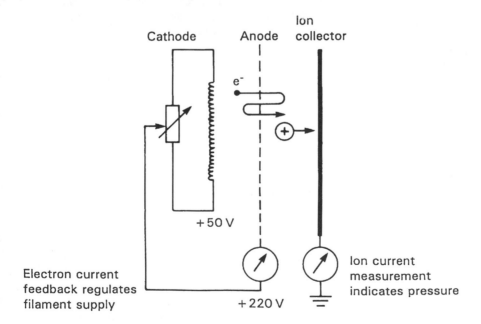

Figure 2.12. Schematic of a hot cathode ionisation gauge showing typical operating voltages

a high fraction of the emitted electrons pass through it. The electrical potentials applied to the anode and the cathode are such that the electrons have sufficient energy to ionise gas particles in the system on collision. Gas particles which are ionised on the far side of the anode from the filament will be attracted towards the ion collector which is at a negative potential with respect to the anode. It is this ion current collected that is proportional to the number density of particles in the system and is therefore expressed as a pressure. The abundance of electrons from the filament source means that no magnetic fields are required in these types of gauges and therefore the electrode assembly of the gauge head can be place directly in the vacuum system generally without causing interference with other components in the vacuum system. Only where the technique being used is sensitive to stray electrons or light is this not the case.

Except for the case of specially designed gauge heads, the upper limit on pressure measurement from hot cathode ionisation gauges is $\approx 10^{-2}$ mbar, above this pressure a glow discharge will form in the region of the electrodes which prevents operation of the gauge. Operation above this pressure will also ultimately result in filament burn out. The limit on the low pressure range of measurement of $\approx 10^{-12}$ mbar is due to two effects, the so-called X-ray and ion desorption effects.

The X-ray effect is caused by the electrons which impinge on the anode. This will emit soft X-ray photons, which may then strike the ion collector causing the emission of further electrons from it. This electron current from the ion collector will be indistinguishable from an ion current flowing to the collector and will simulate a high pressure reading. Also the emitted photons can collide with the walls of the vacuum chamber surrounding the gauge head and so produce electrons. If the electrical potentials in the system are such that these electrons can travel to the ion collector, then they will cause an electron current to flow which will simulate a lower pressure reading. The scale of these effects will be dependent on the anode voltage and ion collector voltages with respect to their surroundings and the surface area of the ion collector.

When electrons impinge on the anode they can cause the desorption of gas species from the surface of the anode often as positive ions. These emitted ions will travel to the ion collector leading to a falsely high pressure reading, this is the so-called ion desorption effect. The magnitude of this effect will generally be independent of pressure but will to a point increase with increasing emission current. At small emission currents the effect will increase proportionally with current but as the current increases further the process will have the effect of cleaning up the anode and therefore further increases of current will result in a decrease in the effect.

A schematic representation of an ion gauge head is shown in Figure 2.12. The cathode (or filament) is generally made of tungsten. The electrons oscillate through the anode grid giving them long flight paths increasing the probability for ions being created.

To ensure a linear relation between ion current and pressure the X-ray effect must be minimised. To this end Bayard and Alpert designed a gauge head where the electrode assembly is such that the hot tungsten cathode lies outside the cylindrical anode grid, with the ion collector being a thin wire, with therefore minimum surface area, on the axis of the electrode system. With this design the X-ray effect is reduced by two or three orders of magnitude over that for the early types of gauge due to the great reduction in the surface area of the collector. These gauges can read pressures down into the ultra-high vacuum region of order of 10^{-10} mbar.

Other ion gauges have been designed for specialist areas of application such as the 'Bayard-Alpert gauge with modulator' and the 'Extractor' type gauge. The coverage of these, however, is beyond the scope of this book but a more thorough description can be found in [1].

The pressure measurements from all types of ionisation gauge will again be a function of the gas being measured because the ionisation cross-section will change from species to species. These gauges are again calibrated for air or nitrogen and relative sensitivity factors are needed for accurate pressure determination for other gases. Although these correction factors are to a certain extent dependent on the specific gauge type used, a table of typical correction factors or different gases is presented in Table 2.2. If a gas other than nitrogen or air is predominant in the vessel, then the pressure reading has to be multiplied by the corresponding factor to give the correct pressure reading.

Hot cathode ion gauges exhibit a pumping action, but this is a considerably smaller effect than that for cold cathode gauges.

4.3 PARTIAL PRESSURE MEASURING INSTRUMENTS

In various vacuum processes it is important to know the composition gas or vapour mixture, i.e. the partial pressures of the gases in the system. The different

Table 2.2. Correction factors for ionisation gauge head readings for different gases

Predominantly present:	Factors for gauge reading related to:	
	N_2	Air
He	6.9	6.04
Ne	4.35	3.73
Ar	0.83	0.713
Kr	0.59	0.504
Xe	0.33	0.326
Hg	0.303	0.27
H_2	2.4	1.83
CO	0.92	0.85
CO_2	0.69	0.59
CH_4	0.8	0.7
Higher hydrocarbons	0.1–0.4	0.1–0.4

gases can essentially be distinguished from each other by their molecular masses. A partial pressure measuring device is therefore a sensitive mass analyser in which the measuring system has dimensions which are small enough for it to fit easily into a vacuum system. Also in the case of the measurement of partial pressures in the high and ultra-high vacuum regions they can be baked with the vacuum system they are installed within.

A typical partial pressure measuring instrument (as represented schematically in Figure 2.13) generally consists of three components:

i)   an ion source where the gas particles in the system are ionised, so that they can then be mass analysed and detected,

ii) an ion separation system so that ions of different masses can be distinguished easily, and

iii) an ion collector to measure the ion current at each mass.

These devices most commonly take the form of a small quadrupole mass filter fitted with a hot cathode ion source and a 'Faraday cup' detector, or when more sensitivity is required, a secondary electron multiplier detector. The mass analyser will typically have an operating range up to 100 a.m.u. and certainly no more than 300 a.m.u. and will allow the resolution of peaks 1 a.m.u. apart over this whole range.

The output of these devices is shown in spectral form representing a direct measure of the collected ion current or an equivalent as the mass analyser scans through the mass range required. In the interpretation of these spectra, it must be taken into account that different species of gas can have different detection

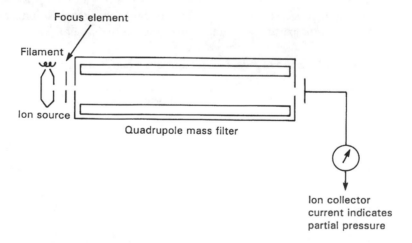

Figure 2.13. Schematic of a simple mass spectrometer for partial pressure measurement

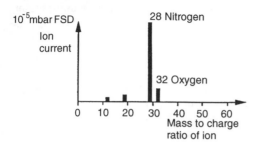

The presence of the oxygen peak indicates an air leak.

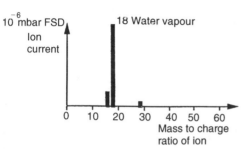

The high water vapour content in the spectrum indicates that the system needs baking.

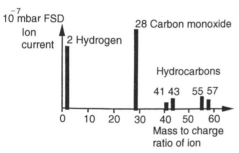

The presence of high mass hydrocarbons in the spectrum indicates backstreaming of oil vapours from rotary pumps or diffusion pumps, a foreline trap or cold trap should be fitted to prevent this.

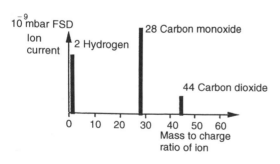

A clean high vacuum system.

Figure 2.14. Typical mass spectra of the four different stages to arrive at ultra-high vacuum conditions

probabilities. This is due not only to their different ionisation probabilities, but also to variations in the transmission of the mass analyser as a function of mass and, in the case of an electron multiplier, the detectability as a function of mass. To add to this, molecular species, especially the higher mass hydrocarbons, can dissociate in the ionisation process giving several peaks from one gas species.

Typical partial pressure spectral for the four different stages passed through, in achieving a clean ultra-high vacuum environment, are shown in Figure 2.14.

ACKNOWLEDGEMENT

Helpful discussions with Dr J. Gordon, Dr N. Aas and many other colleagues are gratefully acknowledged

REFERENCES

[1] *Vacuum Technology its Foundation, Formulae and Tables*, a Leybold-Heraeus Vacuum Products Ltd. publication.
[2] *Vacuum*, A. Chambers.

CHAPTER 3

Electron Spectroscopy for Chemical Analysis

BUDDY D. RATNER AND DAVID G. CASTNER

Department of Chemical Engineering and Center for Bioengineering, University of Washington, Seattle USA

Of all the contemporary surface characterization methods, electron spectroscopy for chemical analysis (ESCA) is the most widely used. ESCA is also called X-ray photoelectron spectroscopy (XPS), and the two acronyms can be used interchangeably. The popularity of ESCA as a surface analysis technique is attributed to its high information content, its flexibility in addressing a wide variety of samples, and its sound theoretical basis. This chapter will introduce the ESCA method and describe its theory, instrumentation, spectral interpretation, and application. The intent of this introduction is to provide a broad overview. Many general reviews on this subject exist and further reading about ESCA theory and applications is encouraged [1–91]. This review is aimed at readers who have had little or no formal introduction to the ESCA method — it should provide an appreciation of the power and limitations of the contemporary surface analytical method. The jargon associated with ESCA will also be introduced and discussed, thereby assisting the reader in digesting the specialist literature.

1 OVERVIEW

ESCA falls in the category of analytical methods referred to as electron spectroscopies, so called because electrons are measured. Other prominent electron spectroscopies include Auger electron spectroscopy (AES, Chapter 4) and high-resolution electron energy loss spectroscopy (HREELS, Chapter 7).

1.1 THE BASIC ESCA EXPERIMENT

The basic ESCA experiment is illustrated in Figure 3.1. The surface to be analyzed is first placed in a vacuum environment and then irradiated with photons. For ESCA, the photon source is in the X-ray energy range. The atoms comprising

---

Surface Analysis — The Principal Techniques. Edited by John C. Vickerman
© 1997 John Wiley & Sons Ltd

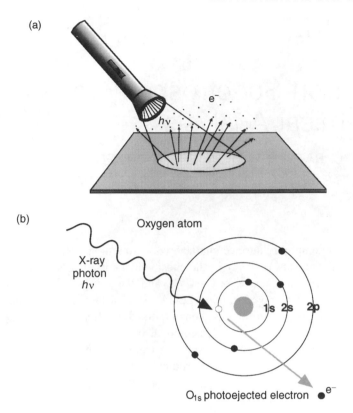

Figure 3.1. (a) A surface irradiated by a photon source of sufficiently high energy will emit electrons. If the light source is in the X-ray energy range, this is the ESCA experiment. (b) The X-ray photon transfers its energy to a core-level electron imparting enough energy for the electron to leave the atom

the surface emit electrons (photoelectrons) after direct transfer of energy from the photon to the core-level electron. These emitted electrons are subsequently separated according to energy and counted. The energy of the photoelectrons is related to the atomic and molecular environment from which they originated. The number of electrons emitted is related to the concentration of the emitting atom in the sample.

1.2 A HISTORY OF THE PHOTOELECTRIC EFFECT AND ESCA

The discovery of the photoelectric effect, its explanation, and the development of the ESCA method are entwined with the revolution in physics that took place in the early years of the twentieth century. This revolution led from classical physics based upon observational mechanics to quantum physics, whose impact is most clearly appreciated at the atomic scale. Some of the developments that took place from the 1880's up to the second half of the twentieth century that are fundamental to the development of ESCA will be briefly reviewed [10–12].

Hertz, in the 1880's, noticed that metal contacts in electrical systems, when exposed to light, exhibit an enhanced ability to spark. Hallwachs, in 1888, observed that a negatively charged zinc plate lost its charge when exposed to ultraviolet (UV) light, but that positively charged zinc plates were not affected. In 1899, J.J. Thompson found that subatomic particles (electrons) were emitted from the zinc plate exposed to light. Finally, in 1905, Einstein, using Planck's 1900 quantization of energy concept, correctly explained all these observations — photons of light directly transferred their energy to electrons within an atom, resulting in the emission of the electrons without energy loss. This process will be clarified in Section 2 of this chapter. Planck received the Nobel Prize for his contribution of the concept of the quantization of energy in 1918. Einstein received the Nobel prize for explaining the photoelectric effect in 1921. To put the revolutionary aspects of these developments in perspective, a quotation from Max Planck, in nominating Einstein for the Prussian Academy in 1913, is illuminating. Planck said of Einstein, 'That he may sometimes have missed the target in his speculations, as for example, in his hypothesis of light quanta, cannot really be held against him.' Of course, history continues to support both the ideas of Planck and Einstein, and these ideas form the foundation for the theoretical understanding of ESCA.

As an analytical method, a more straightforward history can be presented. In 1914, Robinson and Rawlinson studied photoemission from X-ray irradiated gold and, using photographic detection, observed the energy distribution of electrons produced. Although they were hampered by poor vacuum systems and inhomogeneous X-ray sources they were still able to publish a recognizable gold photoemission spectrum. In 1951, Steinhardt and Serfass first applied photoemission as a analytical tool. Throughout the 1950's and 1960's, Kai Siegbahn (son of the 1924 Nobel Prize winner, Manne Siegbahn) developed the instrumentation and theory of ESCA to give us the method we use today. Siegbahn also coined the term 'electron spectroscopy for chemical analysis,' later modified by his group to 'electron spectroscopy for chemical applications'. In 1981, Kai Siegbahn was rewarded for his contributions with the Nobel Prize in physics.

1.3 WHAT INFORMATION IS LEARNED FROM ESCA?

ESCA is an information-rich method (Table 3.1). The most basic ESCA analysis of a surface will provide qualitative and quantitative information on all the elements present (except H and He). More sophisticated application of the method yields much detailed information about the chemistry, organization, and morphology of a surface. Thus, ESCA can be considered one of the most powerful

Table 3.1. Information derived from an ESCA experiment

In the outermost 10 nm of a surface, ESCA can provide:

- Identification of all elements (except H and He) present at concentrations >0.1 atomic %
- Semiquantitative determination of the approximate elemental surface composition (error < ±10%)
- Information about the molecular environment (oxidation state, bonding atoms, etc.)
- Information about aromatic or unsaturated structures from shakeup ($\pi^* \rightarrow \pi$) transitions
- Identification of organic groups using derivatization reactions
- Non-destructive elemental depth profiles 10 nm into the sample and surface heterogeneity assessment using (1) angular-dependent ESCA studies and (2) photoelectrons with differing escape depths
- Destructive elemental depth profiles several hundred nanometers into the sample using ion etching (for inorganics)
- Lateral variations in surface composition (spatial resolution of 8–150 μm, depending upon the instrument)
- 'Fingerprinting' of materials using valence band spectra and identification of bonding orbitals
- Studies on hydrated (frozen) surfaces

analytical tools available. The capabilities of ESCA highlighted in Table 3.1 will be elaborated upon throughout this article.

2 X-RAY INTERACTION WITH MATTER, THE PHOTOELECTRON EFFECT, AND PHOTOEMISSION FROM SOLIDS

An understanding of the photoelectric effect and photoemission is essential in order to appreciate the surface analytical method, ESCA. When a photon impinges upon an atom, one of three events may occur: (1) the photon can pass through with no interaction; (2) the photon can be scattered by an atomic orbital electron leading to partial energy loss; and (3) the photon may interact with an atomic orbital electron with total transfer of the photon energy to the electron, leading to electron emission from the atom. In the first case, no interaction occurs and it is, therefore, not pertinent to this discussion. The second possibility is referred to as 'Compton scattering' and can be important in high-energy processes. The third process accurately describes the photoemission that is the basis of ESCA. Total transfer of the photon energy to the electron is the essential element of photoemission.

Let us examine four observations associated with this photoelectric effect in more detail. First, no electrons will be ejected from an atom regardless of the illumination intensity unless the frequency of excitation is greater than or equal to a threshold level characteristic for each element. Thus, if the frequency (energy) of the excitation photon is too low, no photoemission will be observed. As the energy of this photon is gradually increased, at some value, we will begin to observe the photoemission of electrons from the atom. Second, once the threshold frequency is exceeded, the number of electrons emitted will be proportional to the intensity of the illumination (i.e., once we've used photons of sufficient energy to stimulate electron emission, the more of those photons we bombard the sample with, the more photoelectrons will be produced). Third, the kinetic energy of the emitted electrons is linearly proportional to the frequency of the exciting photons — if we use photons of higher energy than our threshold value, the excess energy of the photons above the threshold value will be transmitted to the emitted electrons. Finally, the photoemission process from excitation to emission is extremely rapid (10^{-16} s). The basic physics of this process can be described by the Einstein equation, simply stated:

$$E_B = h\nu - \mathrm{KE} \tag{1}$$

where E_B is the binding energy of the electron in the atom (a function of the type of atom and its environment), $h\nu$ is the energy of the X-ray source (a known value), and KE is the kinetic energy of the emitted electron that is measured in the ESCA spectrometer. Thus, E_B, the quantity that provides us with valuable information about the photoemitting atom is easily obtained from $h\nu$ (known) and KE (measured). Binding energies are frequently expressed in electron volts (eV; 1 eV $= 1.6 \times 10^{-19}$ joules). More rigorous descriptions of the photoemission process can be found elsewhere [13, 14].

The concept of the binding energy of an electron in an atom requires elaboration. A negatively charged electron will be bound to the atom by the positively charged nucleus. The closer the electron is to the nucleus, the more tightly we can expect it to be bound. Binding energy will vary with the type of atom (i.e., a change in nuclear charge) and the addition of other atoms bound to that atom (bound atoms will alter the electron distribution on the atom of interest). Different isotopes of a given element have different numbers of neutrons in the nucleus, but the same nuclear charge. Changing the isotope will not appreciably affect the binding energy. Weak interactions between atoms such as those associated with crystallization or hydrogen bonding will not alter the electron distribution sufficiently to change the binding energy measurably. Therefore, the variations we see in the binding energy that provide us with the chemical information content of ESCA are associated with covalent or ionic bonds between atoms. These changes in binding energy are called binding energy shifts or chemical shifts and will be elaborated upon in Section 3.

For gases, the binding energy of an electron in a given orbital is identical to the ionization energy or first ionization potential of that electron. In solids, the influence of the surface is felt, and additional energy must be accounted for to remove an electron from the surface. This extra energy is called the work function and will be discussed in Section 3.

Irradiation of a solid by X-rays can also result in emission of Auger electrons (Figure 3.2). Auger electrons, discussed in detail in Chapter 4, differ in a number of respects from the photoelectrons with which this chapter primarily deals. A characteristic of Auger electrons is that their energy is independent of irradiation energy. Photoelectron energy is directly proportional to irradiation energy, according to equation (1).

Much of the basic physics of photoemission, and the background material for other surface analysis methods as well, is couched in the jargon of solid state physics. An excellent 'translation' of this jargon is now available [15].

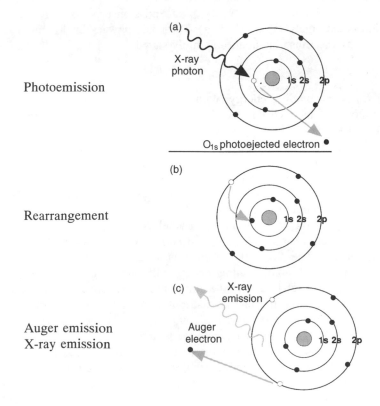

Figure 3.2. (a) The X-ray photon transfers its energy to a core-level electron leading to photoemission from the n-electron initial state. (b) The atom, now in an $(n-1)$-electron state, can reorganize by dropping an electron from a higher energy level to a vacant core hole. (c) Since the electron in (b) dropped to a lower energy state, the atom can rid itself of excess energy by ejecting an electron from a higher energy level. This ejected electron is referred to as an Auger electron. The atom can also shed energy by emitting an X-ray photon, a process called X-ray fluorescence

3 BINDING ENERGY AND THE CHEMICAL SHIFT

The general concept of the electron binding energy and its relationship to the energy of the incident X-ray and the emitted photoelectron was introduced in the previous section. This section will develop this relationship in more detail, with particular emphasis placed on describing the quantities that affect the E_B magnitude and how measurement of the E_B can be used to characterize materials.

3.1 KOOPMANS' THEOREM

The E_B of an emitted photoelectron is simply the energy difference between the $(n - 1)$-electron final state and the n-electron initial state (see Figure 3.2). This is written as:

$$E_B = E_f(n - 1) - E_i(n) \tag{2}$$

where $E_f(n - 1)$ is the final state energy and $E_i(n)$ is the initial state energy. If no rearrangement of other electrons in the atom or material occurred during the photoemission process, then the observed E_B would be just the negative orbital energy, $-\varepsilon_k$ for the ejected photoelectron. This approximation comes from Koopmans' theorem [16] and is written as:

$$E_B \approx -\varepsilon_k \tag{2}$$

The values of ε_k can be calculated using the Hartree–Fock method. These values are typically within 10–30 eV of the actual E_B values. The disagreement between E_B and $-\varepsilon_k$ is because Koopmans' theorem and the Hartree–Fock calculation method do not provide a complete accounting of the quantities that contribute to E_B. In particular, the assumption that other electrons remain 'frozen' during the photoemission process is not valid. During emission of the photoelectron, other electrons in the sample will respond to the creation of the core hole by rearranging to shield, or minimize, the energy of the ionized atom. The energy reduction caused by this rearrangement of electrons is called the 'relaxation energy.' Relaxation occurs for both electrons on the atom containing the core hole (atomic relaxation) and on surrounding atoms (extra-atomic relaxation). Relaxation is a final state effect and will be described in more detail later in this section. In addition to relaxation, quantities such as electron correlation and relativistic effects are neglected by the Koopmans/Hartree–Fock scheme. Thus, a more complete description of E_B is given by:

$$E_B = -\varepsilon_k - E_r(k) - \delta\varepsilon_{corr} - \delta\varepsilon_{rel} \tag{3}$$

where $E_r(k)$ is the relaxation energy and $\delta\varepsilon_{corr}$ and $\delta\varepsilon_{rel}$ are corrections for the differential correlation and relativistic energies. Both the correlation and relativistic terms are typically small and usually can be neglected.

3.2 *INITIAL STATE EFFECTS*

As shown in equation (2), both initial and final state effects contribute to the observed E_B. The initial state is just the ground state of the atom prior to the photoemission process. If the energy of the atom's initial state is changed, for example, by formation of chemical bonds with other atoms, then the E_B of electrons in that atom will change. The change in E_B, ΔE_B, is called the chemical shift. To a first approximation all core-level E_B's for an element will undergo the same chemical shift.

It is usually assumed that initial state effects are responsible for the observed chemical shifts, so that, as the formal oxidation state of an element increases, the E_B of photoelectrons ejected from that element will increase. This assumes that final state effects such as relaxation have similar magnitudes for different oxidation states. For most samples, the interpretation of ΔE_B solely in terms of initial state effects (equation (4)) frequency is adequate.

$$\Delta E_B = -\Delta \varepsilon_k \qquad (4)$$

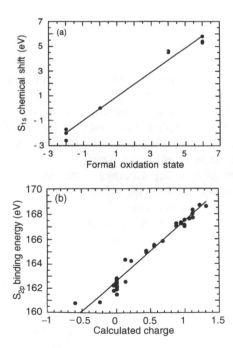

Figure 3.3. (a) The sulfur 1s chemical shifts, versus formal oxidation state for several inorganic sulfur species. (b) The sulfur 2p binding energy, versus calculated charge for several inorganic and organic sulfur species. Date taken from the results of Siegbahn *et al* [2]

Several examples showing a correlation between the initial state of an element and its ΔE_B are given in the classical work of Siegbahn and colleagues [2]. For example, Figure 3.3(a) shows their observation that E_B for the S_{1s} orbital increased by nearly 8 eV as the formal oxidation state of S increased from -2 (Na_2S) to $+6$ (Na_2SO_4). Tables 3.2 and 3.3 list typical C_{1s} and O_{1s} E_B values for functional groups present in polymers. The C_{1s} E_B is observed to increase monotonically as the number of oxygen atoms bonded to carbon increases

Table 3.2. Typical C_{1s} binding energies for organic samples*

Functional group		Binding energy (eV)
hydrocarbon	C−H,C−C	285.0
amine	C−N	286.0
alcohol, ether	C−O−H, C−O−C	286.5
Cl bound to carbon	C−Cl	286.5
F bound to carbon	C−F	287.8
carbonyl	C=O	288.0
amide	N−C=O	288.2
acid, ester	O−C=O	289.0
urea	N−C−N (O double bond)	289.0
carbamate	O−C−N (O double bond)	289.6
carbonate	O−C−O (O double bond)	290.3
2F bound to carbon	$-CH_2CF_2-$	290.6
carbon in PTFE	$-CF_2CF_2-$	292.0
3F bound to carbon	$-CF_3$	293−294

*The observed binding energies will depend on the specific environment where the functional groups are located. Most ranges are ±0.2 eV, but some (e.g., fluorocarbon samples) can be larger

Table 3.3. Typical O_{1s} binding energies for organic samples*

Functional group		Binding energy (eV)
carbonyl	C=O, O−C=O	532.2
alcohol, ether	C−O−H, C−O−C	532.8
ester	C−O−C=O	533.7

*The observed binding energies will depend on the specific environment where the functional groups are located. Most ranges are ± 0.2 eV.

(C–C < C–O < C=O < O–C=O < O–(C=O)O–). This is consistent with an initial state effect, since, as the number of oxygen atoms bonded to a carbon increase, the carbon should become more positively charged, resulting in an increase in the C_{1s} E_B.

A caution must be made against solely using initial state effects for interpreting chemical shifts. There are examples where final state effects can significantly alter the relationship between the formal oxidation state and ΔE_B. Also, it is the changes in the distribution and density of electrons of an atom resulting from changes in its chemical environment that contribute to ΔE_B. These quantities do not necessarily have a straightforward relationship to the formal oxidation state. For example, the full charge implied by the formal oxidation state is only attained when the chemical bonding is completely ionic (no covalent character). The degree of ionic/covalent character can vary with the chemical environment. Thus, it is best to correlate ΔE_B with the charge on the atom, not its formal oxidation state. Siegbahn and colleagues have shown this yields consistent correlations for both inorganic and organic sulfur species [2]. Figure 3.3(b) shows the linear relationship they observed for the S_{2p} E_B versus calculated charge on the sulfur atom.

One approach developed to provide a physical basis for chemical shifts is the charge potential model [2]. This model relates the observed E_B to a reference energy E_B°, the charge q_i on atom i, and the charge q_j on the surrounding atoms j at distances r_{ij}, as follows:

$$E_B = E_B^\circ + kq_i + \sum_{j \neq i}(q_j/r_{ij}) \tag{5}$$

with the constant k. Generally, the reference state is considered to be E_B for the neutral atom. It is then apparent that, as the positive charge on the atom increases by formation of chemical bonds, E_B will increase. The last term on the right-hand side of equation (5) is often called the Madelung potential because of its similarity to the lattice potential of a crystal $V_i = \sum q_j/r_{ij}$). This term represents the fact that the charge q_i removed or added by formation of a chemical bond is not displaced to infinity, but rather to the surrounding atoms. Thus, the second and third term on the right-hand side of equation (5) are of opposite sign. Using equation (5) the chemical shift between states 1 and 2 can now be written as:

$$\Delta E_B = k[q_i(2) - q_i(1)] + V_i(2) - V_i(1) \tag{6}$$

$$\Delta E_B = k\Delta q_i + \Delta V_i$$

where ΔV_i represents the potential change in the surrounding atoms.

3.3 FINAL STATE EFFECTS

As noted in Section 3.1, relaxation effects can have a significant impact on the measured E_B. In all cases the electron rearrangements that occur during photoemission result in the lowering of E_B. If the magnitude of the relaxation energy varies significantly as the chemical environment of an atom is changed, the E_B ranking that would be expected, based on initial state considerations can be altered. For example, the ranking of the Co $2p_{3/2}$ E_B values is Co^0 (778.2 eV) $<$ Co^{+3}(779.6 eV) $<$ Co^{2+}(780.5 eV) [17]. Also, both Cu^0 and Cu^{+1} have $2p_{3/2}$ E_B values of 932.5 eV ($\Delta E_B = 0$) [18]. Thus, for the Co and Cu systems, final state effects cause deviations in the E_B versus oxidation state ranking expected from initial state considerations.

Contributions to the relaxation energy arise from both the atom containing the core hole (atomic relaxation) and its surrounding atoms (extra-atomic relaxation). Most of the atomic relaxation component results from rearrangement of outer shell electrons, which have a smaller E_B than the emitted photoelectron. In contrast, the inner shell electrons (E_B larger than the emitted photoelectron) make only a small contribution to the atomic relaxation energy and can be neglected. The form of extra-atomic relaxation depends on the material being examined. For electrically conducting samples such as metals, valence band electrons can move from one atom to the next to screen the core hole. For ionically bonded solids such as the alkali halides, electrons are not free to move from one atom to the next. The electrons in these materials, however, can be polarized by the presence of a core hole. The magnitude of reduction in E_B produced by extra-atomic relaxation in ionic materials is smaller than the extra-atomic relaxation in metallic samples.

Other types of final state effects such as multiplet splitting and shakeup satellites can contribute to E_B. Multiplet splitting arises from interaction of the core hole with unpaired electrons in the outer shell orbitals. Shakeup satellites arise from the outgoing photoelectron losing part of its kinetic energy to excite a valence electron into an unoccupied orbital (e.g., $\pi \rightarrow \pi^*$ transition). These features and their presence in ESCA spectra are described in Section 6. Detailed discussion of final state effects is given elsewhere [19].

3.4 BINDING ENERGY REFERENCING

As is apparent from the preceding sections, accurate measurement of E_B can provide information about the electronic structure of a sample. As discussed in Section 2, E_B is determined by measuring the KE of the emitted photoelectron. In order to do this properly, a calibrated and suitably referenced ESCA spectrometer is required. The following paragraphs will describe how to set up an ESCA spectrometer to measure accurately the photoelectron KE (and therefore E_B for different types of samples.

Conducting samples such as metals are placed in electrical contact with the spectrometer, typically by grounding both the sample and the spectrometer. This puts the Fermi level (E_f), the highest occupied energy level, of both the sample and spectrometer, at the same energy level. Then the photoelectron KE can be measured as shown in Figure 3.4. As can be seen in this figure, the sum of the

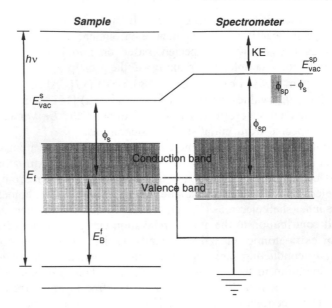

Figure 3.4. The energy level diagram for an electrically conducting sample that is grounded to the spectrometer. The Fermi levels of the sample and spectrometer are aligned ($E_f^s = E_f^{sp}$) so that E_B is referenced with respect to E_f. The measurement of E_B is independent of the sample work function, ϕ_s, but is dependent on the spectrometer work function, ϕ_{sp}

KE and E_B does not exactly equal the X-ray energy, as implied in the Einstein equation. The difference is the work function of the spectrometer (ϕ_{sp}). The work function, ϕ, is related to the E_f and vacuum level (E_{vac}) by:

$$\phi = E_f - E_{vac} \tag{7}$$

Thus, ϕ is the minimum energy required to eject an electron from the highest occupied level into vacuum. The Einstein equation now becomes:

$$E_B^f = h\nu - \text{KE} - \phi_{sp} \tag{8}$$

Therefore, both KE and ϕ_{sp} must be measured to determine E_B^f. The superscript f on E_B means that E_B is referenced to E_f. For conducting samples it is the work function of the spectrometer (ϕ_{sp}) that is important (see Figure 3.4). This can be calibrated by placing a clean Au standard in the spectrometer and adjusting the instrumental settings such that the known E_B values for Au are obtained (e.g., $E_f = 0$ eV, $4f_{7/2} = 83.98$ eV). The linearity of the E_B scale is then calibrated by adjusting the energy difference between two widely spaced lines of a sample (e.g., the 3s and $2p_{3/2}$ peaks of clean Cu) to their known values. The operator continues to iterate between the two calibration procedures until they converge to the accepted values. Further details of the calibration procedure have been described elsewhere [20–22]. Once the spectrometer energy scale has been calibrated, it is assumed to remain constant. This is valid as long as the spectrometer is maintained in an UHV environment. If the pressure of the spectrometer is raised above the UHV range, particularly when exposed to a reactive gas, different species can adsorb to components in the analyzer. This will change the ϕ_{sp} and necessitate recalibration.

3.5 CHARGE COMPENSATION IN INSULATORS

For measuring E_B, the procedure described in the previous section is the one of choice when the electrical conductivity of the sample is higher than the emitted current of photoelectrons. However, some materials do not have sufficient electrical conductivity or cannot be mounted in electrical contact with the ESCA spectrometer. These samples require an additional source of electrons to compensate for the positive charge built up by emission of the photoelectrons. Ideally, this is accomplished by flooding the sample with a monoenergetic source of low-energy (<20 eV) electrons. When the only source of compensating electrons is monoenergetic low-energy electrons, the vacuum level of the sample will be in electrical equilibrium with the energy of the electrons (Figure 3.5). Therefore, the measured E_B of an insulated sample depends on its work function

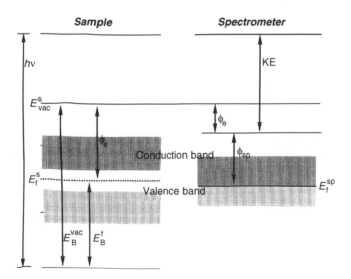

Figure 3.5. The energy level diagram for a sample electrically insulated from the spectrometer. The vacuum level of the sample (E^s_{vac} is aligned with the energy of the charge neutralization electrons (ϕ_s) so that E_B is referenced with respect to ϕ_e. The measurement of E_B is dependent on the sample work function, ϕ_s

(ϕ_s) and the energy of the flooding electrons, $\phi_{e'}$, as shown in equation (9) and Figure 3.5 [23].

$$E_B^{vac} = E_B^f + \phi_s = h\nu - KE + \phi_e \qquad (9)$$

Thus, for insulators E_B is referenced to E_{vac} and ϕ_e. This makes it difficult or impossible to measure absolute E_B values for samples not in electrical contact with the spectrometer. Under these conditions it is best to use an internal reference. For polymer and organic samples, the hydrocarbon component (C−C/C−H) of the C_{1s} peak is typically set to 285.0 eV. For supported catalysts, a major peak of the oxide support (Si_{2p}, Al_{2p}, etc.) is typically used. Internal referencing of the E_B scale allows the accurate measurement of other E_B values in the sample.

Usually the energy of the flooding electrons is varied to obtain the narrowest width of the photoemission peak. It is important to have the entire sample either electrically grounded or fully isolated. A sample in partial electrical contact with the spectrometer can lead to differential charging, which will produce distorted peak shapes and, under extreme conditions, new peaks. These experimental artifacts must be avoided to obtain a proper analysis of the sample. The analyst must be aware of the electrical properties of the sample and how they can affect the ESCA experiment. For example, a conducting metal substrate with a thin (~5 nm or less) insulating overlayer can usually be analyzed with the sample grounded. However, if the insulating overlayer becomes too thick (~10 nm or more), differential charging can occur. Then the entire sample must be electrically isolated from the spectrometer for proper analysis.

3.6 PEAK WIDTHS

The observed width of a given photoelectron peak is determined by the lifetime of the core hole, instrumental resolution, and satellite features. The peak width due to the core hole lifetime can be calculated from Heisenberg's uncertainty relationship:

$$\Gamma = h/\tau \qquad (10)$$

where Γ is the intrinsic peak width in eV, h is Planck's constant in eV-seconds, and τ is the core hole lifetime in seconds. For the C_{1s} orbital, Γ is ~0.1 eV. For a given element, the value of Γ is typically larger for inner shell orbitals versus outer shell orbitals. This is because an inner shell core hole can be filled by electrons from the outer shells. Thus, the deeper the orbital, the shorter the core hole lifetime and the larger the intrinsic peakwidth. For example, the intrinsic peakwidths of Au increase in the order 4f < 4d < 4p < 4s, the order of increasing E_B. Similarly, the value of Γ for a given orbital (e.g., 1s) increases as the atomic number of the element increases, since the valence electron density, and therefore the probability of filling the core hole, increases with increasing atomic number. The lineshape due to the core hole lifetime is Lorenzian.

Instrumental effects can broaden the width of a photoemission peak by the energy spread of the incident X-rays and the resolution of the analyzer. For insulating materials, additional broadening of the peak can occur from the energy spread of the flooding electrons and the resulting energy spread in the surface potential [24]. Typically it is assumed that instrumental contributions to the photoemission peak have a Gaussian lineshape. The contributions that the intrinsic and instrumental effects make to the peakwidth are given, to a first approximation, by:

$$\text{FWHM}_{\text{tot}} = (\text{FWHM}_n^2 + \text{FWHM}_x^2 + \text{FWHM}_a^2 + \text{FWHM}_{\text{ch}}^2 + \cdots)^{1/2} \qquad (11)$$

where FWHM is the full-width at half-maximum of the observed peak (tot), core hole lifetime (n), X-ray source (x), analyzer (a), and charging contribution (ch).

The third contribution to peakwidths is satellite features. These can arise from several sources such as vibrational broadening, multiplet splitting, and shakeup satellites. These features typically have asymmetric lineshapes and, depending on their E_B, may or may not be resolvable from the main photoemission peak. For example, metallic samples have a continuous band of unfilled electron levels above E_f (the conduction band). Upon leaving the sample, a photoelectron can transfer a portion of its KE to excite a valence band electron into the conduction band. Because of the continuous ranges of energies available to this process, an asymmetric tail on the high E_B (low KE) side of the photoemission peak is observed for metallic samples. The degree of peak asymmetry depends on the density of states near E_f [21]. Additional discussion of satellite features will be given in Section 6. Also, further details have been published elsewhere [25].

3.7 PEAK FITTING

To maximize the information extracted from ESCA spectra, the area and E_B of each subpeak for a given orbital (e.g., C_{1s}) must be determined. Typically, the spacing between subpeaks is similar to observed peakwidths ($\sim$1 eV). Thus, it is rare when individual subpeaks are completely separated in an experimental spectrum. This requires the use of a peak-fitting procedure to resolve the desired peak parameters. Quantities used in such procedures include the background, peak shape (Gaussian, Lorenzian, asymmetric, or mixtures thereof), peak position, peak height, and peak width. A computer is usually used for peak fitting or resolution. The most common form used to model the background (inelastic scattering) was developed by Shirley [26]. After the background has been determined, initial guesses are made for each peak parameter and then a least-squares fitting routine is used to iterate to the final values. Caution must be exercised when performing the peak fit since many of the quantities are correlated. This can cause instabilities in the fitting algorithm or non-unique results to be obtained. For spectra containing severely overlapping peaks, the results obtained from peak fitting may depend on the starting parameters chosen (the algorithm converges to a local minimum instead of a global minimum). The experimenter must ensure that the results obtained from the peak fitting procedure agree with all other available information, as described in Section 6. It is best to start with accurate initial peak parameters. Also, additional independent information can be used to constrain peak parameters such as position, width, and area during the initial iterations of the fitting. Once the algorithm is close to convergence, these constraints can be removed. For example, appropriate standards can be used to set peak positions. If the instrumental resolution dominates the peak width, then a 100% Gaussian peak shape can be used. Under these conditions all peaks should have similar widths. As the instrumental resolution is improved, mixtures of Gaussian–Lorenzian-asymmetric tailing must be used. For non-conducting polymer samples, when the C_{1s} peak width is < 1eV, Gaussian–Lorenzian mixtures are required. For narrow peaks of metallic samples, some asymmetric tailing should also be included. The peak fitting results obtained for the polyurethane C_{1s} spectrum are discussed in Section 6. This example shows how the concentration and E_B values for different functional groups determined by peak fitting correlate with information from the survey scan and other high-resolution scans. Some additional ideas on peak fitting are presented in Section 13.1.

4 INELASTIC MEAN FREE PATH AND SAMPLING DEPTH

As illustrated schematically in Figure 3.6, while X-rays can readily travel through solids, electrons exhibit significantly less ability to do so. In fact, for X-rays of 1 KeV (a typical order of magnitude for an ESCA excitation source), the X-rays will penetrate 1000 nm or more into matter while electrons of this energy will only penetrate approximately 10 nm. Because of this difference, ESCA, in which only emitted electrons are measured, is surface sensitive. Electrons emitted from X-ray excitation below the uppermost surface zone cannot penetrate far enough to escape from the sample and reach the detector. Let us examine these relationships more quantitatively.

In the ESCA experiment, we are concerned with only the intensity of the emitted photoelectrons (i.e., the total number emitted) that have lost no energy. If an electron suffers energy loss, but still has sufficient energy to escape from the surface, it will contribute to the background signal, but not to the photoemission peak (Figure 3.6). Therefore, the ESCA sampling depth into a solid refers to a characteristic, average length over which the electron can travel with no loss of energy. The decrease in the number of photoemitted electrons that have suffered no energy loss travelling through matter, where each unit thickness of matter through which the electrons travel will absorb the same fraction of the energy, is described by Beer's law (Figure 3.7(a)). The inelastic mean free path (IMFP) term, λ, in this equation is that thickness of matter through which 63% of the traversing electrons will lose energy. Table 3.4 presents a series of definitions of other terms commonly used in ESCA to describe this decrease in elastic electron intensity associated with transport through matter.

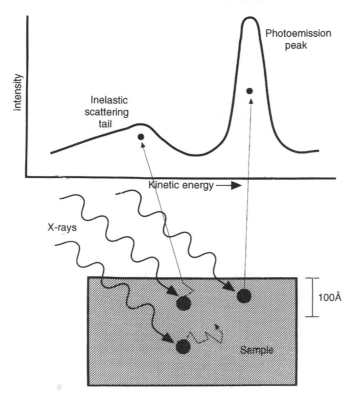

Figure 3.6. X-rays will penetrate deeply into a sample, and stimulate electron emisson throughout the specimen. Only those electrons emitted from the surface zone that have suffered no energy loss will contribute to the photoemission peak. Electrons emitted from the surface zone that have lost some energy due to inelastic interactions will contribute to the scattering background

Table 3.4. Definitions for electron transport in materials*

IMFP (λ) — Inelastic Mean Free Path (ASTM definition). The average distance (in nanometers) that an electron with a given energy travels between successive inelastic collisions.

ED — Escape Depth (ASTM definition). The distance (in nanometers) normal to the surface at which the probability of an electron escaping without significant energy loss due to inelastic processes drops to e^{-1} (38%) of its original value.

AT — Attenuation Length. The average distance (in nanometers) that an electron with a given energy travels between successive inelastic collisions as derived from a particular model in which elastic electron scattering is assumed to be negligible.

ID — Information Depth. The average distance (in nanometers) normal to the surface from which a specified percentage of the detected electrons originates.
Sampling Depth = 3λ (ID where percentage of detected electrons is 95%)

*Definitions from reference [27]

(a) Transmission

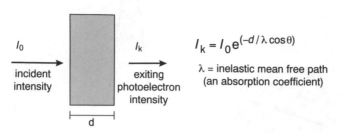

$$I_k = I_0 e^{(-d/\lambda \cos\theta)}$$

I_0 incident intensity

I_k exiting photoelectron intensity

λ = inelastic mean free path (an absorption coefficient)

d

(b) Emission

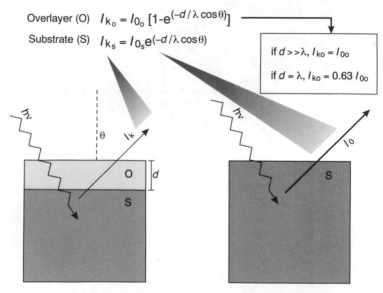

Overlayer (O) $\quad I_{k_o} = I_{0_o}[1-e^{(-d/\lambda \cos\theta)}]$

Substrate (S) $\quad I_{k_s} = I_{0_s}e^{(-d/\lambda \cos\theta)}$

if $d >> \lambda$, $I_{ko} = I_{0o}$

if $d = \lambda$, $I_{ko} = 0.63\, I_{0o}$

Where $d = 3\lambda$, 95% of the signal comes from d

Figure 3.7. (a) For electrons transmitted through a sample, Beer's law of molecular absorption explains the total intensity loss for electrons that lose no energy in traversing the sample. (b) For electron emission from a thick sample, modifications of Beer's law can explain the photoemission intensity from an overlayer, or from the substrate covered by an overlayer

The Beer's law equation, as formulated in Figure 3.7(a), applies to transmission of electrons through a specimen of thickness, d. In ESCA, we generally detect electrons escaping from a solid that is many times thicker than the escape depth of the electrons (Figure 3.6). Since, over the thickness range through which the electron flux is appreciably attenuated, the X-ray flux suffers essentially no diminution, the X-rays can be viewed as stimulating photoemission throughout the bulk of the sample. View the sample, then, as a source of electrons (I_0). We

may now ask, how will the electron flux from this source (the sample) be attenuated, if we cap this electron source with a thin overlayer? This situation, and
the equations that describe it, are presented in Figure 3.7(b). The equations in
Figure 3.7(b) are useful for qualitatively and quantitatively describing the photoelectron emission intensity for many commonly encountered sample types. These
equations will be applied to depth profiling in Section 9.

The actual values for the IMFP of electrons in matter are a function of the
energy of the electrons and nature of the sample through which they travel.
Over the range of electron kinetic energies of most interest in ESCA, the IMFP
increases with electron KE. The form of the dependence of IMFP on KE is
described by KE^n where n has been estimated at 0.54–0.81 (0.7 is frequently
used) [27]. Equations that relate IMFP to electron energy and the type of material through which the electron is traversing have been developed by Seah and
Dench: [28]

$$\text{IMFP} = \lambda = 538KE^{-2} + 0.41(aKE)^{0.5} \quad \text{(for elements)} \tag{12}$$

$$\text{IMFP} = \lambda = 2170KE^{-2} + 0.72(aKE)^{0.5} \quad \text{(for inorganic compounds)} \tag{13}$$

$$\text{IMFP} = \lambda_d = 49KE^{-2} + 0.11KE^{0.5} \quad \text{(for organic compounds)} \tag{14}$$

where: λ is in units of monolayers

   a = monolayer thickness (nm)

   λ_d (in mg m^{-2})

   KE = electron kinetic energy (in eV)

The equations were developed empirically based upon data from a large number
of researchers. As a good fit to experimental data, they provide useful guide-
lines. However, precise values of IMFP in materials have been the subject of
considerable controversy, with little resolution of this issue to date [27, 29–33].
It is reasonable to say that the IMFP for photoelectrons in the range of interest in
ESCA probably falls in the range 1–4 nm. Numbers calculated from the equations
above can be used in most calculations if it is appreciated that, although precise
values cannot be assigned, a reasonable estimate of sampling depths can be made.
The sampling depth, as defined in Table 3.4, is three times the IMFP (i.e., the
depth from which 95% of the photoemission has taken place).

5 QUANTITATION

As discussed in previous sections, the complete ESCA spectrum of a material
contains peaks that can be associated with the various elements (except H and
He) present in the outer 10 nm of that material. The area under these peaks is
related to the amount of each element present. So, by measuring the peak areas
and correcting them for the appropriate instrumental factors, the percentage of

each element detected can be determined. The equation that is commonly used for these calculations is:

$$I_{ij} = KT(\text{KE})L_{ij}(\gamma)\sigma_{ij} \int n_i(z)e^{-z/\lambda(\text{KE})\cos\theta} \, dz \tag{15}$$

where I_{ij} is the area of peak j from element i, K is an instrumental constant, T (KE) is the transmission function of the analyzer, $L_{ij}(\gamma)$ is the angular asymmetry factor for orbital j of element i, σ_{ij} is the photoionization cross-section of peak j from element i, $n_i(z)$ is the concentration of element i at a distance z below the surface, λ (KE) is the inelastic mean free path length, and θ is the take-off angle of the photoelectrons measured with respect to the surface normal. Equation (15) assumes that the sample is amorphous. If the sample is a single crystal, diffraction of the outgoing photoelectrons (see Section 13.3) can cause peak intensities to deviate from values predicted by equation (15). By using a 20–30° solid angle acceptance lens and either amorphous or polycrystalline samples, these diffraction effects can be neglected. Rarely are all of the quantities in equation (15) evaluated. Typically, either elemental ratios (e.g., C/O atomic ratio) or percentages (e.g., atomic percent carbon) are calculated. Thus, it is only necessary to determine the relative relationship, not the absolute values, of the quantities in equation (15).

The instrumental constant, K, contains quantities such as the X-ray flux, area of the sample irradiated, and the solid angle of photoelectrons accepted by the analyzer. It is assumed not to vary over the time period and conditions used to acquire the ESCA spectra for quantitation. Being a constant, it cancels when either elemental ratios or atomic percentages are calculated. The angular asymmetry factor $L_{ij}(\gamma)$ accounts for the type of orbital the photoelectron is emitted from and the angle γ between the incident X-rays and the emitted photoelectrons. The value of $L_{ij}(\gamma)$ for a particular peak can be calculated [34]. If only s orbitals are used for quantitation, $L_{ij}(\gamma)$ will be the same for all peaks and therefore cancel. This situation is frequently encountered with polymeric samples since the 1s orbitals of many elements present in organic polymers (C, N, O, and F) are detectable by ESCA. Even for samples where different types of orbitals are used for quantitation, the variation of $L_{ij}(\gamma)$ is typically small and is usually neglected for solids. However, it is always best to use orbitals of the same symmetry for calculating elemental ratios or atomic percentages.

The transmission function of the analyzer includes the efficiency of the collection lens, the energy analyzer, and detector. Most ESCA instruments are run in the constant-pass energy mode. This means that regardless of the initial KE of the emitted electrons, they will pass through the energy analyzer and strike the detector at a constant energy. This requires the collection lens to reduce the KE of the incoming electrons down to the pass energy. In this case, the only variation in the transmission function with KE of the photoelectrons is due to retardation in the lens system, which can be determined experimentally and usually has the form of KE^n. Most manufactures provide information about the transmission function of their instruments. Published data are also available for many instruments [35, 36].

The photoionization cross-section σ_{ij} is the probability that the incident X-ray will create a photoelectron from the jth orbital of element i. Values for σ_{ij} are typically taken from the calculations of Scofield [37]. Selected values of the Scofield cross-sections are listed in Table 3.5. Empirically determined cross sections are also available [38]. The IMFP, λ (KE), has been discussed in Section 4. For quantitative analysis, the values calculated from the equations published by Seah and Dench [28] (equations (12) (13) and (14)) are commonly used. These equations show that λ depends both on the sample type (elemental, inorganic species, or organic species) and the KE of the photoelectron. Both quantities must be properly accounted for to obtain good quantitative results. The $\cos \theta$ term accounts for the decrease in sampling depth as the surface normal of the sample is rotated away from the axis of the acceptance lens. This is described in detail in Sections 4 and 9 and Figure 3.17.

5.1 QUANTITATION METHODS

The concentration of element i, n_i, is the unknown quantity in equation (15). All other terms in this equation can either be measured (e.g, I_{ij}) or calculated (e.g., σ_{ij}). Therefore, equation (15) can be solved for n_i. Once n_i is known for each element present in the ESCA spectrum, the atomic percentages can be calculated as:

$$\%n_i = 100(n_i / \sum n_i) \qquad (16)$$

where $\%n_i$ is the atomic percent of element i. Atomic ratios (n_i/n_k) can also be calculated. To remove the integral from equation (15) it is usually assumed that the elemental concentrations are homogeneous within the ESCA sampling depth. Equation (15) can then be integrated to obtain:

$$I_{ij} = KT(KE)L_{ij}(\gamma)\sigma_{ij}n_i\lambda(KE)\cos\theta \qquad (17)$$

When n_i is not homogeneous with respect to z, the depth profiling experiments described in Section 9 are required to determine the form of $n_i(z)$.

5.2 QUANTITATION STANDARDS

Standard samples can be used to evaluate the validity of the quantitation equations presented above. There are four criteria for standard samples, they should: have a known composition, be homogeneous with depth, be relatively stable, and be free of contaminants. Two polymer samples that meet these criteria are poly(tetrafluoroethylene) (PTFE) and poly(ethylene glycol) (PEG).

PTFE is composed exclusively of CF_2 chains giving a F/C atomic ratio of 2.0. This polymer is known to be unreactive, so it is relatively easy to prepare

Table 3.5. Select photoelectron binding energies (eV) and-Scofield photoemission cross sections [37]*

Element	1s	2s	2p1/2	2p3/2	3s	3p1/2	3p3/2	3d3/2	3d5/2	4s	4p1/2	4p3/2	4d3/2	4d5/2	4f5/2	4f7/2
C	284 [1.00]															
N	399 [1.80]															
O	532 [2.93]															
F	686 [4.43]															
Al		118 [0.753]	73 [0.181]	73 [0.356]												
Si		149 [0.955]	100 [0.276]	99 [0.541]												
P		189 [1.18]	136 [0.430]	135 [0.789]	16 [0.112]											
S		229 [1.43]	165 [0.567]	164 [1.11]	16 [0.147]											
Ti		564 [3.24]	461 [2.69]	455 [5.22]	59 [0.473]	34 [0.276]	34 [0.537]									
Cu		1096 [5.46]	951 [8.66]	932 [16.73]	120 [0.957]	74 [0.848]	74 [1.63]									
Ag					717 [2.93]	602 [4.03]	571 [8.06]	373 [7.38]	367 [10.66]	95 [0.644]	62 [0.700]	56 [1.36]				
I					1072 [3.53]	931 [5.06]	875 [10.62]	631 [13.77]	620 [19.87]	186 [0.959]	123 [1.11]	123 [2.23]	50 [1.69]	50 [2.44]		
Au										759 [1.92]	644 [2.14]	546 [5.89]	352 [8.06]	334 [11.74]	87 [7.54]	84 [9.58]

*Cross-sections listed in brackets.

a clean PTFE surface. The presence of any oxidation or contamination can be readily determined by examination of the ESCA spectra. Since only F and C are present in PTFE, the detection of O by ESCA indicates the presence of surface oxidation or a surface contaminant. Likewise, any C_{1s} peaks other than the CF_2 peak at 292 eV indicate the presence of a surface contaminant. The most common contaminant is adsorbed hydrocarbon, which has a C_{1s} peak at 285 eV (Figure 3.8). Even small amounts of hydrocarbon contamination can be readily detected. The amount shown in Figure 3.8 represents ~0.3 atomic percent carbon. If the PTFE sample shows significant hydrocarbon contamination, sonicating the PTFE in methanol followed by acetone usually removes the contaminant. The results in Table 3.6 for cleaned PTFE samples were obtained using equation (17). The excellent agreement between the experimental values and stoichiometric values support the use of equations listed above for quantitation.

Like PTFE, PEG is also a good candidate for a standard material. The formula for PEG is $HO-(CH_2-CH_2-O)_n-H$ so only C and O should be present with a

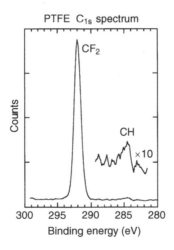

PTFE C_{1s} spectrum

Figure 3.8. The ESCA C_{1s} spectrum of poly(tetrafluoroethylene). The peak at 292 eV corresponds to the CF_2 groups present in this sample. The weak peak at 285 eV corresponds to a small amount (~0.3 atomic percent) of a hydrocarbon contaminant present on the surface of the sample

Table 3.6. Quantitation results for poly(tetrafluoroethylene)

Atomic percent F = 67.1 ± 0.4
Atomic percent C = 32.9 ± 0.5
F/C atomic ratio = 2.04 ± 0.04

(Number of samples = 22)

Table 3.7. Quantitation results
for poly(ethylene glycol)

Atomic percent O $= 33.8 \pm 0.4$
Atomic percent C $= 66.2 \pm 0.4$
C/O atomic ratio $= 1.96 \pm 0.03$

(Number of samples $= 12$)

C/O atomic ratio of 2.0. The PEG C_{1s} spectrum should only have one peak at 286.5 eV (C–O species), so the presence of a hydrocarbon contaminant ($E_B = 285$ eV) can readily be detected. The results for PEG listed in Table 3.7, like the PTFE results, show excellent agreement between experiment and stoichiometry. This provides further support for the use of equation (17) for quantitation experiments.

5.3 QUANTITATION EXAMPLE

The results for a polyurethane sample presented in Tables 3.8 and 3.9 provide an additional example of the accuracy of quantitative analysis of polymeric materials. Further information about the structure and identification of ESCA spectral features of this polyurethane are provided in the next section. In the quantitation experiments only C, N, and O were detected, and the 1s peak areas were used for quantitation. Therefore, $L_{ij}(\gamma)$ can be considered a constant. The spectra were acquired on a Surface Science Instruments X-probe spectrometer with the following characteristics: T(KE) is constant, λ (KE) varies as $KE^{0.7}$, $hv = 1487$ eV, and $\theta = 55°$. Under these conditions equations (16) and (17) can be combined to yield:

$$\%n_i = (I_{ij}/\sigma_{ij}KE^{0.7})/\sum(I_{ij}/\sigma_{ij}KE^{0.7}) \qquad (18)$$

Table 3.8 shows the values of I_{ij} KE, σ_{ij}, and $\%n_i$ for analysis of one polyurethane sample. Table 3.9 summarizes the results from eight different analyses of this material with the calculated standard deviations. The composition expected from the polyurethane stoichiometry is also listed. The results show both good reproducibility and accuracy. As the atomic percentage of the element decreases towards the ESCA detection limits (~ 0.1 atomic percent), the relative standard deviation will increase significantly. Near the detection limit the

Table 3.8. Quantitation results for a polyurethane sample

Element	Orbital	KE (eV)	σ	Peak area	Atomic percent
carbon	1s	1200	1.00	26557	76.9
nitrogen	1s	1085	1.80	4478	7.7
oxygen	1s	955	2.93	13222	15.4

Table 3.9. Quantitation results for a polyurethane sample

Element	Atomic percent	
	ESCA	Stoichiometry
carbon	76.6 ± 1.0	76.0
nitrogen	7.9 ± 0.5	8.0
oxygen	15.5 ± 0.8	16.0

(Number of samples = 8)

magnitude of the standard deviation is typically the same as the magnitude of the atomic percentage. Based upon the results in Table 3.9, the polyurethane sample has similar surface and bulk compositions. This is not always the case, and examples of variation in the surface composition with respect to the bulk composition are shown in Section 9. In addition to the examples in Section 9, the presence of contaminants is also often detected at the surface of samples. Oxidation of the sample and adsorption of hydrocarbons and silicones are common contamination processes.

6 SPECTRAL FEATURES

The understanding and analysis of ESCA spectra require an appreciation of the spectral features that are observed. ESCA analyses are typically performed by first taking a wide scan or survey scan spectrum, often covering a range of 1000 eV, and then looking in more detail over smaller ranges (perhaps 20 eV) at specific features found in the wide scan spectrum. A characteristic wide scan spectrum, energy referenced to compensate for sample charging as described in Section 3, is presented in Figure 3.9. High-resolution spectra of specific features observed in the wide scan spectrum are shown in Figure 3.10. First, let us consider the wide scan spectrum.

The wide scan spectrum of a synthetic polymer, a polyurethane (Figure 3.9), has been annotated specifically for this example. The chemical structure of this polymer is also contained within this figure. First, note the x-axis. This axis is generally labeled 'binding energy.' However, from the Einstein equation, it is apparent that we could also plot it in terms of KE. As discussed in Section 3, the KE of the emitted photoelectron is a precisely measured value. The binding energy is a calculated value computed from the KE, the energy of the X-ray photon, the work function of the surface, and a correction term due to electrical charge accumulation on the surface. Still, when the ESCA instrument is in proper calibration, there is an inverse, linear relationship between KE and binding energy. Since binding energy has meaning for the chemistry and structure of the surface, it is most common to plot ESCA spectra in terms of binding energy. The y axis is typically 'intensity' or 'number of counts.' For the presentation of ESCA data, it is usually linear rather than logarithmic.

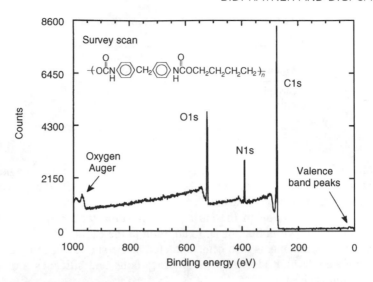

Figure 3.9. The ESCA survey scan of a hard-segment polyurethane

Next, we can observe the background. The number of counts attributed to the background increases with increasing binding energy (decreasing kinetic energy). This is the inelastic scattering, as suggested in Figure 3.6. After each photoemission event, there is a cumulative background signal associated with photoelectrons that have lost energy due to inelastic collisions in the solid, but that still have sufficient energy to escape the work function of the surface. There is a continuum of energies of such electrons ranging between the photoemission KE and zero KE since the collision events reducing the KE of the photoelectron do not have discrete energies.

Rising prominently above the background signal we observe two types of peaks in Figure 3.9. There are photoemission peaks associated with core-level photoionization events and X-ray-induced Auger electron emission peaks. If binding energy referencing has been performed, peaks can be readily identified from their positions using tabulations of binding energy values [7, 39]. Where ambiguity exists as to the identify of a peak, it is useful to look for other photoemission lines from the same element. For example, iridium (irradiated by an aluminum anode X-ray source) should have reasonably strong emissions at 690 eV (4s), 577eV ($4p_{1/2}$), 495 eV ($4p_{3/2}$), 312 eV ($4d_{3/2}$), 295 eV $4d_{5/2}$), 63 eV ($4f_{5/2}$), and 60 eV ($4f_{7/2}$), with the latter five lines particularly intense. If all members of this series of lines (and especially the most intense of the set) are not observed in a spectrum, then iridium is probably not present. Table 3.5 contains the binding energies of a few select photoemission lines produced with Al Kα irradiation (1487 eV).

Auger lines are usually also listed in photoelectron peak tabulations. Auger lines can be readily distinguished from photoemission lines by changing X-ray

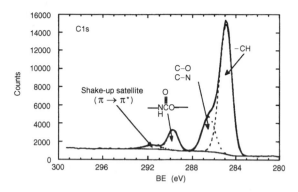

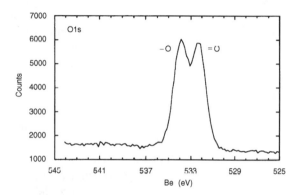

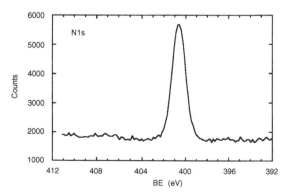

Figure 3.10. (a) The C_{1s} spectrum (resolved into component peaks) for the hard-segment polyurethane; (b) the O_{1s} spectrum for the hard-segment polyurethane; (c) the N_{1s} spectrum for the hard-segment polyurethane

sources (e.g., using a Mg Kα source instead of an Al Kα source). The kinetic energies of all the Auger lines will remain the same, while the kinetic energies of the photoemission lines shift by the difference in energies of the two X-ray sources. Auger peaks can be used analytically, in conjunction with the photoemission peaks, to distinguish between different possible chemical species using the modified Auger parameter [40]:

$$\alpha' = h\nu_{X-ray} + KE_{Auger} - KE_{photoelectron} \qquad (19)$$

The final feature to be discussed in this wide scan spectrum is observed at low binding energies. The low-intensity features seen between 0 and 30 eV are due to photoemission of valence (outer shell) electrons. Interpretation of these spectral features is often more complex than the core-level lines, and has been presented elsewhere [41].

Much additional detail can be observed in the high-resolution ESCA spectra. Consider Figure 3.10(a), the high-resolution C_{1s} spectrum from the polyurethane sample used to generate Figure 3.9. From the peak shape, it is apparent that this spectrum is composed of a number of subpeaks. These subpeaks, attributed to chemical shifts from atoms and groups bound to the carbons (see Section 3), are identified in the figure. Methods and the rationale for resolving a peak envelope into subpeaks have been described [8, 21].

As well as peaks for each of the major types of carbon species, another feature is noted at 6.6 eV from the lowest binding energy (hydrocarbon) peak. This peak is referred to as a shake-up satellite. It represents photoelectrons that have lost energy through promotion of valence electrons from an occupied energy level (e.g., a π level) to an unoccupied higher level (e.g., a π^* level). Shake-up peaks (also called 'loss peaks' because intensity is lost from the primary photoemission peak) are most apparent for systems with aromatic structures, unsaturated bonds or transition metal ions. In contrast to the continuum of reduced energies seen in the inelastic scattering tail, shakeup peaks have discrete energies (6.6 eV higher binding energy than the primary peak in C_{1s} spectra of aromatic-containing molecules) because the energy loss is equivalent to a specific quantitized energy transition (i.e., the $\pi \rightarrow \pi^*$ transition). If the departing photoelectron transfers sufficient energy into the valence electron to ionize it into the continuum, the photoemission loss peak is called a 'shake-off' peak. The shakeoff satellite peaks of the photoemission peak can have a wide range of possible energies (of course, always of lower KE than the photoemission peak). This energetically broad feature is typically hidden within the background signal and is usually not detected or used analytically.

Much additional information about the polymer is available by examining the other high-resolution spectra (Figures 3.10(b) and (c)). Specific interesting features are annotated on the spectra. Quantitative information comes from the ratio of the areas under the peaks in Figures 3.10(a), (b), and (c).

As a general note, ESCA is best used by taking advantage of all the information available in the spectra. Thus, the analysis does not end with the acquisition of the wide scan spectrum. High-resolution spectra of each of the features found in the wide scan spectrum are examined to extract maximum information. The information from a complete data set should be corroborative and not contradictory. For example, where significant levels of oxygen are seen in a wide scan spectrum of an organic polymer, subpeaks associated with carbons bound to oxygen should be found in the high-resolution C_{1s} spectrum. Furthermore, the subpeaks in the O_{1s} spectrum should also have binding energies appropriate for oxygen–carbon functionalities. When contradictions are found, further analysis and perhaps re-acquisition of the data are in order. The theoretical and experimental understanding of ESCA spectra are reasonably well developed, and contradictory evidence should have a sound explanation. One explanation that must always be considered is spectral artifact, often due to surface charging. Thus, exercising more care in charge compensation during reacquisition of the data can often resolve such problems.

In inorganic systems a number of other spectral features are observed that must be understood. These include spin–orbit doublets, multiplet splitting, and plasmon losses. Each will be described.

In Figure 3.11, the initial state and final states (after photoemission) for a pair of electrons in a 3p orbital are schematically illustrated. Note that two energetically equivalent final states are possible, 'spin up' or 'spin down.' If there is an open shell (quantum number $1 > 0$, i.e., a p, d, or f orbital) with two states of the same energy (orbital degeneracy), a magnetic interaction between the spin of the electron (up or down) and its orbital angular momentum may lead to a splitting of the degenerate state into two components. This is called spin–orbit coupling or j–j coupling (j quantum number $= 1 + s$). Figure 3.11 also shows common spin–orbit pairs. The ratio of their respective degeneracies, $2j + 1$, determines the intensities of the components. Figure 3.12 illustrates the $f_{5/2}$ and $f_{7/2}$ components of a gold 4f photoemission peak. The total 4f photoemission intensity for gold as used in quantitation is the sum of the two spin–orbit peaks. The trend for the doublet separation is p > d > f within a given atom.

A related phenomenon, referred to as multiplet or electrostatic splitting, is seen for the s orbital photoemission from some transition metal ions (e.g., Mn^{+2}, Cr^{+2}, Cr^{+3}, Fe^{+3}). A requirement for this splitting of the s photoemission peak into a doublet is that there be unpaired orbitals in the valence shells. Complex peak splittings can be observed in transition metal ions and rare earth ions when multiplet splitting occurs in p and d levels. Additional detail on this splitting can be found elsewhere [19, 21].

The conduction electrons in metals have been likened to a 'sea' or continuum. Characteristic collective vibrations have been noted for this continuum of electrons and are referred to as plasmon vibrations. In some cases, the exiting photoelectron can couple with the plasmon vibrations leading to characteristic,

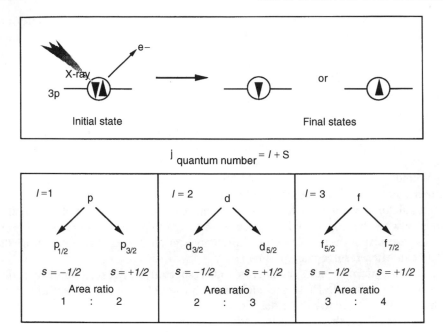

Figure 3.11. After electron from a 3p orbital subshell, the remaining electron can have a spin-up or spin-down state. Magnetic interaction between these electrons and the orbital angular momentum may lead to spin–orbit coupling

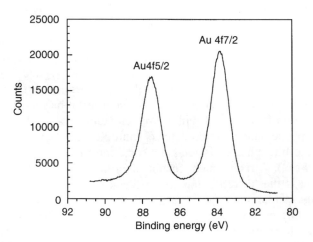

Figure 3.12. Spin–orbit coupling leads to a splitting of the 4f photoemission of gold into two subpeaks

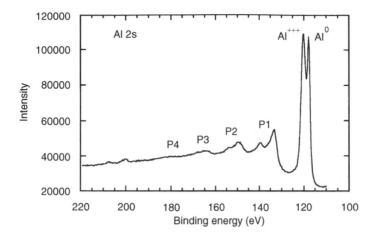

Figure 3.13. Plasmon loss peaks for the aluminum 2s photoemission peak

periodic energy losses. An example of the plasmon loss series of peaks for the aluminum 2s photoemission is presented in Figure 3.13.

The features most often observed in ESCA spectra, independent of the type of instrument used, have been briefly reviewed. A feature that is associated with the type of instrument is the X-ray satellite. Non-monochromatized X-ray sources (see Section 7) may excite the sample with more than one X-ray line. From equation (1), each X-ray line will lead to a distinct photoemission energy. Low-intensity X-ray lines, particularly the $K_{\alpha3,4}$, will produce low-intensity

Table 3.10. Features observed in ESCA spectra

---

1. Photoemission peaks
    - Intense
    - Narrow
    - Nearly symmetric
    - Shifted by chemistry
    - Contain vibrational fine structure
2. X-ray satellite peaks
    - Not observed with a monochromatized source
    - Always the same energy shift from the photoemission peak
*3. Shake-up satellites and shake-off satellites
4. Photon-enduced Auger lines
*5. Inelastic scattering background
6. Valence band features
7. Spin–orbit coupling
8. Multiplet splitting
*9. Plasmon loss peaks

---

*Loss process.

photoemission peaks with approximately 10 eV higher KE than the primary
photoemission peak. These X-ray satellite peaks are not observed with
monochromatic X-ray sources that are described in Section 7.

Table 3.10 summarizes all features that are important both to understand the
spectra obtained, and for enhancing the information content of the ESCA experiment.

7 INSTRUMENTATION

The ESCA experiment is necessarily tied to the complex instrumentation needed
to stimulate photoemission and to measure low fluxes of electrons. A schematic
drawing of a contemporary ESCA instrument is shown in Figure 3.14. The

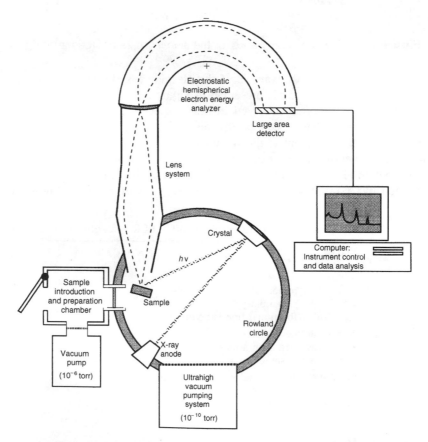

Figure 3.14. A schematic diagram of an ESCA spectrometer. The key components
of a state-of-the-art spectrometer (X-ray anode, monochromator crystal, collection
lens, hemispherical analyzer, and large area detector) are shown

primary components that make up the ESCA instrument are the vacuum system, X-ray source, electron energy analyzer, and data system.

7.1 VACUUM SYSTEMS FOR ESCA EXPERIMENTS

Vacuum systems are described in detail in Chapter 2, so only the aspects of the vacuum system that are pertinent to ESCA will be described here. The heart of the ESCA instrument is the main vacuum chamber where the sample is analyzed. The ESCA experiment must be done under vacuum for three reasons. First, the emitted photoelectrons must be able to travel from the sample through the analyzer to the detector without colliding with gas phase particles. Second, some components such as the X-ray source require vacuum conditions to remain operational. Third, the surface composition of the sample under investigation must not change during the ESCA experiment. Only a modest vacuum (10^{-6}–10^{-7} torr; 1 torr = 133 Pa) is necessary to meet the first two requirements. More stringent vacuum conditions are necessary to avoid contamination of the sample. The actual vacuum required will depend on the reactivity of the sample (e.g., metallic Na will require a better vacuum than PTFE). For most applications a vacuum of 10^{-10} torr is adequate. For studies on polymeric materials, good results can usually be obtained with a vacuum of 10^{-9} torr.

Samples are typically introduced into the main or analytical vacuum chamber via a load-lock or preparation chamber. In its simplest form, the load-lock is a small volume chamber that can be isolated from the analytical chamber and then backfilled to atmospheric pressure. One or more samples are placed in the load-lock chamber, which is then evacuated, typically with a turbomolecular pump. After the samples are pumped down, they are transferred into the analytical chamber. Depending on the vacuum requirements and the type of samples, the pumpdown process can be as short as a few minutes or as long as several hours. In many cases it is desirable to do more elaborate sample processing before introducing the sample into the analytical chamber. For these cases, custom chambers providing an UHV environment with ion guns, deposition sources, sample heating and cooling, sample cleaving, gas dosers, etc., are available. The configuration of these sample preparation chambers depends on their intended use.

After the samples have been placed in the analytical chamber, they must be properly positioned for analysis. This is accomplished with a sample holder/manipulator. Sample manipulators typically have the capability to translate a sample in three directions and to rotate it in one or two directions. Temperature control is also available on most manipulators. For spectrometers used for multisample analysis, the translation and rotation motions are computer controlled, so unattended operation of the instrument is possible. By coupling different sample mounting techniques with the manipulator capabilities and/or adding other components such as ion guns, a range of different ESCA experiments can be done (variable temperature, variable angle, multisample, destructive depth profiling, etc.).

7.2 X-RAY SOURCES

X-rays for an ESCA experiment are usually produced by impinging a high-energy ($\sim$10 keV) electron beam on a target. Core holes are created in the target material or anode, which in turn emits fluorescence X-rays and electrons (see Section 2 and Figure 3.2). It is the fluorescence X-rays that are used in the ESCA experiments. Common anodes along with the energy of their characteristic emission lines are listed in Table 3.11. A specific fluorescence line is used instead of the background emission (Bremsstrahlung) since its intensity is several orders of magnitude higher. Thus, the incident X-ray energy is fixed for each anode. A multi-anode configuration is used to provide a range of X-ray energies. Most spectrometers use only one or two anodes, with Al and Mg the most common. Since most of the incident electron energy is converted into heat, anodes are usually water cooled. This allows operation at higher power loads without significant degradation (e.g., melting).

The emission from the anode can be allowed to strike the sample directly. Although this provides a high X-ray flux, it has several disadvantages. First, the energy resolution of the X-ray source is determined by the natural width of the fluorescence line (typically 1–2 eV). Second, the emission from weaker (satellite) X-ray fluorescence lines will also strike the sample, resulting in the appearance of satellite peaks in the ESCA spectrum. Third, high-energy electrons, Bremsstrahlung, and heat will strike the sample, which can result in sample degradation. The flux of electrons and Bremsstrahlung can be eliminated or minimized by placing a thin, relatively X-ray-transparent foil between the X-ray source and the sample. The presence of the foil will also minimize contamination of the sample by the X-ray source. For Al and Mg anodes, a $\sim$2 μm thick Al foil is commonly used. The best way to optimize single energy production is to use an X-ray monochromator. The most popular monochromatized source combines an Al anode with one or more quartz crystals. The lattice spacing for the $10\bar{1}0$ planes in quartz is 0.425 nm, which is appropriate for the Al Kα wavelength (0.83 nm). For these wavelengths, the Bragg relationship ($n\lambda = 2d\sin\theta$) is satisfied at an angle of 78°. The geometry of a monochromatized X-ray source is illustrated in Figure 3.14. The quartz monochromator crystal and a thin Al foil to isolate the

Table 3.11. Characteristic energies and linewidths for common ESCA anode materials

Anode material	Emission line	Energy (eV)	Width (eV)
Mg	Kα	1253.6	0.7
Al	Kα	1486.6	0.85
Si	Kα	1739.5	1.0
Zr	Lα	2042.4	1.7
Ag	Lα	2984	2.6
Ti	Kα	4510	2.0
Cr	Kα	5415	2.1

source from the sample will prevent electrons, Bremsstrahlung, satellite X-ray lines, and heat radiation from striking the sample. It will also narrow the energy spread of X-rays striking the sample. The disadvantage of a monochromator is the lower X-ray intensity that reaches the sample. This decrease in flux can be compensated for by using an efficient collection lens, energy analyzer, and multi-channel detector system. Such a monochromatized instrument was successfully commercialized in the early 1970s [42]. In the mid 1980s, other manufacturers adopted this approach.

The area of the sample irradiated by the X-source depends on the geometry of the source and the type of electron gun used to stimulate X-ray emission. Most non-monchromatized sources illuminate a spot that is a few centimeters in diameter. In contrast, the monochromatized sources typically illuminate an area that is a few millimeters in diameter. With a focused electron gun and the quartz crystal used as both a monochromator and a focusing element, even smaller spot sizes can be realized [43]. Currently X-ray spot sizes of ~ 50 μm can be obtained.

The above discussion of X-ray sources deals with conventional instrumentation for individual laboratory experiments. The increased availability of synchrotron radiation in the last ten years has opened another avenue for ESCA experiments. The synchrotron provides a broad band of intense radiation (infrared to hard X-rays) that is highly collimated and polarized. When used with a suitable monochromator, synchrotron radiation can provide a tunable source of X-rays for photoemission experiments. However, the number of synchrotron facilities is far less than the number of stand-alone ESCA instruments. This often requires the investigator to travel extended distances to carry out the synchrotron experiments. Further discussion of synchrotron facilities, instrumentation, and capabilities have been presented elsewhere [44–47].

7.3 ANALYZERS

The analyzer system consists of three components: the collection lens, the energy analyzer, and the detector. On some ESCA spectrometers, the lens system can collect photoelectrons over a 30° solid angle. The higher the collection angle, the higher the number of photoelectrons collected per incident X-ray, which is generally advantageous. An efficient lens system can offset, in part, the decreased signal intensity encountered when using monochromatized and focused X-ray sources. The increased collection angle is particularly important for samples that degrade upon exposure to X-rays, since the more efficient the detection system is (e.g., the more photoelectrons collected per X-ray) the more data that can be collected before the sample is damaged. One type of experiment in which a large acceptance angle is a disadvantage, is non-destructive depth profiling. A large acceptance angle, by definition, contains a broad range of photoelectron take-off angles. This degrades the depth resolution obtainable in a variable take-off angle experiment. To improve the depth resolution, an aperture is placed over the entrance to the analyzer lens [48].

In addition to collecting the photoelectrons, the lens system on most spectrometers also retards their KE down to the pass energy of the energy analyzer. The entire ESCA spectrum is acquired by ramping appropriate voltages on the different lens elements. The range and retardation ratio used depends on the pass energy of the energy analyzer and the spectral range to be examined [21]. The lens system also projects the analyzed area a distance away from the entrance of the energy analyzer, which allows the sample to be positioned so that it is more readily accessible to the X-ray source and other components in the vacuum system.

The most common type of energy analyzer used for ESCA experiments is the electrostatic hemispherical analyzer. It consists of two concentric hemispheres of radius R_1 and R_2. A potential of ΔV is placed across the hemispheres such that the outer hemisphere is negative and the inner hemisphere is positive with respect to the potential at the centre line, $R_0 = (R_1 + R_2)/2$. The centre line potential is known as the pass energy. As noted previously, most ESCA experiments are done with a constant pass energy. This will maintain a constant absolute resolution, ΔE, for all photoelectron peaks, since the analyzer resolution is defined as $\Delta E/E$, where E is the energy of the electron as it passes through the analyzer. This ratio is a constant for a given analyzer, so if E is fixed (constant pass energy), ΔE will be fixed. This relationship shows that the lower the pass energy, the smaller ΔE will be. However, the signal intensity will also decrease at smaller pass energies. Typically 5–25 eV pass energies are used to acquire high-resolution ESCA spectra, while 100–200 eV pass energies are used to acquire survey scans.

Hemispherical analyzers are classified as dispersive analyzers, that is, the electrons are deflected by an electrostatic field. There is a range of electron energies that can successfully travel from the entrance to the exit of the analyzer without undergoing a collision with one of the hemispheres. The magnitude of this electron energy range depends on quantities such as the pass energy, the size of the entrance slits, and the angle with which the electrons enter the analyzer. In modern commercial analyzers, this range is ~10% of the pass energy. Further information about hemispherical analyzers has been published elsewhere [21].

The electrons are counted once they have passed through the energy analyzer. Since the electrons arrive at the analyzer exit with a range of energies, the most efficient means of detection is to use a multichannel array to count the number of electrons leaving the analyzer at each energy. One method of accomplishing this is to use a channel plate to magnify the electron current and a resistive strip anode to monitor the position, and therefore energy, of the electrons. A less elegant method is to place a slit at the analyzer exit so that only electrons in a narrow energy range strike the detector. In this case, a device such as a channeltron is used to measure the number of electrons. Compared to a multichannel detection method using N channels, the single-channel detection method takes $N^{1/2}$ times longer to acquire the same spectrum.

Some of the analyzer systems used for ESCA experiments maintain the spatial relationship of the emitted photoelectrons throughout their transmission through the lens and energy analyzer. This means that the position where the photoelectrons strike the detector is related to their emission position from the sample. Thus, a position-sensitive detector can be used to image the sample. Depending on how the analyzer system is designed, spatial imaging can be done in one or two lateral directions. Spatial resolutions near 10 μm have been achieved with imaging detectors [49, 50].

7.4 DATA SYSTEMS

Modern computers provide a powerful means both for controlling instrument operation and performing data analysis. State-of-the-art ESCA spectrometers have virtually all aspects of their operation under computer control. Most accessories and components (ion guns, electron guns, valves, etc.) can be activated by the computer. The power supplies that control the analyzer functions (pass energy, scan rate, E_B range, etc.) are also under computer control. This, along with computer control of the sample positioning system, allows unattended, multi-sample runs to be executed. Since each sample may require several different types of scans, it is useful to be able to pre-select and store the desired scan parameters along with the sample position. Then, execution of these commands, which may take several hours, can be done automatically.

Since most computers are now multi-tasking, data acquisition and data analysis can be done simultaneously. Current software programs contain a wide range of data analysis capabilities. Complex peak shapes can be fit in seconds. Automatic peak finding, identification, and quantitation for survey scans can also be accomplished in seconds. A whole array of data scaling, smoothing, plotting, transferring, and transforming are readily available. Images, X–Y maps, and depth profiles can also be generated. Some software programs even include mathematical analysis packages (multivariant statistics, pattern recognition, etc.). Other software programs allow ESCA data to be directly transferred into word processing packages. In general, as the speed and power of computer systems increase, so do the capabilities for ESCA data acquisition and analysis. Over the last ten years this, along with improvements in the ESCA hardware, have dramatically increased the number of samples that can be run in a day.

7.5 ACCESSORIES

The types of accessories that can be added to an ESCA spectrometer are almost limitless. Common accessories include ion guns, electron guns, gas dosers, and quadrupole mass spectrometers. The accessories selected for a given system depend on the applications that are planned for the system. In many cases the ESCA instrument is part of a multicomponent surface analysis system that has one or more additional techniques (Auger, ion scattering, SIMS, LEED, EELS, etc.) mounted on the same vacuum chamber. In some cases, the ESCA analyzer is used in several different techniques (Auger, ion scattering, etc.). For monochromatized ESCA systems, the most important accessory is the low-energy electron flood gun, which is required to obtain high-quality spectra from insulating materials (described in Section 3.5).

8 SPECTRAL QUALITY

Like all other spectroscopic techniques, the signal-to-noise (S/N) ratio and resolution are the most important properties to consider when evaluating spectral quality. Usually manufacturers advertise the count rates for their ESCA spectrometers, which is not a useful specification. More important is how noise-free the spectrum is, which determines the counting time needed to acquire a high-quality spectrum. Specifically, the length of time it takes to reach a given S/N ratio at a given energy resolution is the important criterion. As resolution is increased, the S/N will decrease. There are several ways to evaluate the S/N ratio. One convenient method is the peak-to-peak S/N ratio shown in Figure 3.15. A spectrum containing a photoemission peak is acquired for a known time period, then the channels with the highest and lowest number of counts are noted. The difference between these two channels (16738–48) is the peak-to-peak signal. The peak-to-peak noise is determined by using identical scan parameters (same

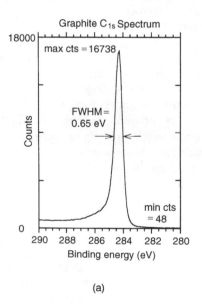
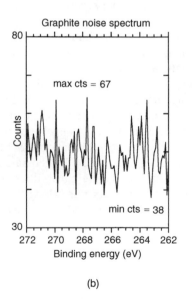

(a)   (b)

Figure 3.15. The C_{1s} and noise spectra from a graphite sample. The data in each region were acquired with the same scan parameters (time, eV/point, window, pass energy, etc.). These data, obtained after three minutes of scanning, have a peak-to-peak signal-to-noise of 575 at an energy resolution of 0.65 eV

acquisition time, number of data points, eV/point, pass energy, X-ray setting, etc.), except the scan window is shifted below the photoemission peak. The peak-to-peak noise, like the peak-to-peak signal, is the difference between the channels with the highest and lowest number of counts (67–38). The peak-to-peak S/N ratio is simply the ratio of the peak-to-peak signal to the peak-to-peak noise (16690/29 = 575). The energy resolution can be determined by measuring the FWHM of the peak. Thus, for the graphite sample shown in Figure 3.15, it was determined that three minutes of scanning produced a peak with an energy resolution of 0.65 eV and an S/N ratio of 575. The S/N of a spectrum can be increased by either increasing the scan time or decreasing the energy resolution. For a properly designed spectrometer, the S/N ratio will increase as $t^{1/2}$ where t is the scan time.

9 DEPTH PROFILING

Although ESCA would seem to provide information from a highly surface-localized zone, in fact the surface zone has a finite thickness and often is composed of a vertical composition gradient. If we estimate that the sampling depth of ESCA is 8 nm and atomic dimensions are 0.3 nm, then the surface region could be composed of 20–30 atomic layers. Each of these layers may have a different composition. The ESCA spectrum we obtain will be composed of a convolution of the information from all the layers. This problem is illustrated schematically in Figure 3.16. Depth profiling methods are used to

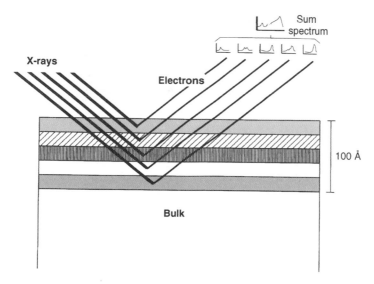

Figure 3.16. ESCA spectra are convolutions of the information from each depth within the sampling depth

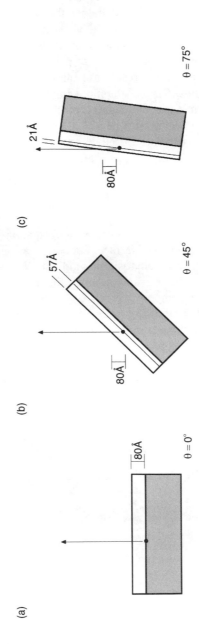

Figure 3.17. As the sample is rotated, maintaining the X-ray source and detector in fixed positions, the effective sampling depth decreases by a factor of $\cos\theta$. The sample angle, θ, is defined relative to the normal to the surface

deconvolute from the ESCA signal the composition as a function of depth. Three sampling depths are of interest here: 0–8 nm (the sampling depth of ESCA using conventional X-ray sources), 0–20 nm (the sampling depth of ESCA using special X-ray sources), and 0 to ∼1000 nm using destructive depth profiling methods. Each of these zones will be separately described.

The information from the outermost ∼8 nm of a surface is converted into a depth profile using data acquired in an angular dependent ESCA experiment. As the sample angle to the analyzer entrance is increased, with the X-ray source and the analyzer kept in a fixed position, the photoelectrons originate from an increasingly surface localized zone (Figure 3.17). If data are acquired at angles of, for example, 0°, 50°, and 80° from the surface normal, three sets of ESCA data can be obtained that contain information about the composition as a function of depth. Qualitatively, the shape of this composition versus take-off angle curve can reveal much about the compositional organization of a surface (Figure 3.18). To convert from a plot of 'angle versus composition' to a plot of 'depth into the surface versus composition' necessitates a deconvolution of the data set. The mathematical functions upon which the deconvolutions are based are forms of the equations in Figure 3.7. A number of algorithms for performing such a deconvolution have been published [51–56]. Table 3.12 contains an ESCA data set taken at five electron take-off angles for a sample of a cast film of a fluorine-containing polyurethane [51]. Figure 3.19 shows the deconvolution of this data set using the algorithm developed by Tyler et al [51]. Surface depletion of fluorine and nitrogen (hard segment components) is evident. For accurate and meaningful depth profiles from angular dependent ESCA data, we assume that the surfaces and interfaces studied are molecularly smooth and that overlayers are of uniform thickness. Effectively, an aspect ratio of 10:1 (length to height) for surface irregularities (i.e., 'gently rolling hills') can be tolerated [57]. Other assumptions also apply for interpreting angular dependent ESCA data [51, 52, 57].

Non-destructive ESCA depth profiling can also be performed using X-ray sources of different energies. According to equation (1) a higher energy X-ray source will liberate higher KE photoelectrons. These more energetic photoelectrons have a greater IMFP and, consequently, an increased sampling depth. If Al Kα (1487 eV), Ag Lα (2984 eV), and Cr Kα (5415 eV) X-ray sources are each used to generate ESCA spectra of the same sample, the C_{1s} electron sampling depths, using equation (12) developed by Seah and Dench [28], can be estimated at 10.8, 16.2, and 22.4 nm, respectively. Thus, the information needed for a depth profile is acquired.

Depth profiles deep into the sample surface (to a micron or more) can be generated by ion etching the surface, and then analyzing the bottom of the etching crater at regular time intervals using ESCA. However, for organic material, structural information will be lost due to the damaging effects of the ion beam. Also, the ion beam will induce scrambling and knock-in of atoms at the bottom of the

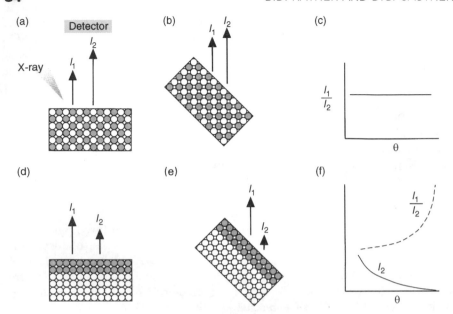

Figure 3.18. The morphology of a specimen will influence the angular dependence of the ESCA signal intensities: (a) For a specimen with homogeneously distributed atoms, note the ratio of the total intensity of the photoemission from the grey atoms and the white atoms; (b) the ratio as described in (a) will be constant at any sample angle; (c) because the ratio of intensities does not change with sample angle, for a sample homogeneous in depth, a plot of the ratio of photoemission intensities (or the ratio of atomic percents) with sample angle will show zero slope; (d) a sample is illustrated with an overlayer of grey atoms on a substrate of white atoms; (e) when this sample is rotated, the photoemission signal will localize closer to the outermost surface. Therefore, the intensity of the signal from the grey atoms will increase relative to the intensity from the white atoms with increasing angle; (f) a plot of the ratio of the grey atom photoemission intensity to the white atom intensity with sample angle will increase in an exponential fashion with sample angle. The photoemission from the white atoms will decrease in intensity with increasing sample angle

Table 3.12. Angular dependent ESCA data from a fluorine-containing polyetherurethane (normalized signal intensities)*

Angle	C	O	N	F
0°	5456	1267	189	236
39°	4341	979	118	157
55°	3498	822	103	126
68°	2736	642	68	70
80°	1706	395	34	39

*Data from reference [51].

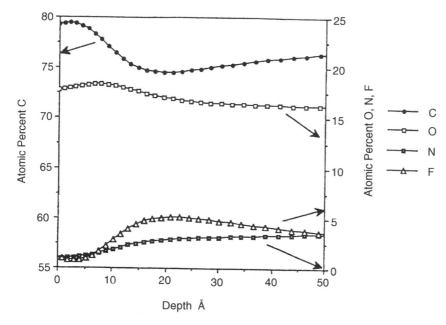

Figure 3.19. A depth profile diagram of a fluorine-containing polyurethane estimated using the regularization method [51] for deconvoluting the angular dependent data set in Table 12 (From [51] used with permission of John Wiley & Sons)

crater reducing the accuracy of the analysis — the longer the etching time (the deeper the crater), the more degraded will be the accuracy of the depth profile.

10 X–Y MAPPING

In recent years there have been marked improvements made in the spatial resolution of ESCA. Historically, spatial resolutions achieved with ESCA have been poorer than spatial resolutions achieved with other surface analysis techniques such as Auger and SIMS. This is because it is harder to focus an X-ray beam than an electron or ion beam. However, much can be gained from improving the spatial resolution of ESCA. One benefit would be an improved ability to do spot analyses. As features on microelectronic chips become smaller, improved spatial resolution is needed. Likewise, the analysis of surface defects requires good spatial resolution. The second benefit would be the improved ability to construct images of a sample. By using the chemical specificity of ESCA, both elemental and functional group maps of a surface can be constructed.

As discussed in Section 7, there are two methods for obtaining spatial resolution in stand-alone ESCA spectrometers. One method is to focus the X-ray

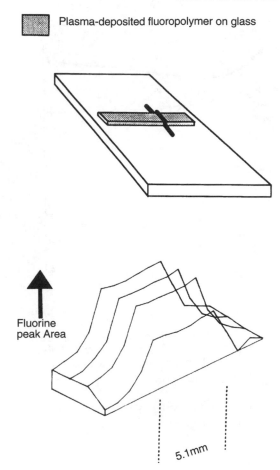

□ Plasma-deposited fluoropolymer on glass

Figure 3.20. An X–Y map of the fluorine signal from a RF glow discharge deposited film of tetrafluoroethylene. A mask was placed over the glass substrate during the deposition. The highest ESCA F concentrations correspond to the location of a rectangular opening in the mask

source [43]. The best spatial resolution currently obtained with this method is ~50 μm. The second method is to use a position-sensitive detector to image the surface [49, 50]. The best spatial resolution currently obtained with this method is ~10 μm. The imaging detector has the advantage in spatial resolution and does not require a focused X-ray source. However, there are advantages to using focused X-rays. First, only the area being analyzed is irradiated. This is particularly important when working with samples that are prone to X-ray degradation. With a large-spot imaging detector system, parts of the sample can be degraded before they are even analyzed. Second, positioning the X-ray beam for analysis is straightforward with focused X-rays. After calibrating the X-ray position with a microscope, the sample is moved to the appropriate position and the analysis is started.

Spatial resolutions of 1 μm or better are desirable for many applications, requiring at least an order of magnitude increase from currently obtainable spatial resolutions. Some synchrotron-based techniques are being developed to meet these needs. One is a superconducting ESCA analyzer, in which the sample is placed in a superconducting magnet and then irradiated with X-rays [58]. The emitted photoelectrons follow the magnetic field lines as they leave the sample. A position-sensitive detector is then placed outside of the magnet to obtain a magnified image. With this system, spatial resolution of a few microns have been obtained. Photoelectron microscopes are also being developed at several synchrotron facilities and should allow spatial resolutions of at least 1 μm to be obtained. A review of the progress made in this field has recently been published [59].

Even with the limited spatial resolution of commercial laboratory ESCA spectrometers, useful information can be obtained. Figure 3.20 shows an X-Y map made from an RF glow discharge deposited film of tetrafluoroethylene. A mask with a rectangular opening was placed over the glass substrate during the deposition. The fluorine map shows the F concentration was significantly higher in the unmasked area.

11 CHEMICAL DERIVATIZATION

The binding energy shifts in ESCA associated with specific functional groups often do not permit precise identification of the group. For example, carbons bound in an ether environment ($\underline{C}$−O−C) and carbons in a hydroxyl environment ($\underline{C}$−OH) are both observed at 286.5 eV. Also, carbons in a carboxylic acid environment and carbons in an ester environment are not readily distinguishable based upon binding energy shifts. There are many other examples of functional groups that are difficult to identify based upon their binding energies. To assist with the precise identification of chemical groups, a chemical reaction specific to only the functional group of interest can be performed. This reaction should uniquely identify the functional group in the ESCA spectrum by altering its binding energy position or by adding a tag atom not present in the specimen prior to reaction. This idea is illustrated schematically in Figure 3.21. There are many derivatization reactions that have been explored for this purpose [8, 60]. A few common derivatization reactions are illustrated in Table 3.13. There are also many control studies that must be performed so derivatization reactions can be used with confidence. The concerns that must be addressed for derivatization studies are outlined in Table 3.14. After performing appropriate control studies, derivatization reactions can be used to enhance the understanding of the chemistry of complex surfaces [61–63].

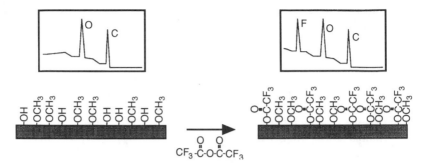

Figure 3.21. Hydroxy groups and ethers on a surface cannot be distinguished based upon ESCA chemical shifts. If the surface is reacted with trifluoroacetic anhydride, only hydroxyl groups will pick up F. The size of the F peak in the ESCA spectrum will be proportional to the number of reacted hydroxyl groups

Table 3.13. Four surface chemical derivatization reactions

Group	Reaction
Hydroxyl $-OH +$ $(CF_3CO)_2O$ trifluoroacetic anhydride $\longrightarrow$ $-OCOCF_3$	
Carbonyl $-\overset{\mid}{C}=O +$ NH_2NH_2 hydrazine $\longrightarrow$ $-\overset{\mid}{C}=NNH_2$	
Carboxylic acid $-COOH +$ CF_3CH_2OH trifluoroethanol $\xrightarrow{\substack{carbodiimide \\ C_6H_5N}}$ $-CO_2CH_2CF_3$	
Unsaturated $-CH=CH- + Br_2 \longrightarrow -CHBr - CHBr-$	

Table 3.14. Concerns in derivatization studies

- Is the reaction stoichiometric?
- Has the reaction gone to completion (kinetics)?
- Does the derivatizing reagent cross-react with any of the other functional groups in the sample?
- The marker atom should be unique in the sample and have a high photoemission cross-section
- The derivative formed must be stable under vacuum.
- The derivative formed must be stable over time, even under X-ray and electron flux.
- Surface rearrangement leading to migration of the marker atom from the interface is undesirable.
- The derivatizing reagent should not extract components out of the sample.

12 VALENCE BAND

Most ESCA experiments focus on the core level peaks. As discussed in previous sections of this chapter, the core level spectra are comprised of relatively sharp, intense peaks that can be used to identify oxidation states, molecular functional groups, concentrations, etc. of atoms in the surface region of a sample. For quantitation each core level is assumed to have a specific photoionization cross section that does not depend on the sample matrix or bonding. This is a good assumption for core electrons which are not involved directly in bond formation and allows ESCA to be used as a quantitative technique. This assumption is not valid for valence band analysis. The valence electrons are directly involved in bond formation, so the intensity and energy of the valence band peaks depend on their bonding environment. Thus, quantitative interpretation of the valence band spectrum for most materials requires a full molecular orbital calculation. For multi-component materials or those containing a large number of atoms per structural unit (e.g., polymers) this requires lengthy computer calculations [64].

Valence band analysis is worth pursuing because it can provide electronic structure information that can not be obtained from typical core level analysis. Also, it is sometimes possible to extract useful structural information from valence band spectra. In spite of these benefits, valence band spectra are not generally used in typical ESCA experiments. One reason is that the ESCA cross-sections for valence band peaks are significantly lower than core level cross-sections, which dictates that long analysis times (up to several hours, depending on the instrument used) must be used to acquire valence band spectra with good signal-to-noise characteristics. This can be a problem for organic samples that degrade during analysis.

Although polymer valence band studies are not abundant, they do show the power of valence band analysis. Early studies showed that the combination of a monochromatized X-ray source, to minimize sample degradation, and theoretical calculations, to aid interpretation, resulted in an enhanced understanding of the polymer surface structure [41]. Small molecule models were also found to be valuable aids in interpreting valence band spectra, examples being linear alkanes for polyethylene and benzene for aromatic containing polymers (polyphenyl, polystyrene, etc.) [41]. The sensitivity of the valence band to polymer structure was highlighted in studies with polymers having the same elemental composition and core level spectra. These polymers were easily differentiated based on their valence band spectra. For example, poly(propylene oxide) and poly(vinyl methyl ether) both have C_3H_6O monomer units, but exhibit different valence band spectra [41]. Isomeric effects in pure hydrocarbon components can also be distinguished with valence band spectra. Examples include poly(3-methyl 1-butene) and poly(1-pentene), both having a C_5H_{10} monomer repeat unit [41], methyl substituted polystyrenes [65], and the normal, iso, and tertiary butyl side chains in methacrylates (C_4H_9 units) [66, 67]. Further details such as head-to-head versus head-to-tail linking of monomer units and the tacticity of the monomer units can also be differentiated with valence band spectra [41]. Until recently, it required several hours of acquisition with a monochromatic X-ray source to acquire suitable polymer valence band spectra. With the recent improvement in instrument performance, this acquisition time has been shortened considerably, resulting in the appearance of a handbook that contains an extensive collection of polymer valence band spectra [68].

In contrast to polymeric materials, more extensive use of valence band spectra has been made for metallic and semiconductor materials [69–71]. This is probably due to several reasons. First, experimental metallic and semiconducting samples are easier to study since they are less susceptible to degradation and charging problems than organic materials. Second, single crystalline samples are readily available for a wide range of metals and semiconductors. This allows one to perform a detailed investigation of the valence band electronic structure through angle-resolved photoemission experiments [70]. To overcome the low cross-section limitation in ESCA valence band experiments, many experimenters use synchrotron radiation for the incident photon source [72, 73]. By using a synchrotron, the energy of the incident photons can be tuned to maximize a given valence band peak, since the photoemission cross-section depends on the energy of the excitation source. Thus, the combination of a tuneable X-ray source, a single crystal sample, and angle-resolved detection provides a powerful method for obtaining detailed information about the electronic structure of a material (work functions, band gap energies, band dispersions, band bending, etc). Valence band experiments can also be used to probe the development of the electronic structure of metal clusters as a function of cluster size [74] and the electronic interactions in metallic alloys [75].

13 PERSPECTIVES

New instrumentation, technique development, and enhanced data analysis continue to expand the utility of ESCA for the analysis of surfaces. Five frontier areas are described.

13.1 ENERGY RESOLUTION AND SPECTRAL INTERPRETATION

The energy resolution of ESCA instruments continues to improve, largely because of improvements in monochromator design and charge compensation [76]. With improvements in energy resolution, new information can be extracted from ESCA spectra [76–79]. For example, where a C_{1s} spectrum of poly(2-chloroethyl methacrylate) was traditionally resolved into three peak components using high-resolution spectrometers, it now becomes apparent that five peaks can be accurately fitted to this peak envelope [77]. The enhanced resolution of the newest instruments may even reveal vibrational components that broaden the ESCA peaks [21, 80].

13.2 LOW-TEMPERATURE ESCA STUDIES

Many surfaces molecularly re-orient in contact with certain liquid phases [81]. It would be useful to be able to examine the liquid–solid interface, rather than the vacuum–solid interface usually studied in UHV ESCA instruments. Such experiments can now be performed by rapidly freezing a specimen under liquid, subliming the ice phase from the surface (while maintaining the solid at a temperature sufficiently low that molecular mobility is effectively halted), and then examining the surface (still at low temperature) by ESCA. With experiments of this type, the surface of the solid phase as it exists at the liquid–solid interface is 'locked in place' for study. Examples showing the power of this method for examining the effect of water on re-orientation of the surface structure have been published [82–84].

13.3 PHOTOELECTRON DIFFRACTION

For clean single-crystal surfaces, oscillations can be observed in the absolute core photoemission intensity as electron take-off angle is varied. These oscillations are attributed to constructive and destructive interferences between the coherent components of the photoelectron wave. More specifically, forward scattering is often observed where the photoemission intensity is enhanced along nuclear axes. Analysis of the photoelectron diffraction plots can offer much insight into the interatomic spacings in the uppermost few atomic layers. Instrumentally, high signal to noise and a narrow electron acceptance angle into the analyzer are needed to obtain good data. Reviews are available on photoelectron diffraction [85].

13.4 LOW-COST ESCA

A typical ESCA system costs $350 000–$700 000. This high price, and the need for highly trained personnel for instrument operation, have prevented the migration of ESCA into routine analysis and quality control associated with commodity products. Recent development of a compact, self-contained, highly automated ESCA instrument that sells for approximately $150 000 may open up new areas for ESCA analysis [86].

13.5 IMAGING ESCA

Advances in the ability of ESCA systems to image chemical features with fine spatial detail are being made [49, 50, 58, 77, 87]. With spatial resolutions approaching 1 μm, a revolution in the application of ESCA to surface analytical problems involving the distribution of chemical features at surfaces is now taking place. Instrumental strategies for spatially resolved ESCA are described in Section 10.

14 CONCLUSIONS

The ESCA technique has been a commercially available method since the late 1960's. In 25 years it has gone from a physicist's experiment to a practical and widely available surface analysis tool with thousands of published applications. The advantages of ESCA are its simplicity, flexibility in sample handling, and high information content. The heightened interest in materials science, biotechnology and surface phenomena in general, coupled with advances in ESCA technique and instrumentation, make it probable that ESCA will remain the predominant surface analysis technique in the foreseeable future. When used in conjunction with other surface analysis methods, ESCA will play a pivotal role in expanding our understanding of the chemistry, morphology, and reactivity of surfaces.

ACKNOWLEDGEMENTS

Support was received from NIH grants RR01296 and HL25951 during the writing of this article, and for some of the experiments described in it. We thank Deborah Leach-Scampavia for acquiring some of the ESCA data used in this article, and Thomas Menduni for editorial assistance.

REFERENCES

[1] R.S. Swingle and W.M. Riggs, ESCA, *CRC Crit. Rev. Anal. Chem.*, **5**, 267–321 (1975).
[2] K. Siegbahn, C. Nordling, A. Fahlman, R. Nordberg, K. Hamrin, J. Hedman, G. Johansson, T. Bergmark, S.E. Karlsson, I. Lindgren, and B. Lindberg, ESCA: atomic, molecular and solid state structure studied by means of electron spectroscopy, *Nova Acta Regiae Societatis Scientiarum Upsaliensis, Ser. IV*, **20**, 5–282 (1967).
[3] K. Siegbahn, Electron spectroscopy for atoms, molecules, and condensed matter, *Science*, **217**, 111–121 (1982).
[4] A. Dilks, X-ray photoelectron spectroscopy for the investigation of polymeric materials, in *Electron Spectroscopy: Theory, Techniques, and Applications, Vol. 4*, A.D. Baker and C.R. Brundle, (Eds) Academic Press, London, (1981) pp. 277–359.
[5] P.K. Ghosh, *Introduction to Photoelectron Spectroscopy*, John Wiley & Sons, New York, (1983).
[6] B.D. Ratner and B.J. McElroy, Electron spectroscopy for chemical analysis: applications in the biomedical sciences, in *Spectroscopy in the Biomedical Sciences*, R.M. Gendreau, (Ed.) CRC Press, Boca Raton, FL, (1986) pp. 107–140.
[7] T.A. Carlson, *Photoelectron and Auger Spectroscopy*, Plenum Press, New York, (1975).
[8] J.D. Andrade, X-ray photoelectron spectroscopy (XPS), in *Surface and Interfacial Aspects of Biomedical Polymers, Vol. 1: Surface Chemistry and Physics*, J.D. Andrade, (Ed.) Plenum Press, New York, (1985) pp. 105–195.
[9] D.T. Clark, Some experimental and theoretical aspects of structure, bonding and reactivity of organic and polymeric systems as revealed by ESCA, *Physica Scripta*, **16**, 307–328 (1977).

[10] M.J. Klein, The beginnings of the quantum theory, in *History of Twentieth Century Physics*, C. Weiner, (Ed.) Academic Press, New York, (1977) pp. 1-39.

[11] A. Pais, *Inward Bound*, Oxford University Press, Oxford, (1986).

[12] E. Segre, *From X-rays to Quarks*, W.H. Reeman and Co., San Francisco, (1980).

[13] J. Berkowitz, *Photoabsorption, Photoionization, and Photoelectron Spectroscopy*, Academic Press, New York, (1979).

[14] L.C. Feldman and J.W. Mayer, *Fundamentals of Surface and Thin Film Analysis*, North-Holland, New York, (1986).

[15] R. Hoffmann, *Solids and surfaces. A Chemists View of Bonding in Extended Structures, Vol. 1*, VCH Publishers, New York, (1988).

[16] T.S. Koopmans, Über die Zuordnung von Wellenfunktionen und Eigenwerten zu den Einzelnen Elektronen eines Atoms, *Physica*, **1**, 104-113 (1934).

[17] C.R. Brundle, T.J. Chuang, and D.W. Rice, X-ray photoemission study of the interaction of oxygen and air with clean cobalt surfaces, *Surf. Sci.*, **60**, 286-300 (1976).

[18] N.S. McIntyre and M.G. Cook, X-ray photoelectron studies on some oxides and hydroxides of cobalt, nickel, and copper, *Anal. Chem.*, **47**, 2208-2213 (1975).

[19] D.A. Shirley, Many-electron and final-state effects: beyond the one-electron picture, in *Photoemission in Solids. I. General Principles*, M. Cardona and L. Ley, (Eds) Springer-Verlag, Berlin, (1978) pp. 165-195.

[20] M.P. Seah, Post-1989 calibration energies for X-ray photoelectron spectrometers and the 1990 Josephson constant, *Surf. Interface Anal.*, **14**, 488 (1989).

[21] D. Briggs and M.P. Seah, *Practical Surface Analysis, Vol. 1*, John Wiley & Sons, Chichester, (1990).

[22] M.T. Anthony and M.P. Seah, XPS: energy calibration of electron spectrometers 1- An absolute, traceable energy calibration and the provision of atomic reference line energies, *Surf. Interface Anal.*, **6**, 95-106 (1984).

[23] R.T. Lewis and M.A. Kelly, Binding-energy reference in X-ray photoelectron spectroscopy of insulators, *J. Elect. Spectrosc. Rel. Phenom.*, **20**, 105-115 (1980).

[24] C.E. Bryson III, Surface potential control in XPS, *Surf. Sci.*, **189/190**, 50-58 (1987).

[25] M. Cardona and L. Ley (Eds.) *Photoemission in Solids. I. General Principles*, Springer-Verlag, Berlin, (1978).

[26] D.A. Shirley, High-resolution x-ray photoemission spectrum of the valence bands of gold, *Physical Review B*, **5**, 4709-4714 (1972).

[27] C.J. Powell, The quest for universal curves to describe the surface sensitivity of electron spectroscopies, *J. Elect. Spectrosc. Rel. Phenom.*, **47**, 197-214 (1988).

[28] M.P. Seah and W.A. Dench, Quantitative electron spectroscopy of surfaces: A standard data base for electron inelastic mean free paths in solids, *Surf. Interface Anal.*, **1**, 2-11 (1979).

[29] P. Cadman, S. Evans, G. Gossedge, and J.M. Thomas, Electron inelastic mean free paths in polymers: comments on the arguments of Clark and Thomas, *J. Polym. Sci., Polym. Lett. Ed.*, **16**, 461-464 (1978).

[30] D.T. Clark, H.R. Thomas, and D. Shuttleworth, Electron mean free paths in polymers: A critique of the current state of the art, *J. Polym. Sci., Polym. Lett. Ed.*, **16**, 465-471 (1978).

[31] R.F. Roberts, D.L. Allara, C.A. Pryde, D.N.E. Buchanan, and N.D. Hobbins, Mean free path for inelastic scattering of 1.2 keV electrons in thin poly(methyl methacrylate) films, *Surf. Interface Anal.*, **2**, 5-10 (1980).

[32] C.D. Wagner, L.E. Davis, and W.M. Riggs, The energy dependence of the electron mean free path, *Surf. Interface Anal.*, **2**, 53-55 (1980).

[33] C.R. Brundle, H. Hopster, and J.D. Swalen, Electron mean-free path lengths through monolayers of cadmium arachidate, *J. Chem. Phys.*, **70**, 5190–5196 (1979).

[34] R.F. Reilman, A. Msezane, and S.T. Manson, Relative intensities in photoelectron spectroscopy of atoms and molecules, *J. Elect. Spectrosc. Rel. Phenom.*, **8**, 389–394 (1976).

[35] M.P. Seah, M.E. Jones, and M.T. Anthony, Quantitative XPS: The calibration of spectrometer intensity-energy response functions. 2. Results of interlaboratory measurements for commercial instruments, *Surf. Interface Anal.*, **6**, 242–252 (1984).

[36] M.P. Seah and M.T. Anthony, Quantitative XPS: The calibration of spectrometer intensity-energy response functions. 1. The establishment of reference procedures and instrument behaviour, *Surf. Interface Anal.*, **6**, 230–241 (1984).

[37] J.H. Scofield, Hartree-Slater subshell photoionization cross-sections at 1254 and 1487 eV, *J. Elect. Spectrosc. Rel. Phenom.*, **8**, 129–137 (1976).

[38] C.D. Wagner, L.E. Davis, M.V. Zeller, J.A. Taylor, R.H. Raymond, and L.H. Gale, Empirical atomic sensitivity factors for chemical analysis by electron spectroscopy for chemical analysis, *Surf. Interface Anal.*, **3**, 211–225 (1981).

[39] C.D. Wagner, W.M. Riggs, L.E. Davis, J.F. Moulder, and G.E. Muilenberg, *Handbook of X-ray Photoelectron Spectroscopy*, Perkin-Elmer Corporation, Eden Prairie, MN, (1979).

[40] C.D. Wagner and A. Joshi, The Auger parameter, its utility and advantages: A review, *J. Elect. Spectrosc. Rel. Phenom.*, **47**, 283–313 (1988).

[41] J.J. Pireaux, J. Riga, R. Caudano, and J. Verbist, Electronic structure of polymers, *ACS Symp. Ser.*, **162**, 169–201 (1981).

[42] M.A. Kelly and C.E. Tyler, *Hewlett-Packard Journal*, **24** (1972).

[43] R.L. Chaney, Recent developments in spatially resolved ESCA, *Surf. Interface Anal.*, **10**, 36–47 (1987).

[44] H. Winick and S. Doniach, *Synchrotron Radiation Research*, Plenum, New York, (1980).

[45] J.C. Schuchman, Vacuum systems for synchrotron light sources, *MRS Bulletin*, **15**, 35–41 (1990).

[46] G. Margaritondo, Synchrotron radiation photoemission spectroscopy of semiconductor surfaces and interfaces, *Annual Rev. Mater. Sci.*, **14**, 67–93 (1984).

[47] D.A. King, Looking at solid surfaces with a bright light, *Chem. Brit.*, **22**, 819–822 (1986).

[48] B.J. Tyler, D.G. Castner, and B.D. Ratner, Determining depth profiles from angle dependent X-ray photoelectron spectroscopy: The effects of analyzer lens aperture size and geometry, *J. Vac. Sci. Technol: A*, **7**, 1646–1654 (1989).

[49] I.W. Drummond, F.J. Street, L.P. Ogden, and D.J. Surman, AXIS: An imaging X-ray photoelectron spectrometer, *Scanning*, **13**, 149–163 (1991).

[50] P. Coxon, J. Krizek, M. Humpherson, and I.R.M. Wardell, ESCASCOPE — a new imaging photoelectron spectrometer, *J. Elect. Spectrosc. Rel. Phenom.*, **52**, 821–836 (1990).

[51] B.J. Tyler, D.G. Castner, and B.D. Ratner, Regularization: A stable and accurate method for generating depth profiles from angle dependent XPS data, *Surf. Interface Anal.*, **14**, 443–450 (1989).

[52] R.S. Yih and B.D. Ratner, A comparison of two angular dependent ESCA algorithms useful for constructing depth profiles of surfaces, *J. Elect. Spectrosc. Rel. Phenom.*, **43**, 61–82 (1987).

[53] H. Iwasaki, R. Nishitani, and S. Nakamura, Determination of depth profiles by angular dependent X-ray photoelectron spectra, *Japan. J. Appl. Phys.*, **17**, 1519-1523 (1978).

[54] O.A. Baschenko and V.I. Nefedov, Depth profiling of elements in surface layers of solids based on angular resolved X-ray photoelectron spectroscopy, *J. Elect. Spectrosc. Rel. Phenom.*, **53**, 1-18 (1990).

[55] T.D. Bussing and P.H. Holloway, Deconvolution of concentration depth profiles from angle resolved X-ray photoelectron spectroscopy data, *J. Vac. Sci. Technol. A*, **3**, 1973-1981 (1985).

[56] R.W. Paynter, Modification of the Beer-Lambert equation for application to concentration gradients, *Surf. Interface Anal.*, **3**, 186-187 (1981).

[57] C.S. Fadley, Solid state-and surface-analysis by means of angular-dependent X-ray photoelectron spectroscopy, *Prog. Sol. State Chem.*, **11**, 265-343 (1976).

[58] P.L. King, R. Browning, P. Pianetta, I. Lindau, M. Keenlyside, and G. Knapp, Image processing of multispectral X-ray photoelectron spectroscopy images, *J. Vac. Sci. Technol. A*, **7**, 3301-3304 (1989).

[59] B.P. Tonner, Photoemission spectromicroscopy of surfaces in materials science, *Synchrotron Radiation News*, **4**, 27-32 (1991).

[60] C.D. Batich, Chemical derivatization and surface analysis, *Applied Surface Science*, **32**, 57-73 (1988).

[61] A. Chilkoti and B.D. Ratner, An X-ray photoelectron spectroscopic investigation of the selectivity of hydroxyl derivatization reactions, *Surf. Interface Anal.*, **17**, 567-574 (1991).

[62] A. Chilkoti, B.D. Ratner, and D. Briggs, Plasma-deposited polymeric films prepared from carbonyl-containing volatile precursors: XPS chemical derivatization and static SIMS surface characterization, *Chem. Mater.*, **3**, 51-61 (1991).

[63] D. Briggs and C.R. Kendall, Derivatization of discharge-treated LDPE: an extension of XPS analysis and a probe of specific interactions in adhesion, *Int. J. Adhesion and Adhesives*, **2**, 13-17 (1982).

[64] J.-M. Andre, J. Dehalle, and J.J. Pireaux, Band structure calculations and their relations to photoelectron spectroscopy, *ACS Symp. Ser.*, **162**, 151-168 (1981).

[65] A. Chilkoti, D.G. Castner, and B.D. Ratner, Static SIMS and XPS of deuterium-and methyl- substituted polystyrene, *Applied Spectroscopy*, **45**, 209-217 (1991).

[66] D.T. Clark and H.R. Thomas, Applications of ESCA to polymer chemistry. XI. Core and valence energy levels of a series of polymethacrylates, *J. Poly. Sci.: Poly. Chem. Ed.*, **14**, 1701-1713 (1976).

[67] D.G. Castner and B.D. Ratner, Surface characterization of butyl methacrylates by XPS and Static SIMS, *Surf. Interface Anal.*, **15**, 479-486 (1990).

[68] G. Beamson and D. Briggs, *High Resolution XPS of Organic Polymers*, John Wiley and Sons, Chichester, (1992).

[69] S. Huefner, *Photoelectron spectroscopy. Principles and applications*, Springer, Berlin, (1995).

[70] K.E. Smith and S.D. Kevan, The electronic structure of solids studied using angle resolved photoemission spectroscopy, *Progress in Solid State Chemistry*, **21**, 49-131 (1991).

[71] C.S. Fadley, Elastic and inelastic scattering in core and valence emission from solids: some new directions, *AIP Conference Proceedings*, **215**, 796-813 (1990).

[72] P.O. Nilsson, Photoelectron spectroscopy by synchrotron radiation, *Acta Physica Polonica A*, **82**, 201-219 (1992).

[73] C.G. Olson and D.W. Lynch, An optimized undulator beamline for high-resolution photoemission valence band spectroscopy, *Nuclear Instruments and Methods in Physics Research, Section A*, **347**, 278–281, (1994).

[74] W. Eberhardt, P. Fayet, D.M. Cox, Z. Fu, A. Caldor, R. Sherwood, and D. Sondericker, Photoemission from mass-selected monodispersed Pt clusters, *Phys. Rev. Lett.*, **64**, 780–783, (1990).

[75] R.I.R. Blyth, A.B. Andrews, A.J. Arko, J.J. Joyce, P.C. Canfield, B.I. Bennett, and P. Weinberger, Valence band photoemission and Auger line shape study of Au_xPd_{1-x}, *Phys. Rev. B*, **49**, 16149–16155, (1994).

[76] G. Beamson, A. Bunn, and D. Briggs, High-resolution monochromated XPS of poly(methyl methacrylate) thin films on a conducting substrate, *Surf. Interface Anal.*, **17**, 105–115 (1991).

[77] B.D. Ratner, The surface characterization of biomedical materials: how finely can we resolve surface structure? in *Surface Characterization of Biomaterials*, B.D. Ratner, (Ed.) Elsevier, Amsterdam, (1988) pp. 13–36.

[78] A.P. Pijpers and W.A.B. Donners, Quantitative determination of the surface composition of acrylate copolymer latex films by XPS, *J. Polym. Sci., Polym. Chem. Ed.*, **23**, 453–462 (1985).

[79] R.J. Meier and A.P. Pijpers, Oxygen-induced next-nearest neighbour effects on the C1s-levels in polymer XPS-spectra, *Theoret. Chim. Acta*, **75**, 261–270 (1989).

[80] U. Gelius, S. Svensson, H. Siegbahn, E. Basilier, A. Faxalv, and K. Siegbahn, Vibrational and lifetime line broadenings in ESCA, *Chem. Phys. Lett.*, **28**, 1–7 (1974).

[81] F.J. Holly and M.F. Refojo, Water wettability of hydrogels, *ACS Symp. Ser.*, **31**, 252–266 (1976).

[82] B.D. Ratner, P.K. Weathersby, A.S. Hoffman, M.A. Kelly, and L.H. Scharpen, Radiation-grafted hydrogels for biomaterial applications as studied by the ESCA technique, *J. Appl. Polym. Sci.*, **22**, 643–664 (1978).

[83] A. Takahara, N.-J. Jo, K. Takamori, and T. Kajiyama, Influence of aqueous environment on surface molecular mobility and surface microphase separated structure of segmented poly(ether urethanes) and segmented poly(ether urethane ureas), in *Progress in Biomedical Polymers*, C.G. Gebelein and R.L. Dunn, (Eds.) Plenum Press, New York, (1990) pp. 217–228.

[84] K.B. Lewis, B.D. Ratner, L.A. Klumb, and S.I. Ertel, Surface restructuring of biomedical polymers, *Trans. Soc. Biomat.*, **14**, 176 (1991).

[85] W.F. Egelhoff, X-ray photoelectron and Auger electron forward scattering: a new tool for surface crystallography, *CRC Crit. Rev. Solid St. Mat. Sci.*, **16**, 213–235 (1990).

[86] Inspector ESCA System Data Sheet, Fisons Instruments, San Carlos, CA, (1991).

[87] M.P. Seah and G.C. Smith, Concept of an imaging XPS system, *Surf. Interface Anal.*, **11**, 69–79 (1988).

QUESTIONS

1. The observed E_B values for the carbonyl C_{1s} and O_{1s} subpeaks in the high-resolution ESCA spectra of polyacrylamide, polyurea, and polyurethane samples are listed below. The atoms bound to the carbonyl group in each sample are also shown. All E_B values have been referenced by setting the hydrocarbon C_{1s} peak of each polymer to 285.0 eV. Based on the different structures of these functional groups, explain their different C_{1s} and O_{1s} E_B

values. Make sure an explanation is given that is consistent for both the C_{1s} and O_{1s} spectra.

| Sample | Functional group | E_B values (eV) | |
		C_{1s}	O_{1s}
polyacrylamide	$\begin{array}{c} NH_2 \\ \mid \quad \mid \\ -CH-C\!=\!O \end{array}$	288.2	531.4
polyurea	$\begin{array}{c} O \\ \parallel \\ -NH-C-NH- \end{array}$	289.2	531.7
polyurethane	$\begin{array}{c} O \\ \parallel \\ -O-C-NH- \end{array}$	289.6	532.2

2. An ESCA survey scan of a material detected the presence of carbon, nitrogen, and oxygen. High-resolution C_{1s}, N_{1s}, and O_{1s} scans of this material showed the presence of two, one, and one sub-peaks, respectively. Assuming that the transmission function does not vary with KE, λ varies as $KE^{0.7}$, and that there is an Al $K\alpha$ X-ray source, use the data provided below to calculate the percentage of each component present in this sample. Also, propose a chemical structure for this sample and provide a consistent assignment of functional groups for the sub-peaks. Remember, ESCA does not detect hydrogen. Correct the E_B values for sample charging by referencing C_{1s} hydrocarbon peak to 285.0 eV.

Peak	E_B (eV)	Area
C_{1s}	276.8	4000
C_{1s}	279.9	2000
N_{1s}	391.6	3355
O_{1s}	523.2	4995

3. An ESCA survey scan of a material detected the presence of only carbon and oxygen. High-resolution C_{1s} and O_{1s} scans of this material showed the presence of four and three sub-peaks, respectively. Assuming the transmission function does not vary with KE, λ varies as $KE^{0.7}$, and an Al $K\alpha$ X-ray source, use the data provided below to calculate the C/O atomic ratio and the percentage of each component present in this sample. Also, propose a chemical structure for this sample and provide a consistent assignment of functional groups for the subpeaks. The E_B values have been corrected for sample charging.

Peak	E_B (eV)	Area
C_{1s}	285.0	1925
C_{1s}	286.6	675
C_{1s}	289.0	675
C_{1s}	291.6	100

O_{1s}	532.1	1600
O_{1s}	533.7	1685
O_{1s}	538.7	85

4. There is much interest in special properties of buckminsterfullerenes. The buckminsterfullerene consists of 60 carbon atoms arranged in an icosahedron (a soccer ball-like shape). The diameter of the icosahedral 'sphere' is 0.71 nm. Consider the cases where we have three close-packed layers, one close-packed layer, and a partial monolayer (covering 70% of the surface area) of 'buckyballs' on molecularly smooth, contamination-free gold substrates. These specimens are examined by ESCA. Roughly sketch plots of the anticipated relative gold photoemission signal intensity (I) as a function of photoelectron (sample) take-off angle (θ) (*note: in this example, θ is measured to the normal to the specimen*) for the three specimens. Since the icosahedra pack closer than spheres, you can assume that the buckminsterfullerene molecules can be modelled as solid cubes, 0.71 nm on a side. The plots you sketch should approximately represent the anticipated characteristics of the signal intensity variation without the need for accurate x- or y-axis numbers — the functional relationship is being explored here. The y-axis should have two numbers on it: 100% signal (relative to clean gold) and 0% signal (no gold signal). The x-axis should have $0°$ and $80°$ on it.

5. Consider a molecularly smooth Teflon $[-(CF_2CF_2)_n-]$ overlayer deposited on a molecularly smooth, contamination-free platinum surface. Using ESCA (the sample analysis angle relative to the surface normal is specified in parentheses) to study this type of specimen, arrange the following cases in order of decreasing relative platinum signal: [a] a 7.0 nm Teflon overlayer and an Al Kα X-ray source ($0°$), [b] a 0.5 nm Teflon overlayer and an Al Kα X-ray source ($0°$), [c] a 0.5 nm Teflon overlayer and a Ag Lα X-ray source ($0°$), [d] a 7.0 nm Teflon overlayer and an Al Kα X-ray source ($80°$), and [e] a 0.5 nm Teflon overlayer and a Ti Kα X-ray source ($0°$).

6. A polymeric surface rich in hydroxyl groups is derivatized with a vapour-phase reagent that converts the $-OH$ groups to $-OCF_3$ groups. The specimen is studied in an ESCA instrument with a non-monochromatized Mg Kα X-ray source. The relative fluorine signal is observed to decrease with increasing time under analysis. Suggest three reasons why this might occur. Suggest two instrumentation strategies to make this derivatization analysis useful for analytically comparing specimens as to $-OH$ content.

CHAPTER 4

Auger Electron Spectroscopy

HANS JÖRG MATHIEU
EPFL, Lausanne, Switzerland

1 INTRODUCTION

Auger Electron Spectroscopy (AES) represents today the most important chemical surface analysis tool for conducting samples. The method is based on the excitation of so-called 'Auger electrons'. Already in 1923 Pierre Auger [1] had described the β emission of electrons due to ionization of a gas under bombardment by X-rays. This ionization process can be provoked either by electrons — commonly known as the Auger process — or by photons as used by P. Auger. In the latter case we will call this method photon induced Auger Electron Spectroscopy (see also the chapter on ESCA/XPS). Today's AES is based on the use of primary electrons with typical energies between 3 and 30 keV and the possibility to focus and scan the primary electron beam in the nanometer and micrometer range analyzing the top-most atomic layers of matter. The emitted Auger electrons are part of the secondary electron spectrum obtained under electron bombardment with a characteristic energy allowing one to identify the emitting elements. The experimental setup is very similar to that of a Scanning Electron Microscope — with the difference that the electrons are not only used for imaging but also for chemical identification of the surface atoms.

Auger electrons render information essentially on the elemental composition of the first 2–10 atomic layers. Figure 4.1 shows schematically the distribution of electrons, i.e. primary, backscattered and Auger electrons together with the emitted characteristic X-rays under electron bombardment. We notice that under typical experimental conditions the latter have a larger escape depth due to a much smaller ionization cross section with matter, i.e. a higher probability to escape matter. Auger electrons with energies up to 2000 eV, however, have a high probability to escape only from the first few monolayers because of their restricted kinetic energy. Consequently, they are much better suited for surface analysis. A second important detail is shown in Figure 4.1 revealing that the diameter of the analyzed zone can be larger than the diameter of the primary beam due to scattering of electrons.

Surface Analysis — The Principal Techniques. Edited by John C. Vickerman
© 1997 John Wiley & Sons Ltd

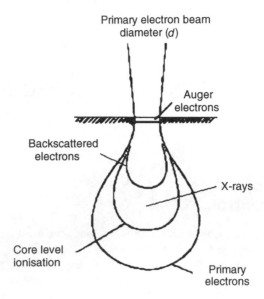

Figure 4.1. Distribution (schematic) of primary, backscattered and Auger electrons together with X-rays

2 PRINCIPLE OF THE AUGER PROCESS

Before determining the kinetic energy of Auger electrons let us have a quick look at quantum numbers and nomenclature. A given energy state is characterized by four quantum numbers, i.e. n (principal quantum number), l (orbital), s (spin) and j (spin-orbit coupling with $j = l + s$). The latter can only have values with $j = 1 \pm 1/2$ except $j = -1/2$. The energy E (nlj) of a given electronic state can therefore be characterized by these three numbers as indicated in Table 4.1 for certain elements.

Table 4.1. Nomenclature of AES and XPS peaks

n	l	j	Index	AES notation	XPS notation
1	0	1/2	1	K	$1s_{1/2}$
2	0	1/2	1	L_1	$2s_{1/2}$
2	1	1/2	2	L_2	$2p_{1/2}$
2	1	3/2	3	L_3	$2p_{3/2}$
3	0	1/2	1	M_1	$3s_{1/2}$
3	1	1/2	2	M_2	$3p_{1/2}$
3	1	3/2	3	M_3	$3p_{3/2}$
3	2	3/2	4	M_4	$3d_{3/2}$
3	2	5/2	5	M_5	$3d_{5/2}$
etc.				etc.	etc.

2.1 KINETIC ENERGIES OF AUGER PEAKS

Figure 4.2 shows schematically the Auger process. The primary beam energy has to be sufficiently high to ionize a core level W (i.e. K, L, ...) with energy E_W. The empty electron position will be filled by an electron from a level E_X closer to the Fermi level. The transition of the electron between levels W and X liberates an energy corresponding to $\Delta E = E_W - E_X$ which in turn is transferred to a third electron of the same atom at level E_Y. The kinetic energy of this third electron corresponds therefore to the difference of energy between the three electronic levels involved minus the sample work function, Φ_e. If the sample is in good contact with the sample holder (i.e Fermi levels of sample and instrument are

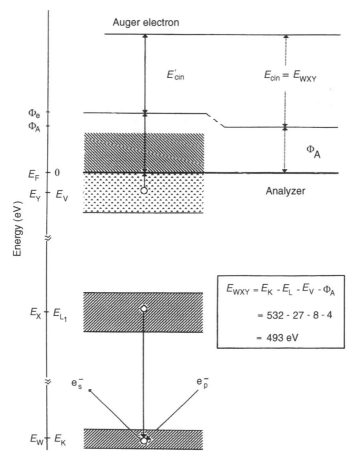

Figure 4.2. Auger process: E_F is the Fermi level (zero atomic energy level for binding energies of electrons. Φ_e and Φ_A are the work function of sample (e) and analyzer (A), respectively

Table 4.2. Binding energies of some elements

Z	El	$1s_{1/2}$ K	$2s_{1/2}$ L_1	$2p_{1/2}$ L_2	$2p_{3/2}$ L_3	$3s_{1/2}$ M_1	$3p_{1/2}$ M_2	$3p_{1/2}$ M_3	$3d_{3/2}$ M_4	$3d_{5/2}$ M_5
1	H	14								
2	He	25								
3	Li	55								
4	Be	111								
5	B	188			5					
6	C	284			6					
7	N	399			9					
8	O	532	24		7					
9	F	686	31		9					
10	Ne	867	45		18					
11	Na	1072	63		31	1				
12	Mg	1305	89		52	2				
13	Ml	1560	118	74	73	1				
14	Si	1839	149	100	99	8				
15	P	2149	189	136	135	16	10			
16	S	2472	229	165	164	16	8			
17	Cl	2823	270	202	200	18	7			
18	Ar	3202	320	247	245	25	12			
19	K	3608	377	297	294	34	18			
20	Ca	4038	438	350	347	44	26		5	
21	Sc	4493	500	407	402	54	32		7	
22	Ti	4965	564	461	455	59	34		3	
23	V	5465	628	520	513	66	38		2	
24	Cr	5989	695	584	757	74	43		2	
25	Mn	6539	769	652	641	84	49		4	
26	Fe	7114	846	723	710	95	56		6	
27	Co	7709	926	794	779	101	60		3	
28	Ni	8333	1008	872	855	112	68		4	
29	Cu	8979	1096	951	932	120	74		2	
30	Zn	9659	1194	1044	1021	137	90		9	
31	Ga	10367	1299	1144	1117	160	106		20	
42	Mo	20000	2866	2625	2520	505	410	393	208	205
46	Pd	24350	36304	3330	3173	670	559	531	340	335
48	Ag	25514	3806	3523	3351	718	602	571	373	367
73	Ta*	67416	11681	11136	11544	*566	*464	*403	*24	*22
79	Au*	80724	14352	13733	14208	*763	*643	*547	*88	*84

* 4s, 4p et 4f levels indicated, respectively

identical) we can determine the kinetic energy of an element with atomic number Z and an Auger transition between level W, X and Y as follows:

$$E_{WXY} = E_W(Z) - E_X(Z + \Delta) - E_Y(Z + \Delta) - \Phi_A \qquad (1)$$

where Φ_A represents the work function of the analyzer and W, X and Y the three energy levels of the Auger process involved (i.e. KLL, LMM, MNN — omitting to note the sub-levels). The term Δ (between 0 and 1) denotes the displacement

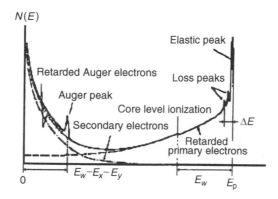

Figure 4.3. Schematic representation of an Auger spectrum

of an electronic level towards higher binding energies after the ionization of the atom by the primary electron. $\Delta = 0.5$ represents a fair approximation for an estimate of the kinetic energy. The work function of the analyzer detector is typically 4 eV. Taking such values (see Table 4.2) we obtain for the transition of oxygen O_{KLL} an energy $E_{KLL} = 493$ eV as indicated by Figure 4.2. Experimentally one finds energy values between 493 eV and 512 eV depending on the oxide.

Figure 4.3 shows schematically an Auger spectrum in which the number of emitted electrons N is given as a function of the kinetic energy E.

We observe that the Auger peaks are superimposed on the spectrum of the secondary electrons. The elastic peak E_p represents the primary energy applied. We further notice on the tail of the elastic peak characteristic loss peaks from the ionization levels (E_W, E_X, etc.) and on the low kinetic energy side of the Auger peaks tails which are due to uncharacteristic energy losses. Auger transitions have been calculated and can be found in the literature [2–6]. Figure 4.4 gives the principal Auger transitions of all elements starting form Li. Since for an Auger transition a minimum of three electrons is required, only elements with $Z \geqslant 3$ can be analyzed. Table 4.3 gives numerical values of the principal transitions together with other useful parameters in AES.

2.2 IONIZATION CROSS-SECTION

The probability of an Auger transition is determined by the probability of the ionization of the core level W and its de-excitation process involving the emission of an Auger electron or a photon. Primary electrons with a given energy E arriving at the surface will ionize the atoms starting at the surface of the sample. The cross-section, $\sigma_W(E)$, calculated by quantum mechanics for the Auger process (A) at an energy core level W can be estimated by

$$\sigma_W = \text{const} \times \frac{C(E_p/E_W)}{E_W^2} \qquad (2)$$

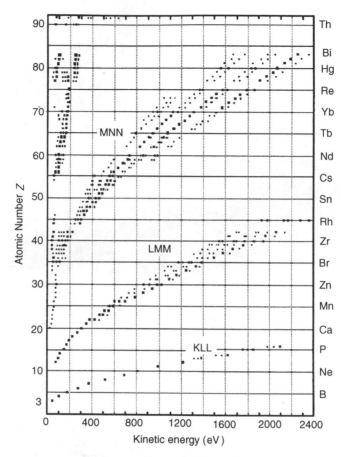

Figure 4.4. Principal transitions of AES

where the constant depends on the core level W (= K, L, M). σ_W is a function of the primary energy E_P and the core level E_W. Figure 4.5 shows experimental results together with the calculated σ_W according to equation (2) as a function of the ratio E_P/E_W. One observes that the ionization cross-section passes through a maximum at approximatively $E_P/E_W = 3$. Typical values for σ_W are 10^{-3}–10^{-4}. This means that the probability of an ionization followed by an Auger de-excitation is 1 in 10^4. Thus one finds experimentally Auger electron transition superimposed on a high secondary electron spectrum as indicated in Figure 4.3.

2.3 COMPARISON OF AUGER AND PHOTON EMISSION

Figure 4.2 indicated schematically the Auger process. We have already learned that after ionization of the core level W the de-excitation takes place by an

Table 4.3. AES transitions and their relative sensitivity factors

Legend (cell key):

Atomic Number	Z
Atomic Volume	A/ρ* [× 10⁻⁶ m³ / mol] — shown as A/ρ^*
Element	
S(5)	AES rel. sensitivity at 5 keV**
S(10)	AES rel. sensitivity at 10 keV**
LMM KE	AES transition, Kinetic Energy of AES transition (eV)

Cell layout:

```
Z   Element   A/ρ
    S(5)
    S(10)
    LMM  KE
```

* atomic weight A – mass density ρ
** valid for CMA only

Periodic-table data (Z, Element, A/ρ, S(5), S(10), AES transition, KE in eV):

Z	Element	A/ρ	S(5)	S(10)	Transition	KE
1	H	14.1				
2	He	31.8				
3	Li	13.1	0.160		KLL	43
4	Be	5.0	0.10	0.045	KLL	104
5	B	4.6	0.120	0.055	KLL	179
6	C	5.3	0.14	0.08	KLL	272
7	N	17.3	0.230	0.160	KLL	379
8	O	14	0.400	0.350	KLL	508
9	F	17.1	0.48	0.45	KLL	647
10	Ne	16.8			KLL	805
11	Na	23.7	0.25	0.23	KLL	990
12	Mg	14.0	0.13	0.13	KLL	1186
13	Al	16.0	0.19	0.15	LMM	68
14	Si	12.1	0.28	0.15	LMM	92
15	P	17.0	0.47	0.30	LMM	120
16	S	15.5	0.75	0.57	LMM	152
17	Cl	18.7	1.05	0.69	LMM	181
18	Ar	24.2			KLL	215
19	K	45.3	0.90	0.37	KLL	252
20	Ca	29.9	0.40	0.22	LMM	291
21	Sc	15.0	0.28	0.20	LMM	340
22	Ti	10.6	0.34	0.23	LMM	418
23	V	8.35	0.38	0.29	LMM	473
24	Cr	7.23	0.31	0.28	LMM	529
25	Mn	7.39	0.193	0.160	LMM	589
26	Fe	7.1	0.22	0.15	LMM	703
27	Co	6.7	0.23	0.19	LMM	775
28	Ni	6.6	0.27	0.22	LMM	848
29	Cu	7.1	0.23	0.20	LMM	920
30	Zn	9.2	0.19	0.18	LMM	994
31	Ga	11.8	0.16	0.14	LMM	1070
32	Ge	13.6	0.130	0.125	LMM	1147
33	As	13.1	0.12	0.11	LMM	1228
34	Se	16.5	0.092	0.088	LMM	1315
35	Br	23.5	0.075	0.074	LMM	1376
36	Kr	32.2			MNN	53
37	Rb	55.9	0.052	0.053	LMM	1565
38	Sr	33.7	0.043	0.045	LMM	1649
39	Y	19.8	0.11	0.01	MNN	127
40	Zr	14.1	0.16	0.15	MNN	147
41	Nb	10.8	0.21	0.18	MNN	167
42	Mo	9.4	0.28	0.28	MNN	186
43	Tc					
44	Ru	8.3	0.50	0.37	MNN	273
45	Rh	8.3	0.68	0.47	MNN	302
46	Pd	8.9	0.89	0.50	MNN	330
47	Ag	10.3	0.97	0.67	MNN	356
48	Cd	13.1	0.99	0.68	MNN	376
49	In	15.7	0.97	0.65	MNN	404
50	Sn	16.3	0.90	0.53	MNN	430
51	Sb	18.4	0.65	0.40	MNN	454
52	Te	20.5	0.47	0.28	MNN	483
53	I	25.7	0.34	0.21	MNN	511
54	Xe	42.9	0.24	0.15	MNN	532
55	Cs	70	0.17	0.12	MNN	563
56	Ba	39	0.12	0.08	MNN	584
57	La	22.5	0.88	0.60	MNN	625
58	Ce	21.0	0.068	0.045	MNN	661
59	Pr	20.8	0.055	0.038	MNN	699
60	Nd	20.6	0.047	0.032	MNN	730
61	Pm					
62	Sm	19.9	0.033	0.026	MNN	814
63	Eu	28.9	0.029	0.025	MNN	858
64	Gd	19.9	0.027	0.024	MNN	895
65	Tb	19.2	0.025	0.025	MNN	1073
66	Dy	19.0	0.027	0.027	MNN	
67	Ho	18.7	0.030	0.030	MNN	1126
68	Er	18.4	0.036	0.035	MNN	1175
69	Tm	18.1	0.042	0.040	MNN	1393
70	Yb	24.8	0.051	0.048	MNN	1449
71	Lu	17.8	0.062	0.058	MNN	1514
72	Hf	13.6	0.141		NNN	185
73	Ta	10.9	0.136	0.093	NNN	179
74	W	9.53	0.115	0.079	NNN	179
75	Re	8.85	0.096		NNN	176
76	Os	8.43				
77	Ir	8.54	0.046		NOO	54
78	Pt	9.10	0.28		NOO	64
79	Au	10.2	0.34	0.21	NOO	69
80	Hg	14.8	0.030		NOO	76
81	Tl	17.2	0.42		NOO	84
82	Pb	18.3	0.40		NOO	94
83	Bi	21.3	0.37		NOO	101
84	Po	22.7				
85	At					
86	Rn					
87	Fr					
88	Ra	45				
89	Ac					
90	Th	19.9	0.286		OPP	65
91	Pa	15.0			OPP	
92	U	12.5	0.437 (3 keV)		OPP	72

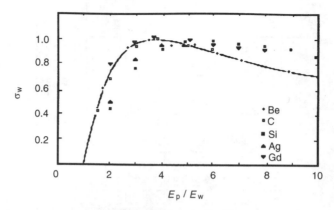

Figure 4.5. Variation of the ionization cross-section with the ratio of primary electron beam energy E_P and core level energy E_W

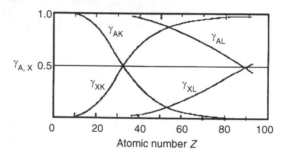

Figure 4.6. Emission probability of an Auger electron (A) or photon (X)

electron filling the place at level W. The liberated energy difference $\Delta E = E_W - E_X$ can either be transferred to an electron of the same atom or a photon with same energy $\Delta E = h\nu$. Again, whether an Auger electron or a photon is emitted is determined by quantum mechanical selection rules. The emission probability varies with the atomic number Z and the type of atomic level involved (K, L, M, etc) leading to cross-sections γ_{AK} and γ_{XK} or γ_{AL} and γ_{XL} for detection via emission of an Auger electron (A) or a photon (X-ray (X)), respectively, as indicated by Figure 4.6. Probability of excitation via an Auger process is very high for light elements and transition of the KLL type, γ_{AK}. However, even for heavy elements one observes a relatively high probability for elements of type LMM (γ_{AL}) or MNN (γ_{AM} — not shown).

2.4 ELECTRON BACKSCATTERING

In Auger Electron Spectroscopy, primary electrons arrive at the sample surface with an energy of 3–30 keV. Monte Carlo calculations indicate that such electrons can penetrate up to a depth of several microns. During their trajectory, these electrons lose a certain amount of energy, change their direction and are also backscattered. They may create secondary electrons, Auger electrons and photons. Some of the backscattered electrons can in turn produce themselves Auger electrons if they have sufficient energy. This way the backscattered electrons contribute to the total Auger current. Since the number of Auger electrons is proportional to the total Auger current one obtains:

$$I_{total} = I_o + I_M = I_o(1 + r_M) \tag{3a}$$

Figure 4.7 shows the backscattering factor, r_M, calculated for various matrices with atomic number Z. One notices that r_M becomes larger for increasing Z, i.e. elements with more free electrons like gold ($Z = 79$), produce more backscattered electrons.

The backscattering factor can be estimated by the following equation:

$$1 + r_M = 1 + 2.8 \left[1 - 0.9 \frac{E_W}{E_p}\right] \eta(Z) \tag{3b}$$

where E_W is the ionization energy of the core level and E_p the primary beam energy with:

$$\eta(Z) = -0.0254 + 0.16Z - 0.00186Z^2 + 8.3 \times 10^{-7}Z^3 \tag{3c}$$

Inspection of Figure 4.7 indicates the importance of the variation of r_M for Auger analysis, especially for very thin films on a substrate producing a large number of backscattered electrons.

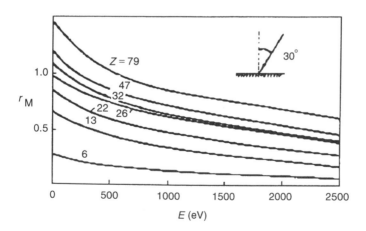

Figure 4.7. Electron backscattering factor r_M as a function of kinetic energy energy for a primary electron energy of 5 keV and an angle of incidence of $\theta = 30°$ (courtesy of M.P. Seach in [2])

2.5   ESCAPE DEPTH

The attenuation length of Auger electrons λ with a kinetic energy E_{kin} determines the escape depth Λ according to:

$$\Lambda = \lambda \cos \theta \tag{4a}$$

where θ is the emission angle of the Auger electrons with respect to surface normal. The probability for an electron to travel over a distance x without any collision is proportional to $\exp(-x/\Lambda)$ 95% of the Auger intensity comes from within 3Λ of the surface. A rough estimate of λ is obtained for the elements by [2]:

$$\lambda = 0.41a^{1.5}E_{cin}^{0.5} \tag{4b}$$

where a (in nm) is the monolayer thickness of a cubic crystal calculated by:

$$\rho N\, a^3 = A \tag{4c}$$

with ρ the density (in kg/m^3), N Avogadro's number ($N = 6.023 \times 10^{23}$ /mol, a (in m) and A the molecular mass (kg/mol) of the matrix in which the Auger electron is created. The ratio A/ρ (atomic volume) is given in Table 4.3. Figure 4.8 shows λ as a function of the kinetic energy. It reveals that λ varies from 2 to 20 monolayers for typical Auger energies up to 2000 eV. The thickness of a

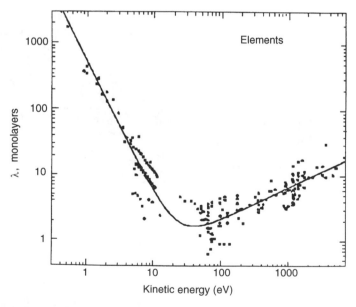

Figure 4.8. Dependence of attenuation length, λ, on kinetic energy. (courtesy of M.P. Seah in [2])

monolayer is approximatively 0.2–0.25 nm for metals. Since the kinetic energy determines the escape depth, a measurement of two peaks of the same element but of different energy can be used as a measure for the variation of composition with depth.

2.6 CHEMICAL SHIFTS

A change of the oxidation state of an element results in a shift of the binding energy of the valence band level. Therefore, in principle, each time a change of a

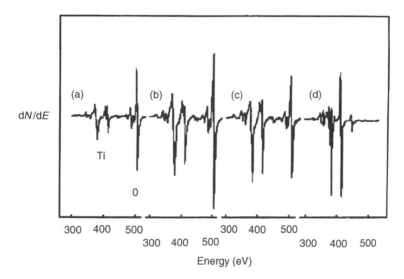

Figure 4.9. Examples of chemical shift in AES (a) TiO_2, (b) Ti_2O_3, (c) TiO and (d) Ti metal

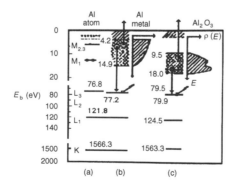

Figure 4.10. Energy levels of (a) Al atom, (b) metal and (c) Al_2O_3

binding energy occurs, one observes also a 'chemical shift' for Auger transitions. The same phenomenon is found in ESCA. However, since three energy levels are involved in an Auger transition such shifts cannot always easily be correlated to a shift of one particular level. Such fine structure of peaks of certain elements is well known (i.e. C, Si, Al) allowing the experimentalist to distinguish between different states of oxidation as indicated in Figure 4.9. Figure 4.10 illustrates an example of the variation of the levels of the different peaks of aluminum and indicates schematically differences of the density of electrons $\rho(E)$ of the M-level. Chemical shifts during AES profiling will be discussed in Section 4.6.

3 INSTRUMENTATION

The main parts of an Auger spectrometer are the electron gun and the electrostatic energy analyzer. Both are placed in an ultra-high vacuum chamber with base pressures between 10^{-9} and 10^{-11} mbar. Such low pressures are necessary to guarantee a contamination-free surface to keep adsorption of residual gases below 10^{-3} monolayers/second. This is achieved for pressures of 10^{-9} mbar or below. Essential accessories of spectrometers are vacuum gauges for total pressure reading, a partial pressure analyzer controlling the rest gas, a fast introduction lock and a differentially pumped ion gun for sample cleaning or in-depth thin film analysis, together with a secondary electron collector for imaging. Figure 4.11 shows an example of a simple Auger spectrometer using a cylindrical mirror analyzer (CMA) with a variable potential applied between an inner and outer cylinder and resulting in a signal which is proportional to the number of detected electrons N at kinetic energy E. The other type of analyzer used in AES, i.e. an hemispherical analyzer (HPA), is shown in Figure 4.12, is often used in XPS analysis. In general, HPA's give a better energy resolution. The electron beam can be static (static AES) or scanned (Scanning Auger Microprobe (SAM)). The

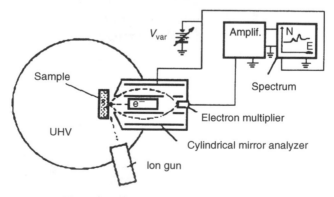

Figure 4.11. Cylindrical mirror analyzer

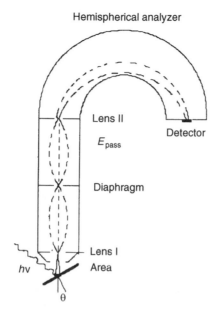

Figure 4.12. Hemispherical analyzer

lateral resolution depends on the electron optics applied (electrostatic or electro-magnetic lenses). Achievable lateral resolution of spectrometers in the Auger detection mode is approx. 0.05 μm.

3.1 ELECTRON SOURCES

Todays scanning Auger systems use three types of electron source with decreasing lateral resolution: (a) tungsten filament, (b) LaB_6 crystal or (c) a field emission gun (FEG). The classical W-filament reaches a minimum beam diameter of 3–5 μm. Only LaB_6 or FEG sources give beam diameters ≤20 nm and their primary electron beam energy has to be increased to 20–30 keV. Lowest beam diameters for a given primary beam current are obtained by a field emission gun (see Figure 4.13) which in turn is more delicate and demands a better control of the vacuum.

The two types of electron sources, thermionic [a,b] and field emission [c] are based on rather different physical principles. The former more common ones apply a certain thermal energy to remove an electron. This energy is called the work function, which represents the barrier at the material surface necessary to free the electron. Typical work function energies are around 4–5 eV. For thermionic sources the material is heated by passing a certain current to obtain a sufficiently high temperature to allow the electrons to reach the vacuum. Field emission is based on the 'tunnelling' process of electrons which is probable, if

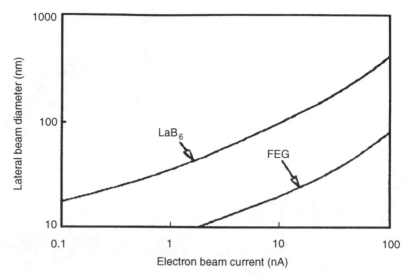

Figure 4.13. Comparison of electron sources (LaB$_6$ and Field Emission Gun (FEG))

a sufficiently high electrical field between the emitter and an extraction electrode is applied. Sharp needle-like points of typically 20–50 nm radii and short distances between emitter and extraction electrode (nm) are needed. In the end the ultimate limit of the lateral resolution is determined by the focussing lenses. Purely electrostatic electron guns allow focussing to 0.2 µm, whereas electromagnetic focussing allows one to decrease spot sizes down to 0.02 µm for LaB$_6$ or tungsten field emitters, respectively. Such field emitters are used in scanning electron microscopes as well. However, focussing of the electron beam may lead to beam damage, particularly for sample areas of low conductivity. To avoid beam damage, beam current densities above 1 mA cm^{-2} corresponding to 1 nA into a spot of 10 µm should be applied. Unfortunately, such limits cannot always be met, particularly in high lateral resolution work, leading in certain cases to local decomposition.

For beam currents $\geqslant$ 10 nA the LaB$_6$ source is superior to both thermionic and field emitting tungsten in terms of spot size obtainable and signal-to-noise ratio. However, for a better lateral resolution at beam currents below 1 nA, the W field emitter is preferred.

3.2 SPECTROMETERS

As mentioned already, two types of analyzers are used in AES, either a CMA or an HPA. The CMA (Figure 4.11) has a larger electron transmission than the HPA (Figure 4.12). The transmission is defined as the ratio of emitted to detected Auger electrons. A scanning electron gun is built coaxially into the CMA avoiding, in certain cases, shadowing effects since the analyzer and electron gun axis are identical. The CMA derives its name from the fact that the electron emitting spot on the sample surface is imaged by the CMA at the detector surface. Primary electrons of known energy which are reflected from the sample surface are used to optimize the signal intensity to find the analyzed spot and calibrate the analyzer.

In HPA's, the primary electron is off-axis allowing a simpler geometry and a better definition of the angle of emitted electrons (compare Figure 4.12). The working distance between sample and analyzer is generally larger for HPA's (approx. 10 mm). At the entrance of the analyzer a system of electrostatic lenses is placed to define the accepted analysis area. In the cylindrical part of the analyzer a diaphragm limits the analyzed area and a second electrostatic lens controls the pass energy of the electrons. A potential is applied to this second lens to reduce the kinetic energy of the Auger electrons allowing one to operate the analyzer at constant pass energy. The hemispherical part of the analyzer focuses the electrons in the plane of a detector which is an arrangement of different channeltron or channelplate electron multipliers. The detecting system measures directly the number of electrons at a certain kinetic energy $N(E)$. Such an analyzer can be used in two detection modes:

1. $\Delta E = $ const (FAT = fixed Analyzer Transmission mode) applying a constant pass energy by controlling lens II.
2. $\Delta E/E = $ const (FRR Fixed Relative Resolution) applying a constant energy ratio where ΔE is the FWHM (Full Width at Half Maximum) of a given peak and E its kinetic energy.

Better electron detectors lead to a better signal-to-noise ratio (800:1 for Cu LMM line) and allow one to decrease the detection limit (1% of a monolayer) at a given spatial resolution (0.1 μm) and fixed primary beam current (10 nA). For better radiation shielding each analyzer is made out of stainless steel and/or completely of mμ-metal.

3.3 MODES OF ACQUISITION

There are four modes of operation in Auger Electron Spectroscopy:

1. point analysis
2. line scan
3. mapping
4. profiling

Figure 4.14 shows a typical survey spectrum as a result of a point analysis indicating the number of detected electrons $N(E)/E$ as a function of kinetic energy (compare also Figure 4.3). One observes a large number of secondary electrons on which the Auger electrons are superimposed. Transitions of oxygen

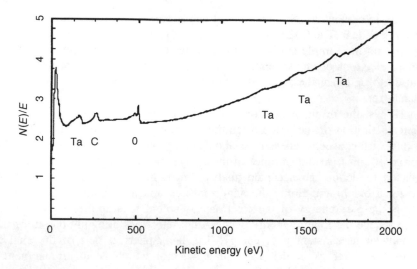

Figure 4.14. Typical survey spectrum Ta_2O_5 as a result of a point analysis

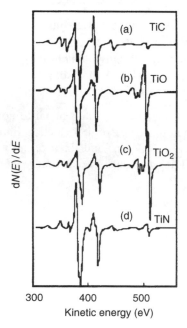

Figure 4.15. Differentiated AES spectra of Titanium as function of their kinetic energy for: (a) TiC; (b) TiO; (c) TiO_2 and (d) TiN

and tantalum have been labelled. Carbon is found as a surface contamination, often observed on samples introduced into the UHV. Peaks can be represented in their differentiated form after background subtraction, i.e. $dN(E)/dE$ as illustrated by Figure 4.15 where different chemical states of titanium are shown. This mode of representation is often applied to point out subtle differences and fine structure of the spectrum. However, as indicated already above, identification of oxidation states is generally easier in ESCA because of the involvement of three energy levels in the Auger process. The advantage of AES is the small spot size and the shorter acquisition time of the measurement.

As we noticed already above, the escape depth of Auger electrons is limited to a few nm. Many practical problems require determination of the variation of an element with depth. The AES systems are equipped to perform different types of depth analysis as illustrated by Figure 4.16. It is evident that the principles shown in Figure 4.16 apply also to other methods like XPS or SIMS. For layers of thicknesses of a few nm one measures the detected intensity as a function of the angle θ making use of equation (4) and illustrated by Figure 4.17. Such angular resolved analysis (AREAS) is limited to a very shallow depth because λ is typically only a few nm as discussed above (Section 2.5)

Composition of thicker layers up to 0.2–1 µm can be determined by combining Auger analysis with Ar^+ (or Kr^+) sputtering by observation of the Auger signal at the bottom of the sputtered crater, either simultaneously or alternately. Sputter depth profiling will be discussed in a separate section (5) below, in more detail.

Layers of even larger thickness (i.e. a few µm) should be analyzed by other methods, i.e. electron microprobe analysis, or — if light elements are to be detected, by scanning the electron beam across a mechanically prepared ball

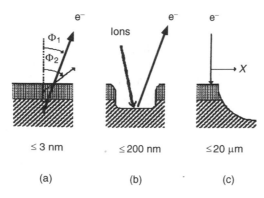

Figure 4.16. Principle of different types of in-depth measurements. (a) non-destructive measurement for layers $\leqslant$ 2–3 nm by variation of the angle of emission; (b) for layers $\leqslant$200–1000 nm by combining AES analysis with destructive sputter erosion and (c) line scan over a creater edge produced by ball cratering or taper sectioning under a small angle applied for layers $\leqslant$ 20 µm.

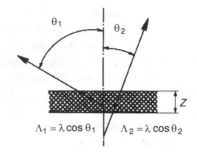

Figure 4.17. Variation of escape depth with angle of emission

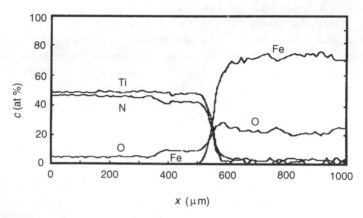

Figure 4.18. Example of a line scan over the crater edge produced by ball cratering showing the atomic concentration as a function of the displacement of the electron beam. The crater edge is located approximatively at $x = 500$ μm

crater or a tapered section. Figure 4.18 gives an example of the line scan mode. A section of a sample of stainless steel is covered with a layer of TiN as shown in Figure 4.19. The crater has been prepared by mechanical abrasion of the layer with a stainless steel sphere — for more details of the method see Lea [7]. The electron beam is scanned from left to right over the TiN layer over the crater edge before reaching the substrate. The displacement x can be correlated to the thickness d of the layer and the depth z by equation (5)

$$z = \tfrac{1}{2}(4R^2 - d^2 - 4x^2 + 4dx)^{1/2} - \tfrac{1}{2}(4R^2 - d^2)^{1/2} \tag{5}$$

where R is the radius of the sphere used during polishing and, d the diameter of the crater produced at the surface.

An application of the third mode of scanning Auger analysis is illustrated by Figure 4.20 showing the lateral distribution of a ternary AlSiMg alloy after a chemical treatment in sulphuric acid and subsequent oxidation in air. In this

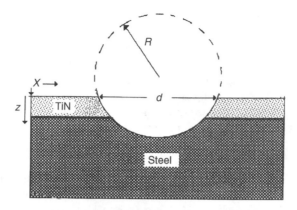

Figure 4.19. Section of a stainless steel sample covered with a layer of TiN of thickness d. The vertical arrows indicate the limits of displacement of the electron beam

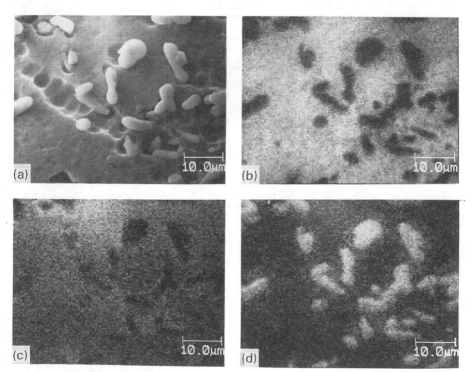

Figure 4.20. Example of an Auger map of AlSiMg alloy showing the secondary electron image. White pixels indicate the presence of an element. (a) SEM; (b) Al; (c) S; (d) Si

mode the electron beam is scanned over a selected area of the sample. The Auger intensity is measured at each point of the area by keeping the analyzer pass energy constant at the peak maximum and minimum of an elemental peak, respectively. The image displayed shows the peak intensity (maximum minus minimum) of each pixel. Inspection of Figure 4.20 indicates a non-uniform distribution of the elements revealing a complementary distribution of the elements. Such mapping allows one to correlate qualitatively, elemental distributions at the surface. Quantification is simpler and faster by selected point analysis.

3.4 DETECTION LIMITS

Identification of Auger peaks is often easier for light elements than for heavier elements because of the interference of peaks of heavier elements with a larger number of transitions. Peaks with higher kinetic energy have a larger width (FWHM is typically 3–10 eV) and therefore peak overlap is more likely. The sensitivity of elements varies only by one order of magnitude, where silver is the most sensitive and yttrium one of the least sensitive elements. The detection limits are set by the signal/noise ratio. Typical limits are:

concentration:	0.1–1% of a monolayer
mass (volume of 1 μm × 1 μm of 1 nm thickness):	10^{-16}–10^{-15} g
atoms:	10^{12}–10^{13} atoms/cm^2

Scanning Auger analysis allows one to decrease the area of detection below the micron level. However, a finely focussed electron beam may provoke a change in composition if the power dissipated into a small area is too large. One should avoid exceeding a limit of 10^4W cm^{-2}. In addition, the detection limit is drastically lower in the mapping mode as illustrated by Figure 4.21 for a pure Cu

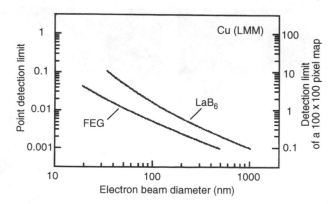

Figure 4.21. Detection limit (monolayer) as a function of lateral resolution for point analysis (left ordinate) and mapping (right ordinate) for the most common electron sources used in AES, i.e. LaB$_6$ crystal sources or field emission (FEG)

sample. Inspection reveals that a static measurement with a lateral resolution of 50 nm gives a detection limit between 0.1 and 0.01 of a monolayer depending on the kind of electron source used. As already mentioned earlier, the field emission gun has a higher brightness and therefore a better detection limit. However, for mapping, the detection limit deteriorates by approximatively a factor of 100 compared to the point analysis, because of the shorter acquisition time per pixel.

4 QUANTITATIVE ANALYSIS

The Auger peak intensity of an element A can be correlated to its atomic concentration $c_A(z)$. Supposing the signal comes from a layer of thickness dz and depth z analyzed at an emission angle θ with respect to surface normal, one obtains the intensity I_A of an Auger peak by:

$$I_A = g \int_0^\infty c_A(Z) \exp \left(-\frac{z}{\lambda \cos \theta} \right) dz \qquad (6)$$

where the attenuation length λ is calculated by equation (4 a–c). g is given by:

$$g = T(E)D(E)I_o \sigma \gamma \, (1 + r_M) \qquad (7)$$

neglecting the influence of the roughness R, where:

$c_A(z)$ concentration of element A which has a function of depth z

λ attenuation length of the Auger electron

θ emission angle with respect to surface normal

$T(E)$ transmission factor which is a function of the kinetic energy E of the Auger electron

$D(E)$ detection efficiency of the electron multiplier — a factor which may vary with time

I_o primary current

σ cross-section of the Auger process

γ probability of an Auger transition (to be compared to the emission of a photon during the de-excitation process)

r_M electron backscatter factor which is matrix (M) dependent (see Section 2.4).

Assuming that we have a flat surface and a homogeneous depth distribution of element A in a matrix M, integration of equation (6) gives

$$I_{A,M} = D(E)T(E)I_o \sigma_A \gamma_A (1 + r_{A,M}) \lambda_{A,M} c_{A,M} \qquad (8)$$

Applying equation (8) to a binary alloy (or an oxide MeO) one obtains for element A or B in a matrix A-B:

$$\frac{I_{A,AB}}{I_{B,AB}} = \frac{\sigma_{A,A}}{\sigma_{B,B}} K \frac{x_A}{x_B} \tag{9}$$

where $\sigma_{A,A}$ and $\sigma_{B,B}$ are the Auger cross-sections of the pure elements and x_A and x_B the mole fractions of A and B, respectively. Instead of Auger cross-sections Auger sensitivity factors, s_i, can be applied for quantification, if relative elemental concentrations are calculated. The constant K is defined by:

$$K = k_r k_\lambda k_c \tag{10}$$

and

$$k_r = \frac{1 + r_{A,AB}}{1 + r_{B,AB}} x \frac{1 + r_{B,B}}{1 + r_{A,A}} \tag{11}$$

$$k_\lambda = \frac{\lambda_{A,AB}}{\lambda_{B,AB}} x \frac{\lambda_{B,B}}{\lambda_{A,A}} \tag{12}$$

$$k_c = \frac{c_{B,B}}{c_{A,A}} \tag{13}$$

In equations (10–13) the first index indicates the element and the second index the matrix (pure A or B or mixed). Unfortunately matrix factors are often not known. However, as above, discussion of λ and r_M has shown, both parameters as well as the atomic concentration depend and vary with Z. Assuming that their ratio is one ($K = 1$) may introduce errors of 10–50% on the absolute value of the calculated atomic concentration. Table 4.4 illustrates the application of matrix factors to binary alloys. E_A and E_B are the kinetic energies of elements A and B, respectively. Inspection of Table 4.4 reveals that for Ni-Pd, Cr-Fe and Mo-Fe alloys K is found close to one, but for Ag-Pd and in particular for Ni-Mg alloys, a matrix correction should be applied. In quantification, an additional error may be introduced by application of elemental Auger sensitivity factors, s_i, neglecting the oxidation state of an element and/or the influence of the transmission function of the analyzer for a given transition.

Table 4.4. Application of AES matrix factors to binary alloys

Alloy	E_A (eV)	E_B (eV)	k_c	k_λ	k_r	K	s_{AA}/s_{BB}	$K(s_{AA}/s_{BB})$	Exp
Ag-Pd	351	330	1.15	0.93	1.00	1.07	1.09	1.17	1.31
Ni-Pd	848	330	0.74	1.16	1.15	0.99	0.30	0.30	0.27
Cr-Fe	527	703	1.02	0.99	1.02	1.03	1.64	1.69	1.74
Mo-Fe	186	703	1.33	0.87	0.87	1.01	1.09	1.10	1.29
Ni-Mg	61	45	0.47	1.42	0.76	0.51	0.81	0.31	0.39
Ni-Mg	848	1186	0.47	1.45	0.80	0.55	3.11	1.71	1.69

The mole fraction in % at is defined by

$$x_A = \frac{I_A}{s_A} \qquad (14)$$

where the intensity I as well as the sensitivity factor s are measured either by taking the peak area in the direct counting mode or by using the peak-to-peak value after differentiation. The variation of the relative sensitivity factors is shown for a CMA analyzer in Figure 4.22. Numerical values of s_i of the main Auger transitions are found also in Table 4.3.

The composition of a sample of n elements can be calculated semi-quantitatively by

$$x_A = \frac{I_A/s_A}{\sum\limits_{i}^{n} I_i/s_i} \qquad (15)$$

neglecting the matrix correction factors in equation (15). Table 4.5 illustrates the result of an application of equation (15) to the analysis of two steels (316 and

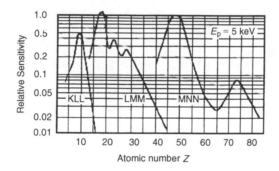

Figure 4.22. Relative Auger sensitivity factors. All factors are defined relative to Ag with: $s_{Ag} = 1$ for a primary electron beam energy of 5 keV. Reproduced with permission from ref. [6]

Table 4.5. Comparison of concentrations (in % at) obtained by Atomic Absorption Spectroscopy (AA) and by AES (after P.W. Palmberg in [5]) – as received (asr) and after sputter cleaning (asp) at a primary electron beam energy of 3 keV

Alloy	Element	σ_A	AA	AES asr	AES asp
316 inox	Cr	0.31	0.20	0.18	0.22
	Fe	0.21	0.66	0.71	0.67
	Ni	0.26	0.13	0.10	0.10
	Mo	0.25	0.02	0.01	0.02
600T Inconel	Cr	0.31	0.18	0.20	0.21
	Fe	0.21	0.08	0.07	0.07
	Ni	0.26	0.74	0.72	0.71
	Ti	0.44	0.00	0.01	0.01

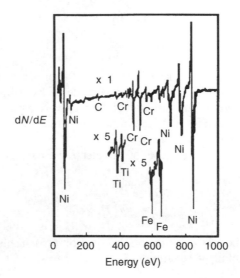

Figure 4.23. Auger spectrum of an Inconel surface fractured in situ (courtesy of Palmberg in [5])

Inconel, respectively) by comparing Auger to bulk analysis. We observe that, although matrix factors are neglected, good agreement for the steel samples between the two measurements is found. The difference between 'asr' and 'asp' is due to sputtering leading to an enrichment of the element with the lower sputter yield. The corresponding Auger spectra are shown in Figure 4.23. Intensities are determined by using the peak-to-peak values of the amplitudes in the dN/dE mode.

5   DEPTH PROFILE ANALYSIS

The fourth mode of data acquisition combines AES with ion beam sputtering yielding in-depth information beyond the escape depth limit of a few nm of the Auger electrons as discussed above. Sputtering is done either by simultaneous or alternating ion bombardment of a raster scanned noble ion beam of known beam energy and current over the sample surface. The ion beam has to be well aligned with the electron beam in order to avoid crater edge effects. Auger analysis should be performed in the centre of the ion crater. Preferably the sputtered area should exceed the analyzed area by a factor of 3–10.

5.1   THIN FILM CALIBRATION STANDARD

The aim of depth profile analysis is to convert the sputter time into depth and the measured intensities into elemental concentrations. The latter can be performed as outlined in Section 4. Depth calibration is often performed by use of a thin film standard which consists of an anodic Ta_2O_5 film on metallic Ta [8]. A typical profile of Ta_2O_5 on Ta is shown in Figure 4.24 reporting in (a) the peak-to-peak amplitude of the oxygen and tantalum peak as a function of Ar^+ sputter time. The thickness of the anodic film has been calibrated separately by Nuclear Reaction Analysis (NRA). Such a depth profile is typical for thin films with an internal film/substrate interface. Four zones can be identified: in zone I the sample is cleaned from surface contamination (adsorbed species like C, CO, C_xH_y). Interaction of the ion with the oxide leads to preferential sputtering, i.e. a change of composition. In the present case oxygen is preferentially removed with respect to the metal. Actually, a chemical reduction of the oxide due to sputtering takes place. A steady state (zone II) is reached after a certain time depending on the sputtered material, crystallinity, ion beam energy, angle of incidence and ion beam current. The composition of zone II depends on the sputtered sample and on the experimental conditions of ion and electron beam applied. In zone III the interface between the oxide film and metal substrate is reached characterized by the decrease of the oxygen and increase of the tantalum signal. In praxi one measures the time to reach the interface, determines the steady state amplitudes in the film and converts the sputter time into a depth provided the film thickness is known from an independent measurement. Zone IV represents the substrate. Due to an interaction of the sputter ions with the surface atoms at the crater bottom atoms are not only removed from the surface but also knocked into the sample. This leads to atomic mixing and broadening of the interface. More details are discussed elsewhere [9]. For very thin films (<10 nm) zone I and II are often not separated since no steady state is reached. In such cases it is difficult and almost impossible to determine the composition and stoichiometry within the film together with the width of the interface. The physical limit of depth resolution of approx. $1-2$ nm is determined by the ion beam energy and doses as well as by the escape depth of the electrons. Figure 4.24(b) shows an example of the interface between an oxide film and the metal substrate. It further underlines the repeatability obtainable in AES thin film analysis even in different laboratories.

5.2 DEPTH RESOLUTION

The resolution Δz of the interface is defined by the width which corresponds to the time necessary to reach 84% and 16% of the steady state value taken as 100% of the oxygen amplitude as shown in Figure 4.24(b). For convenience, the 50% point is taken as measure for the position of the interface. Figure 4.25 illustrates the dependence of Δz with thickness for various films sputtered. In particular, the resolution for amorphous Ta_2O_5 films formed anodically on Ta is given. For many samples, in particular for crystalline samples, the depth resolution Δz decreases proportionally with the square root of the film thickness z. Amorphous films such as Ta_2O_5/Ta (+) or SiO_2/Si exhibit a much better depth resolution because of a smaller sputter induced roughness at a given angle of incidence of the ion beam.

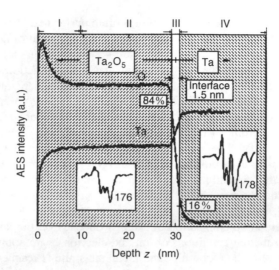

Figure 4.24a. (a) Auger profile of a standard 30 nm Ta_2O_5/Ta film showing the variation of the Auger amplitude of oxygen and the metal as a function of sputter time. The thickness of zone III is $\Delta l = v\Delta t_i$ where v is the sputter rate of the oxide and t_i the time necessary to reach the centre of the interface defined at 50% of the amplitude in the steady state region (II). The shape of the peaks changes from the oxide to the metal as shown in the insert [11]

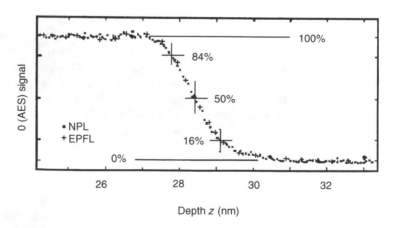

Figure 4.24b. (b) Interlaboratory comparison of the interface of a profile of a 30 nm Ta_2O_5/Ta standard indicating the precision which which an interface can be measured under favourable conditions. Measurements were performed at two laboratories (NPL, Teddington, UK and EPF-Lausanne, Switzerland) [11]

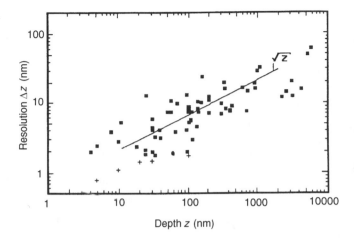

Figure 4.25. Depth resolution Δz as a function of thickness for different crystalline films and Ta_2O_5 films on Ta (+) (courtesy of M.P. Seah [10])

By rotating the sample during analysis one can reduce the effect of induced roughness since the angle of incidence of the primary ions varies, thus avoiding cone formation during sputtering (for further details see [2]). Other important factors which influence the depth resolution are the original roughness and the purity of the the rest gas of the analysis chamber.

5.3 SPUTTER RATES

The objective of a depth profile is to determine the elemental concentration as a function of depth and to convert the time axis into a depth axis. As discussed above, equation (15) can be used as a first-order approximation to calculate the atomic concentration of the elements measured. To convert the time axis into depth we apply the general relation between depth z and sputter time t given by:

$$z(t) = \int_0^\infty v\,\mathrm{d}t \tag{16}$$

where the sputter rate v of elements is obtained by

$$v = \frac{JYA}{\rho\,e\,Nn} \tag{17}$$

with

J ion current density (A m^{-2})
Y sputter yield (atoms/primary ion)
A molecular weight (kg mol^{-1})

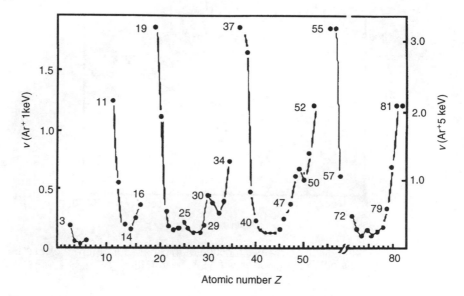

Figure 4.26. Normalized Ar+ sputter rate in nm/s per μA/mm² for elements with Z between 3 and 82 for ion beam energies of 1 and 5 keV

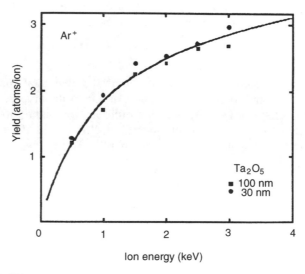

Figure 4.27. Ion sputter yield of Ta_2O_5 for typical sputter energies applied in AES profiles (courtesy of M.P. Seah [10])

ρ mass density (kg m^{-3})
e electron charge (= 1.602 × 10^{-19} A s)
N Avogadro's constant (= 6.023 × 10^{23} mol^{-1})
n number of molecules in a molecule (i.e. 5 in Ta$_2$O$_5$).

The ratio A/ρ (atomic volume) and the sputter yield S for 1 keV Ar$^+$ ions can be found in Table 4.3 and Table 4.6, respectively. The elemental sputter rate (in nm/s) is calculated from the elemental yield after normalizing by the ion current density (in µA/mm^2) according to:

$$\frac{v}{J} = \frac{YA}{100\rho} \left(\frac{\text{nm/s}}{\text{µA/mm}^2} \right) \tag{18}$$

Application of equation (18) results in sputter rates shown in Figure 4.26 for 1 keV and 5 keV Ar$^+$ ions.

The sputter yield, Y, of elements and components depends on several parameters such as ion beam energy E_{ion} and angle of incidence, θ. This dependence is shown as an example in the next two figures for Ta$_2$O$_5$ on Ta. Inspection of Figure 4.27 reveals that Y varies approximatively with log E_{ion} for low ion beam energies. Therefore, knowledge of a sputter yield at a certain ion beam energy E_1 allows one to estimate the sputter yield at E_2. Figure 4.28 indicates that Y varies approximatively with 1/cosθ for angles between $\theta = 0°$ and 45°. Consequently, one concludes that the ion sputter rate v becomes independent of the angle θ below 45° because the primary ion current varies with cos θ as experimentally shown (Figure 4.29). As a rule of thumb one retains that a sputter ion current

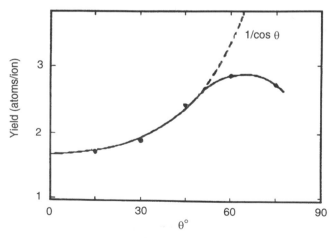

Figure 4.28. Variation of sputter yield of Ta$_2$O$_5$ with the angle of incidence θ defined with respect to surface normal (courtesy of M.P. Seah [10])

Table 4.6. Sputter parameters of elements

Legend:

Atomic Number —	Z / A — Element
Ionisation Potential (eV) —	P / S — Sputter yield for Ar⁺ 1 keV
Electron Affinity (eV) —	EA

(Each cell below is shown as: **Z Symbol** / P S / EA, where P = ionisation potential, S = sputter yield, EA = electron affinity.)

1	2	3	4	5	6	7	8	9	10	11	12	13	14	15	16	17	18
1 H 13.6 0.75																	2 He 24.59 0
3 Li 5.39 0.62	4 Be 9.32 0.9 <0.5											5 B 8.30 0.58 0.28	6 C 11.26 0.5 1.26	7 N 14.53 0.83 0	8 O 13.62 1.87 1.46	9 F 17.42 3.4	10 Ne 21.56 0
11 Na 5.14 4.9 0.55	12 Mg 7.65 3.8 0											13 Al 5.99 1.34 0.44	14 Si 8.15 1.47 1.39	15 P 10.49 2.0 0.75	16 S 10.36 2.34 2.08	17 Cl 12.97 3.62	18 Ar 15.76 0
19 K 4.34 8.2 0.5	20 Ca 6.11 4.13 <0.5	21 Sc 6.54 2.05 0.19	22 Ti 6.82 1.67 0.08	23 V 6.74 1.55 0.53	24 Cr 6.77 2.05 0.67	25 Mn 7.44 2.05 	26 Fe 7.87 2.88 0.16	27 Co 7.86 2.0 0.66	28 Ni 7.64 2.08 1.16	29 Cu 7.73 2.52 1.23	30 Zn 9.39 0	31 Ga 6.0 3.43 0.3	32 Ge 7.9 2.42 1.2	33 As 9.81 3.1 0.81	34 Se 9.75 4.48 2.02	35 Br 11.81 3.36	36 Kr 14.00 0
37 Rb 4.18 12.2 0.49	38 Sr 5.70 <0	39 Y 6.38 2.4 0.31	40 Zr 6.84 1.7 0.43	41 Nb 6.88 1.45 0.89	42 Mo 7.10 1.45 0.75	43 Tc 7.28 1.45 0.55	44 Ru 7.37 1.55 1.14	45 Rh 7.46 1.9 1.14	46 Pd 8.34 2.75 0.56	47 Ag 7.58 3.7 1.30	48 Cd 8.99 9.6 0	49 In 5.79 4.4 0.30	50 Sn 7.34 3.55 1.15	51 Sb 7.34 4.55 1.15	52 Te 9.01 4.77 1.97	53 I 10.45 3.06	54 Xe 12.13 0
55 Cs 3.89 15.3 0.47	56 Ba 5.39 <0	57 La 5.58 2.63 0.5	72 Hf 7.0 2.05 0	73 Ta 7.89 1.65 0.32	74 W 7.98 1.5 0.82	75 Re 7.88 1.5 0.12	76 Os 8.7 1.6 1.12	77 Ir 9.1 1.55 1.57	78 Pt 9.0 1.82 2.13	79 Au 9.23 2.17 2.31	80 Hg 10.44 3.3 0	81 Tl 6.11 0.30	82 Pb 7.42 6.4 0.37	83 Bi 7.29 5.78 0.95	84 Po 8.42 1.9	85 At 9.5 2.8	86 Rn 10.75 0
87 Fr	88 Ra 5.20	89 Ac 6.9															

Lanthanides:

58 Ce	59 Pr	60 Nd	61 Pm	62 Sm	63 Eu	64 Gd	65 Tb	66 Dy	67 Ho	68 Er	69 Tm	70 Yb	71 Lu
5.47 2.72 <0.5	5.42 3.25 <0.5	5.49 3.55 <0.5	5.55 3d	5.63 <0.5	5.67 5.79 <0.5	6.14 <0.5	5.85 3.03 <0.5	5.93 4.1 <0.5	6.02 3.9 <0.5	6.10 3.75 <0.5	6.18 <0.5	6.25 <0.5	5.43 2.9 <0.5

Actinides:

90 Th	91 Pa	92 U
7.0 2.12	2.12	6.08 2.4

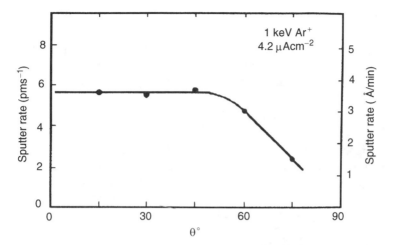

Figure 4.29. Ion sputter rate v as a function of angle of incidence for 1 keV Ar^+ and an ion current density of 4.2 $\mu A/cm^2$ determined for the Ta_2O_5/Ta standard (courtesy of M.P. Seah [10])

density of a few $\mu A/cm^2$ leads to a sputter rate of a few Å/min for 1 keV Ar^+ sputtering for many elements and compounds.

5.4 PREFERENTIAL SPUTTERING

In the previous chapter we have seen that sputter rate and yield vary from element to element. Consequently, while sputtering a multi-element sample one observes a change of composition because of a difference in the elemental sputter yields. This phenomenon is called preferential sputtering. Continued sputtering leads to a steady state as illustrated by Figure 24(a) for Ta_2O_5 on Ta. For a binary alloy we obtain by application of equation (9), use of elemental sensitivity factors and introduction of elemental sputter yields:

$$\frac{I_{A,AB}}{I_{B,AB}} = \frac{s_{A,A}}{s_{B,B}} K \frac{x_A}{x_B} \frac{\kappa_B}{\kappa_A} \tag{19}$$

where κ_A/κ_B is the ratio of the sputter yields of A and B divided by their atomic density $n_M = \rho/M$ (M atomic weight of A or B):

$$\frac{\kappa_B}{\kappa_A} = \frac{Y_B/n_B}{Y_A/n_A} \tag{20}$$

Assuming that for a system A-B, Y_A and Y_B of the alloy are identical to the yields of the pure element, one observes that the ratio of the sensitivity factors, s, is multiplied by the inverse of the sputter yield ratio. This explains why an element with a lower yield is enriched at the surface during profiling.

5.5 λ - CORRECTION

The influence of the attenuation length of Auger electrons on the signal inten-
sity is very important for thin films $\leqslant 3\Lambda$ with $\Lambda = \lambda\cos\theta$ where Λ is of the
same magnitude as the film thickness. To correct for its influence, the measured
intensity $I_A(z)$ given by equation (6) can be replaced by $F_A(z)$ applying, [2],

$$F_A(z) = I_A(z) - \lambda\cos\theta\frac{dI_A(z)}{dz}. \tag{21}$$

The transformation $F_A(z)$ is called the λ-correction. Figure 4.30 gives an
example of the effectiveness of such a transformation showing for a binary alloy
Fe-Cr the enrichment of chromium underneath the surface within the oxide film
which without λ-correction would have passed almost unnoticed. In addition, the
distribution of iron is totally different.

5.6 CHEMICAL SHIFTS IN AES PROFILES

A chemical shift of peak of element A is defined as the displacement of the
kinetic or binding energy of that peak following a change in the state of oxidation (compare Section 2.6). Such displacements are also observed in AES, but
they are less evident than in ESCA, because of the interaction of three different
electron levels of the Auger process. In AES profiling, such shifts are observed
for either a variation of the state of oxidation with depth or during sputtering,
in particular for oxides. However, elemental intensity plots — often used in AES

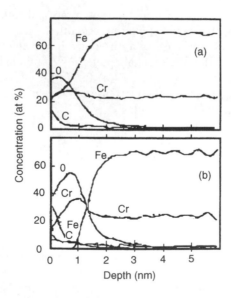

Figure 4.30. λ-correction applied to an oxide layer on an Fe–Cr alloy

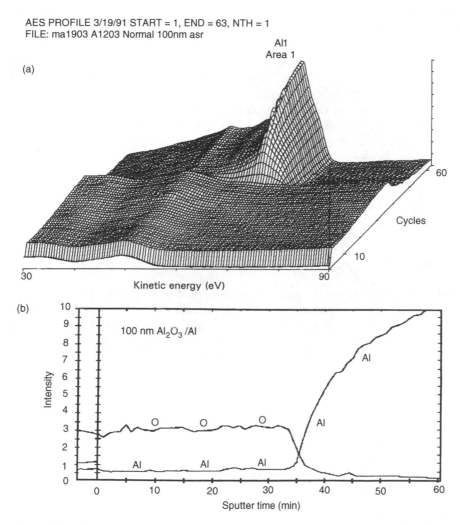

AES PROFILE 3/19/91 START = 1, END = 63, NTH = 1
FILE: ma1903 A1203 Normal 100nm asr

Figure 4.31. Example of an anodic film of 100 nm of Al_2O_3 on Al: (a) spectra of the Al_{LMM} peak during Ar^+ sputtering. They correspond to the data of (b) where the Al_{LMM} intensities are shown together with the O_{KLL} peak

profiling — do not directly allow one to identify such peak shifts because only the total intensity of a peak is reported. As an illustration for a chemical shift in AES the evolution of the Al_{LMM} peak is shown in Figure 4.31(a,b): (a) indicates the measured intensity N as a function of its kinetic energy and sputter cycles for an 100 nm Al_2O_3 film formed anodically on Al. Figure 4.31(b) illustrates the measured intensity (peak to background) of the same Al_{LMM} peak and the O_{KLL} peak as a function of sputter time. The oxygen peak serves as an independent indicator for the oxide/metal interface. For low cycles, numbers corresponding to low sputter times, the typical peak shape of the oxidized Al (Al^{3+}) peak is observed. At the interface (cyc. no. 44 in (a) and $t = 35$ min in (b), respectively) a shift and a change in shape of the Al_{LMM} peak typical for the transition Al^{3+}/Al(metal) is observed. Other similar changes are known for many elements, in particular for light elements. Figure 4.31 suggests that in AES profiling it is necessary not only to use elemental peak intensities, but also to inspect peak shape evolution for chemical oxidation state identification. Needless to say the in cases where a change of shape of a given peak during profiling is observed, standard elemental sensitivity factors become invalid. Peak separation for different oxidation states, redefinition of energy windows and deconvolution routines are necessary to avoid errors of up to 50%. Many software routines of modern AES equipments offer such procedures which, unfortunately however, exclude automatic quantification of the raw data.

6 SUMMARY

Qualitative AES is an elemental analysis detecting all elements except H and He by measuring a characteristic kinetic electron energy specific for a given elemental Auger electron transition on conducting samples. The number of characteristic Auger electrons allows one to quantify data by application of experimentally determined elemental sensitivity factors yielding in general a precision of 10–50 % due to the influence of matrix factors and influence of the state of oxidation on the peak shape. Detection limit of point analysis is about 0.1 to 1 % of a monolayer corresponding to 10^{12}–10^{13} particles/cm^2. This limit may be increased for line scans or elemental mapping depending on the data acquisition time. Spatial Auger resolution is determined by the focus of the primary electron beam and the backscattering of the electrons within the analyzed matrix. Specifications of modern Scanning Auger Microprobes are below 50 nm.

Escape depth and electron attenuation lengths of most elements is 1–3 nm for kinetic energies below 2 keV, typical for Auger transitions. State of the art instruments use primary electron beams with energies up to 30 keV focussed down to 20 nm secondary electron spatial resolution.

In-depth information to a depth of 0.1–1 μm and resolution of 1–20 nm is obtained by combining AES with ion beam sputtering. Despite the fact that ion beam sputtering may change the composition of the layer analyzed, sputter depth profiling is one of the most important applications of AES, because it gives access to internal interfaces and allows one to identify relative composition changes with depth. The quantification of depth profiles is often limited by the absence of either precise sensitivity factors adapted for the measured oxidation state or well defined sputter yield ratios. A Ta_2O_5 film standard on Ta of known thickness serves as a welcome tool to calibrate sputter conditions.

Applications of AES cover all fields of materials science, namely characterization and thin film preparation of surfaces and thin films in micro-electronics, semi- and supraconductors, physics and chemistry, corrosion and electrochemistry, catalysis as well as metallurgy and tribology.

REFERENCES

[1] P. Auger, *J. Phys. Radium* **6**, 205 (1925)
[2] D. Briggs, M.P. Seah, *Practical Surface Analysis by Auger and X-ray Photoelectron Spectroscopy, 2nd ed.* John Wiley, Chichester (1990)
[3] D. David and R. Caplain, *Méthodes usuelles de caractérisation des surfaces*, Eyrolles, Paris (1988)
[4] J.P. Eberhart, *Analyse structurale et chimique des matériaux*, Dunod, Paris (1989)
[5] A.W. Czanderna, *Methodes of Surface Analysis*, Elsevier, New York (1975)
[6] L.E. Davies, N.C. Mac Donald, P.W. Palmberg, G.E. Riach and R.E. Weber (eds.) *Handbook of Auger Electron Spectroscopy, 2nd. ed.*, Perkin Elmer Corp., Phys. Electronics Div, Eden Prairie, Minn./USA (1976)
[7] C. Lea, Metal Sci. **17**, 357 (1983)
[8] BCR ref. material no. CRM 261R availabe from Joint Research Centre, Inst. for Reference Materials and Measurements (IRMM), Retieseweg, B-2440 Geel, Belgium, atten. MRM Unit
[9] H.J. Mathieu in H. Oechsner, (ed). Thin Film and Depth Profile Analysis, *Topics in Current Physics*, Vol. 37 (1984) Springer (Heidelberg)
[10] C.P. Hunt and M.P. Seah, *Surface and Interface Analysis*, **5**, 199 (1983)
[11] C.P. Hunt, H.J. Mathieu and M.P. Seah, *Surface Sci* **139**, 549 (1984)

QUESTIONS

1. Identify the O_{KLL} and C_{KLL} AES transitions in Figure 4.14 and indicate the electronic energy levels involved.

2. Determine the Fe/Cr ratio in at% using the data of Figure 4.23 and the sensitivity of Table 4.5 (a) with and (b) without matrix correction.

3. Explain how to convert measured amplitudes into concentrations. Discuss the limiting factors in AES quantification.

4. Discuss how to measure and interpret AES line scans.

5. Determine the escape depth for Si, Cr and their oxides ($SiO_2 \rho = 2.26$ gcm^{-3}, $A = 60.08$ gmol^{-1} $Cr_2O_3 \rho = 5.21$ gcm^{-3}, $A = 151.99$ gmol^{-1}). A/ρ values of the elements are found in Table 4.3

6. Determine the AES escape depth of Ta for its NNN-peak at 179 eV as a function of the emission angle at $\theta = 10°$ and $90°$.

7. Calculate (a) sputter rate and (b) yield of Si, Fe and Au for 1 keV Ar$^+$ using Table 4.6.

8. Calculate (a) the sputter rate and (b) the depth resolution of the 100 nm Al_2O_3/Al profile shown in Figure 4.31 by using the oxygen and aluminum peak. Discuss their difference.

CHAPTER 5

Secondary Ion Mass Spectrometry – the Surface Mass Spectrometry

JOHN C VICKERMAN*, AND ANDREW J SWIFT[†]

*UMIST, Manchester
[†]CSMA Ltd, Manchester

1 INTRODUCTION

Secondary ion mass spectrometry, SIMS, is the mass spectrometry of ionised particles which are emitted when a surface, usually a solid, is bombarded by energetic primary particles which may be electrons, ions, neutrals or photons. The emitted or 'secondary' particles will be electrons; neutrals species atoms or molecules or atomic and cluster ions. The vast majority of species emitted are neutral but it is the secondary ions which are detected and analysed by a mass spectrometer. It is this process which provides a mass spectrum of a surface and enables a detailed chemical analysis of a surface or solid to be performed.

At first sight the process is conceptually very simple. A pictorial representation of the process is shown in Figure 5.1. Basically when a high energy (normally between 1 and 15 keV) beam of ions or neutrals bombards a surface, the particle energy is transferred to the atoms of the solid by a billiard-ball-type collision process. A 'cascade' of collisions occurs between the atoms in the solid; some collisions return to the surface and result in the emission of atoms and atom clusters some of which are ionised in the course of leaving the surface. The point of emission of low energy secondary particles is remote (up to 10 nm) from the point of primary impact; the final collision resulting in secondary particle emission is of low energy (ca 20 eV); over 95% of the secondary particles originate from the top two layers of the solid. Thus the possibility of a soft ionisation mass spectrometry of the surface layer emerges.

Although the emission of secondary ions from surfaces was observed some 75 years ago [1], *surface* mass spectrometry is a 'young' technique. By far the widest application of SIMS up to the early 1980s was to exploit its destructive capability to analyse the elemental composition of materials as a function of

Surface Analysis — The Principal Techniques. Edited by John C. Vickerman
© 1997 John Wiley & Sons Ltd

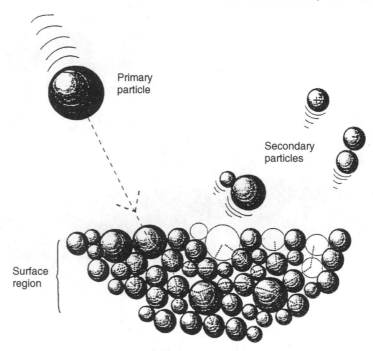

Figure 5.1. A schematic representation of the SIMS process. Reproduced with permission from N. Lockyer, PhD Thenin, UMIST, 1996

depth. The technique is known as *dynamic* SIMS. Dynamic SIMS found extensive application throughout the semiconductor industry where the technique had a unique capability to identify chemically the ultra-low levels of charge carriers in semiconductor materials and to characterise the layer structure of devices. Indeed, after instruments first appeared (Herzog [2] and Leibl [3] and Castang and Slodzian [4]) in the early 1950's and 1960's the technique developed rapidly under the impetus of this industry for the next two decades. Whilst of great importance, this variant of SIMS is clearly not a true surface analysis technique.

Static SIMS emerged as a technique of potential importance in surface science in the late 1960s and early 1970s as a consequence of the work of Benninghoven and his group in Münster [5]. Whilst the SIMS technique is basically destructive, the Münster group demonstrated that using a very low primary particle flux density (<1 nA cm^{-2}) spectral data could be generated in a timescale which was very short compared to the lifetime of the surface layer. The information so derived would be characteristic of the chemistry of the surface layer because statistically no point on the surface would be impacted more than once by a primary particle during an analysis. The surface could be said to be essentially *static*. Obviously, the use of very low primary flux density resulted in a very low yield of secondary particles and this imposed requirements of high sensitivity on the detection equipment. The fact that these experimental conditions could be used was due to advances in single particle detection equipment. Benninghoven and his team first demonstrated the surface analytical possibilities of SSIMS in a series of studies of the initial oxidation of metals [6].

2 BASIC CONCEPTS

2.1 THE BASIC EQUATION

A more extensive introduction can be found in references [7] and [8]. SSIMS is concerned with the analysis of secondary *ions*. Ionisation occurs at, or close to, emission of the particles from the surface with the consequence that the matrix participates in the electronic processes involved. This means that the yield of secondary ions is strongly influenced by the electronic state of the material being analysed with consequent complications for quantitative analysis. The basic SIMS equation is

$$I^m{}_s = I_p y_m \alpha^+ \theta_m \eta \tag{1}$$

where $I^m{}_s$ is the secondary ion current of species m, I_p is the primary particle flux, y_m is the sputter yield, α^+ is the ionisation probability to *positive ions*, θ_m is the fractional concentration of m in the surface layer and η is the transmission of the analysis system.

The two fundamental parameters are y_m and α^+. y_m is the total yield of sputtered particles of species m, neutral and ionic, per primary particle impact. It increases linearly with primary flux. It also increases with primary particle mass, charge and energy, although not linearly [4]. Figure 5.2 shows the variation of y for aluminium. The crystallinity and topography of the bombarded material will also affect the yield. y tends to maximise with energy at around 10 keV. At a given bombardment energy the sputter yield varies by a factor of 3–5 through the periodic table.

Since ionisation is influenced by the electronic state of the surface, the yields of secondary ions vary by several orders of magnitude across the periodic table, Figure 5.3, and are very dependent on the chemical state of the surface. Thus the ion yield for a particular element will vary dramatically, for example, for a metal as compared to its oxide, Table 5.1. It can be seen that oxidation changes the elemental ion yields to differing extents resulting in significant complications when absolute quantitative data is required.

2.2 MONOLAYER LIFETIME AND THE STATIC LIMIT

Static conditions are those which maintain the integrity of the surface layer within the timescale of the analytical experiment. This implies that a very low primary beam dose is used during analysis. The distinction between dynamic and static conditions can be understood by computing the lifetime, t_m, of the topmost atomic

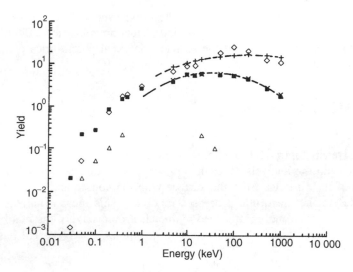

Figure 5.2. Experimental sputter yield data for aluminium as a function of primary ion energy for a number of different primary ions: Δ, He; ◊, Xe; ▢, Ar; +, Xe (theoretical); x, Ar (theoretical)

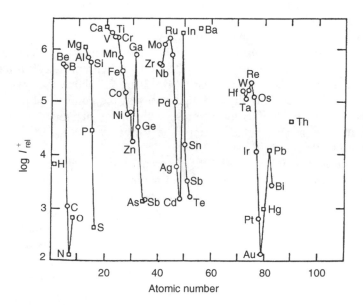

Figure 5.3. The variation of positive ion yield as a function of atomic number for 1 nA 13.5 keV O-bombardment: o from elements; ▢, from compounds. Reproduced with permission from H.A. Storms, K.F. Brown, and J.D. Stein, *Anal. Chem.*, **49**, 2023 (1977). (Copyright (1977) American Chemical Society

Table 5.1. Secondary ion yields from clean and oxidised metal surfaces

Metal	Clean metals M^+ yield	Oxide M^+ yield
Mg	0.01	0.9
Al	0.007	0.7
Si	0.0084	0.58
Ti	0.0013	0.4
V	0.001	0.3
Cr	0.0012	1.2
Mn	0.0006	0.3
Fe	0.0015	0.35
Ni	0.0006	0.045
Cu	0.0003	0.007
Ge	0.0044	0.02
Sr	0.0002	0.16
Nb	0.0006	0.05
Mo	0.00065	0.4
Ba	0.0002	0.03
Ta	0.00007	0.02
W	0.00009	0.035

layer as a function of the primary beam flux at the sample surface.

$$t_m = \frac{10^{15}}{I_p} \times \frac{A}{y} \tag{2}$$

where A cm^2 of the surface (surface layer atom density of 10^{15} atoms cm^{-2}) is bombarded by a primary beam of flux density I_p particles cm^{-2} and the sputter yield is y. Usually the primary beam flux is measured in Amps cm^{-2} (1 amp is equivalent to 6.2×10^{18} charged particles s^{-1}). Using this equation, assuming a sputter yield of 1 Table 5.2 has been assembled.

If an analysis requires say 20 minutes (1200 s) then static conditions can only be safely attained for primary beam currents of about 1 nA cm^{-2} or less. An alternative approach takes account of the fact that each primary particle colliding

Table 5.2. The surface monolayer lifetime as a function of primary beam flux density

I_p (A cm^{-2})	t_m (s)
10^{-5}	16
10^{-7}	1600
10^{-9}	1.6×10^5
10^{-11}	1.6×10^7

with the surface disturbs an area of 10 nm^2, thus it would only require 10^{13} impacts cm^{-2} to influence all the atoms in the surface. This is generally accepted as the static primary particle dose limit.

For dynamic SIMS, high elemental sensitivity and rapid erosion rates are required, so high primary flux densities of 1 µA cm^{-2} or greater are desirable. The time to complete a depth profile will be of principle interest.

Clearly the magnitude of y is significant in these calculations. Whilst these are well known for a wide range of primary particle mass, energy and angle of incidence for metallic targets, data on the sputtering characteristics of organic and other molecular materials are not extensive.

2.3 SURFACE CHARGING

Many of the important technological materials requiring surface analysis are insulators. When an insulating sample is bombarded by a positive ion beam the surface potential rises due to the input of positive charge and the emission of secondary electrons. The potential can rise very rapidly by several hundred volts in a few minutes, such that the kinetic energy of the emitted positive ions rises well beyond the acceptance window of the analyser [9]. The result is the loss of the SSIMS spectrum. There are two linked solutions to this problem. The first, widely used method, is to irradiate the sample surface with a beam of relatively low energy electrons. One disadvantage of this approach is that electron bombardment can also give rise to sample degradation and electron stimulated ion emission. The second solution is to use fast atoms as primary beams in place of ions, fast atom bombardment, FAB. The surface potential does rise under atom bombardment due to the emission of secondary electrons, but it very quickly reaches a plateau of about 20 V. It is straightforward to adjust the ion acceptance optics to cope with ions in this energy range, so positive ion acquisition is very easy even from insulating materials using FABSSIMS. Clearly, even under atom bombardment, negative ion collection still requires the surface potential to be brought negative. Some electron input is required, although considerably less than with ion bombardment.

3 EXPERIMENTAL REQUIREMENTS

The basic arrangement for the SIMS experiment is shown in Figure 5.4. There are three main components: the primary particle source, the mass spectrometer and, since the secondary ions are emitted with a range of kinetic energies, an ion optical system which selects ions within a defined energy band compatible with the capability of the mass analyser.

3.1 PRIMARY BEAM

The range of types of primary beam source which have been used in SIMS can be classified into four basic types according to their mechanisms of primary

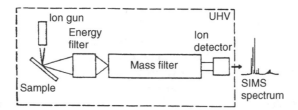

Figure 5.4. A schematic representation of a SIMS instrument. Reproduced by permission from J.C Vickerman, *Chem. Brit.* 969(1987)

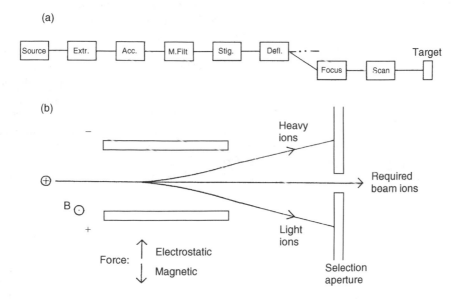

Figure 5.5. A schematic diagram of (a) the main components of a scanning, focused ion gun: Source = ion source; Extr = extractor; Acc = accelerator; M.Filt = mass/energy filter; Stig = stimator; Defl = neutral elimination bend; Focus = focusing lens; scan = xy raster. (b) Operation of a Wien filter for mass selected ion beam. Balance condition is $m/z = 2V(B/E)^2$. Reproduced by permission from [7], *Secondary Ion Mass Spectrometry Principles and Applications*, Oxford Science Publications (1989)

beam production: (i) electron bombardment; (ii) plasma; (iii) surface ionisation; (iv) field ionisation. Each type of source offers different performance in terms of spatial resolution, ease/speed of use, sensitivity, coping with insulating materials, beam induced damage, etc. and all types have been pulsed successfully for time-of-light (ToF) mass spectrometer systems (see later).

The basic components of most types of ion beam source, Figure 5.5, are the source region/extraction zone, focusing and collimating regions, a mass (Wien) filter for beam purification, a pulsing mechanism for systems with time-of-flight mass analysers, stigmation/focus lenses and finally scan rods. The mode of operation of the primary beam largely defines the type of SIMS information accessible. In static and dynamic SIMS it is common practice, to raster scan the primary ion beam across the surface region of interest, see Figure 5.5. For static (scanning) SIMS this enables sensitivity to be optimised by matching the analysed area to the field of view of the collection optics of the analyser. In dynamic SIMS the dimensions of the raster define the crater edges. Scanning SIMS with the mass spectrometer preset to detect certain masses also offers the possibility of mapping the distribution of secondary ions over the area of interest. An instrument with this capability is known as a scanning *SIMS microprobe* and using computer image storage and a colour coded graphics system it is possible to produce colour coded maps of elements and molecules in the surface.

A mapping of the distribution of surface elements can also be derived using an unscanned primary ion beam and a mass spectrometer with an ion optical arrangement such that the positional sense of the ions is retained throughout mass analysis process. This mode of mapping is known as *ion microscopy* and the ion image can be displayed directly onto a fluorescent screen or registered on a position sensitive detector for subsequent computer storage/manipulation.

The next section provides an overview of the basic aspects of some of the more popular primary beam sources. Critical parameters are the brightness and energy spread of the primary beam and the stability and reliability of the device.

3.1.1 Electron Bombardment

These ion sources are based upon the principle of using a high current density of electrons to ionise the primary beam gas, usually argon or xenon. Many types of source arrangement exist. For inert source gases generally a hot cathodic source of electrons (usually tungsten or iridium, frequently treated to increase its electron emission) is used. The electrons are accelerated towards the anode to give them the required energy to ionise the source gas. Cross-sections for gas interaction (and therefore ionisation yield or source efficiency) can be increased using electrostatic or magnetic fields to increase the path length of the travelling electrons. The beam is extracted from the ion source and accelerated and focused to produce a beam at the sample surface of 2 keV–12 keV energy.

A neutral beam can be generated by charge exchanging, say, an argon ion beam by passing it through a chamber (the Wien filter region is used in a number of cases) in the ion beam system, containing a pressure of gaseous argon (10^4 mbar) [10]. A proportion of the ions (10%–30%) lose their charge by capturing an electron from the atoms randomly moving in the chamber. Although these ions have lost their charge, they retain their velocity and direction and a fast atom beam is formed. The residual ion beam can then be deflected away, Figure 5.6.

Most electron bombardment sources are versatile, easy to use and comparatively reliable. They offer only moderate brightness (this is a measure of the

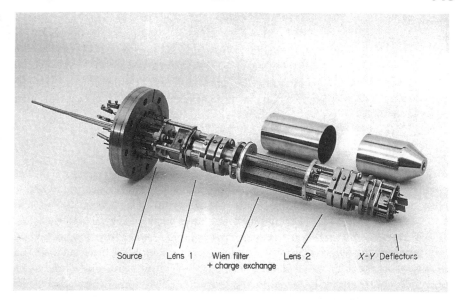

Figure 5.6. A mass filter ion–atom primary beam source. Reproduced by permission from J.C. Vickerman in *Spectroscopy of Surfaces*, (Eds) R.J.H. Clark and R.E Hester, p 174. Copyright John Wiley & Sons Ltd, Chichester (1988)

current density available from the source) of about 10^5 A m^{-2} Sr^{-1}, such that they are most commonly used de-focused over large areas (several mm^2) for static SIMS work and are not preferred for micro-focused studies.

3.1.2 Plasma

Duoplasmatron, RF and hollow cathode sources are all classified under this heading. The output of a simple electron bombardment source may be limited by the density of electrons which can be generated due to space charge effects. If the pressure of the source gas is raised, the ions and neutrals reduce the repulsions between the electrons and a much higher density of electrons can be sustained generating a higher ion beam current. Under such conditions a plasma is formed. The source will have an exit aperture through which the ions are extracted to form the ion beam. The source of electrons may be a hot filament, alternatively electrons may be generated by positive ion bombardment of a cathode to sustain the discharge. Reactive gases, such as oxygen, can be ionised using the cold cathode discharge method. Electric and magnetic fields are used to concentrate the discharge to increase output. The improved ionisation efficiency of this method is reflected in the higher beam brightness ranging from 10^4 to 10^7 A m^{-2} Sr^{-1} attained with this type of source. However, this is achieved partly at the expense of reliability as the violence of the emission process tends to gradually destroy the source components by ion etching. Higher beam brightness renders this source type better suited for dynamic SIMS applications (μA into ≈ 50 μm) and for microfocused scanning analysis (nA into $\leqslant 5$ μm).

As well as inert gases, this source type is used for oxygen (O_2^+) primary ion bombardment, the use of which provides improved sensitivities for the detection of electropositive species and thus is the primary source of choice for many semiconductor depth profiling applications (see Section 7).

3.1.3   Surface Ionisation

Sensitivity to electronegative species is enhanced when an electropositive primary beam is used. The availability of an alkali metal ion source is thus attractive to the depth profiling SIMS analyst and this is offered by the surface ionisation source. In this case, ion emission is thermally stimulated by warming an adsorbed layer of, for example, caesium on the surface of a high work function metal (e.g. iridium) under vacuum conditions. The ionisation potential of the Cs adlayer and the work function of the surface are such that electrons can move freely from adlayer to substrate and upon mild thermal excitation, ions of a very low and uniform energy spread are emitted.

Source brightness depends on the size of the emitting area and $>10^6$ A m^{-2} Sr^{-1} have been attained. There is the drawback of very careful handling and operational requirements of the source metal. However, the benefits they offer in electronegative ion yield for example, are such that this source is built in or retrofitted to nearly all SIMS instruments used for depth profiling semiconductors. Caesium ion sources have been successfully adapted for ToFSIMS analysis where their use in the correspondingly lower ion dose regimes can lead to the generation of cationised secondary ions which are of diagnostic value.

3.1.4   Field Ionisation Sources

These sources operate on the principle of stripping electrons off source atoms situated near to an extremely high local electronic field. A very fine tip ($<$ μm) is used and source energies can be as high as $10-20$ kV. Gas based field ionisation sources have failed yet to produce sufficient beam currents for SIMS work, however, the liquid metal counterpart is widely used. In the latter case (the electrohydrodynamic ion source) a thin 'skin' of liquid metal (typically gallium or indium) is allowed to flow over a fine tungsten tip in a region of very high extraction field, Figure 5.7. This has the effect of distorting the skin towards the exit ring and setting up a (Taylor's) cone and (plasma) ball structure of liquid metal on the probe tip. Primary ions of the metal are stripped away from the plasma ball.

These are the highest brightness sources ($\approx 10^{10}$ A m^{-2} Sr^{-1}) used for surface mass spectrometry and as such are the source of choice for work at the highest spatial resolution. The most commonly used liquid metal ion source is based on

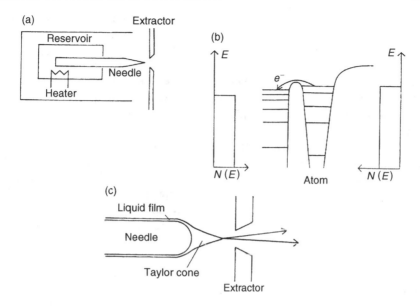

Figure 5.7. Principle of a liquid metal ion source: (a) source schematic; (b) energy diagram; (c) close up of extraction region. Reproduced by permission from [7], *Secondary Ion Mass Spectrometry – Principles and Applications*, Oxford Science Publications (1989)

liquid gallium which relies on field-ionising Ga^+ from a tungsten tip to generate a very bright and highly focusable beam. Spatial resolutions down to the range 200–20 nm have been realised. It is, however, very difficult to maintain *static* conditions and obtain sufficient signal at these levels of spatial analysis.

These beam systems have been adapted for pulsing by the introduction of deflection blanking plates which rapidly sweep the beam across an aperture. When used in conjunction with time-of-flight mass spectrometry (see Section 3.2.3), the pulsed version of this type of source also offers the best prospect for molecular ion imaging at high spatial resolution under static conditions.

3.2 MASS ANALYSERS

The different modes of SIMS have contrasting demands on the mass analyser. For static SIMS it is necessary to maximise the information level achieved per unit of surface damage. The analysis and detection system should be as efficient as possible for the total yield of secondary ions from the surface. In dynamic SIMS the requirement is usually to have the highest sensitivity possible for specific elemental ions. The preservation of surface structure is not important. In scanning or imaging SIMS where the spatial distribution of surface chemical information is being studied the requirements are the same as for static SIMS. If elemental distributions are sought sometimes more dynamic conditions can be tolerated.

The three most widely used mass analysers are the quadrupole RF mass filter, the magnetic sector and the time-of-flight instruments. The quadrupole analyser was widely used in the early work in static SIMS because it was easily incorporated in a UHV system due to its small size. Whilst a great deal of useful information has been obtained using this analyser in SSIMS, it is a low transmission device (less than 1%). Furthermore, it is a scanning instrument so that it only allows the sequential transmission of ions, all other ions being discarded. The information loss is therefore very high. In recent years ToF analysers have been used in SSIMS because of their very high transmission and the fact that they are quasi-parallel detectors — they are not scanning instruments, they collect all the ions generated. Consequently they are about 10^4 more sensitive than a quadrupole instrument. Historically, the magnetic sector analyser has been used for dynamic SIMS because of its high transmission, 10%–50%, and high mass resolution. Although it is usually a scanning device, since detection of only a few specific ions is required using a continuous high flux primary beam, it is usually preferred over the ToF instrument.

3.2.1 Magnetic Sector

This type of mass analyser was first used for conventional mass spectrometry and the principle of operation is well understood. Ions are extracted from the sample using a high extraction potential, circa 4 keV. Upon traversing a magnetic field, a charged particle experiences a field force in a direction orthogonal to the direction of magnetic flux lines and its original axis of travel; it thus adopts a circular path. The extent of force experienced by the particle and hence the radius of its path is directly related to its velocity and therefore since all ions are accelerated to a fixed potential before entering the magnetic field they can be readily separated according to their masses. The radius of curvature, R, for an ion of mass to charge ratio, m/z, travelling through a magnetic field, B, having been accelerated by potential V is given by

$$R = \frac{1}{B} \left(\frac{2mV}{z} \right)^{1/2} \tag{3}$$

The dispersion of adjacent masses, i.e. mass resolution, is proportional to the radius of the magnets used and degrades with increasing mass. Secondary ions are emitted with a spread of kinetic energies. Elemental ions usually have a wider distribution (up to $\approx$100 eV) and peak around 10–20 eV whereas the multi-atomic or molecular ion distribution will peak between 1 and 5 eV and only have a width of a few tens of eV. A wide energy spread can degrade the mass resolution. Double sector instruments incorporating an electrostatic sector are often used in order to combat these resolution degradation effects. The electrostatic sector allows a small energy band of the ions to be selected and focused on the entrance slit of the magnet for analysis, see Figure 5.4 for the principle.

The dispersed ions are commonly scanned across an exit slit of the magnet by scanning the field strength of the electromagnet. In a dynamic SIMS experiment where a few specific elemental ions are to be measured in a depth profile, rapid switching between masses is possible using the electromagnet.

An attractive feature of this form of mass spectrometry for surface analysis is that the positional sense of secondary ions can be retained throughout the analysis process such that secondary ion images can be projected in real time on to a fluorescent screen or directly into computer software via a position sensitive detector. This device can therefore be operated as an *ion microscope*. By irradiating the sample with a large diameter static beam the spatial distribution of the emitted secondary ions can be observed on the viewing screen with a spatial resolution in the $1-5$ μm region.

More normally depth profiling is carried out in the *microprobe* mode. A highly focused ion beam $1-10$ μm diameter is raster scanned across the sample surface to erode a crater with uniform edge and crater bottom, see Section 7.1.1 Emitted secondary ions are collected from the central area of the crater bottom.

Despite impressive performance characteristics, magnetic sector mass spectrometers are not the ideal. They can be large, cumbersome devices and pose considerable difficulties in the generation of true UHV. This is because these instruments cannot be easily baked to desorb chamber wall gases without seriously (irreversibly) modifying the magnetic properties of the magnet. Although transmission is affected, the most serious consequence of non-UHV conditions is in the sample region where interference effects from background gases raise the detection limits for residual gas elements in a matrix such as, for example, carbon in silicon. To minimise this problem instruments with a retractable magnet have been introduced which enables system baking. In another design the introduction of very high pumping rates in the sample region by the use of cryo-pumping reduces the local pressure.

3.2.2 The Quadrupole Mass Analyser

A quadrupole mass spectrometer is so called since it makes use of a combination of a DC and a radio frequency (RF) electric field applied to four parallel rods, in order to separate ions according to their mass-to-charge ratio. A potential consisting of a constant d.c (U) component plus an oscillating r.f component ($V \cos \omega t$) is applied to one pair of rods whilst an equal but opposite voltage is applied to the other pair. The rapid periodic switching of the field sends most ions into unstable oscillations of increasing amplitude until they strike the rods and are hence not transmitted. However, ions with a certain mass to charge ratio, m/z, follow a stable periodic trajectory of limited amplitude and are transmitted to the detector, see Figure 5.8. By increasing the dc and ac fields whilst keeping a constant ratio between them, this resonant condition is satisfied for ions of each ascending m/z ratio in turn. The mass resolution and transmission of this device are interrelated by a complicated series of equations. The ion trajectories are a

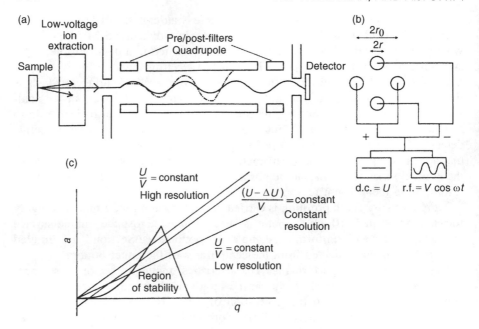

Figure 5.8. Operation of a quadrupole mass filter: (a) longtitudinal cross-section, showing stable and unstable trajectories; (b) radial cross-section, showing applied voltages; (c) ion trajectory stability diagram – ion trajectories are a function of two dimensionless parameters *a* and *b*, (see text). Reproduced by permission from [7], *Secondary Ion Mass Spectrometry-Principles and Applications*, Oxford Science Publications (1989)

function of two dimensionless parameters:

$$a = (8U/r_0^2\omega^2)(z/m)$$

$$q = (4V/r_0^2\omega^2)(z/m) \tag{4}$$

The interested reader is referred to other texts [11] for a more detailed description.

In practice the quadrupole mass analyser is tuned for transmission of secondary ions providing constant resolution $m/\Delta m$ throughout the mass range. Transmission usually falls with increasing mass $\approx m^{-1}$. Ion trajectory aberrations which can degrade performance may also be introduced by fringe field effects at the entrance and exit of the quadrupole rods. These problems can be compensated for by coaxially installing a miniature set of pre- and post-filter quadrupole rods to which is applied a proportion of the rf field.

This is a convenient and widely used device for SIMS and other UHV applications since the electronics can be readily detached and replaced without degradation of performance to facilitate baking of an instrument. The same device can perform under both static and dynamic SIMS conditions and can be readily adapted for postionisation by electron beam SNMS (see Section 8). In static SIMS it is usually operated with low secondary ion extraction fields (10–100 eV). For dynamic SIMS, high fields (>1000 eV) can be used to improve transmission. Quadrupoles are often fitted with an ionising filament attached to the 'front end' of a device which enables residual gas analysis This is particularly convenient for monitoring the quality of UHV in surface analysis (say by SSIMS) but is also capitalised upon in surface science/adsorption studies for parallel thermal desorption studies (TPD) of the system e.g. [12].

The quadrupole is a low transmission device of less than 1%. Furthermore it is a scanning instrument so that it only allows the sequential transmission of ions, all other ions being discarded. The information loss is therefore very high. This is not only undesirable situation from the point of view of inefficiency, but also has ramifications in static SIMS analysis where monolayer lifetime and damage rates can be critical.

3.2.3   Time of Flight Mass Spectrometers

Time-of-flight mass spectrometry is conceptually the simplest means of mass separation used in SIMS. In ToF analysis pulses of secondary ions are accelerated to a given potential (3 to 8 keV) such that all ions possess the same kinetic energy; they are then allowed to drift through a field free space before striking the detector [13]. According to the equation of kinetic energy heavier masses travel more slowly through the 'flight tube' and so the measured flight time, t, of ions of mass-to-charge ratio, m/z, accelerated by a potential V down a flight path of length L provides a simple means of mass analysis.

$$t = L \left(\frac{m}{2zV} \right)^{1/2} \tag{5}$$

The basic experimental requirement is for a precisely pulsed primary ion source, a highly accurate computer clock, a drift tube and considerable computing power for data acquisition. The flight times of all the ions to the detector are electronically measured and related to ion mass. Thus a mass spectrum of all the ions is generated from the flight time spectrum. Mass resolution is critically dependent upon the pulse length of the generated secondary ion pulse which should be precisely defined and very short. This in turn is depen-dent on the pulse length of primary beam which is typically of the order of nanoseconds.

The energy distribution of secondary ions (ca 20–100 eV) will also affect the mass resolution. This initial energy spread will cause ions of the same mass to enter the drift tube with slightly different velocities and thus degrade the resolution in the final spectrum. This is usually compensated for by an energy analyser in the flight tube. The most commonly used device is an *ion mirror* which consists of a series of precisely spaced rings to which is applied a gradually increasing retarding field. The more energetic ions will penetrate further into the mirror before they are reflected, whilst the less energetic ions will take a slightly shorter path. When tuned correctly all ions of the same mass will arrive at the detector at the same time, despite their small energy differences when leaving the sample surface.

Usually detection is by single particle counting using a microchannel plate detector. This is a flat plate device whose surface contains a multitude of miniature channel electron multipliers (10–100 μm diameter, length to diameter 40–100). The inner surface of the channels is a lead/glass matrix which when bombarded secondary ions generates an enormous cascade of electrons. Figure 5.9 shows a common chevron arrangement, with two plates placed with their channels at an angle to each other: 0°/15° or 8°/8° are common arrangements which result in high output gains and suppress ion feedback. In some cases the heaviest ions travel so slowly that they do not register impact on the detector and this has been overcome by the introduction of an acceleration voltage immediately prior to detection.

The transmission of the ToFSIMS system is usually between 10 and 50%, but the great advantage is that, because the analyser is a non-scanning device, none of the ions are discarded in the analysis method. Figure 5.10 shows a schematic outline of the main features of a ToFSIMS system.

Although the input of charged particles in ToFSIMS analysis is very much smaller than with the continuous beam analysers, sample charging is still a problem when analysing insulators. This has been successfully overcome by pulsing electrons on to the sample surface between each primary beam pulse. The superiority of ToFSIMS for static SIMS analysis is illustrated from the

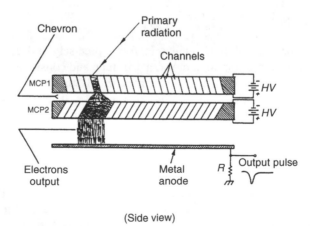

(Side view)

Figure 5.9. A schematic of the operation of a chevron multichannel plate arrangement. Reproduced by permission of Elsevier Science – NL from J.L. Wiza, *Nucl. Instrum. Methods*, **178**, 587 (1979)

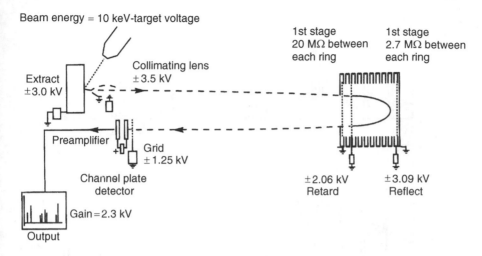

Beam energy = 10 keV-target voltage

Extract ±3.0 kV

Collimating lens ±3.5 kV

1st stage 20 MΩ between each ring

1st stage 2.7 MΩ between each ring

Preamplifier

Grid ±1.25 kV

Channel plate detector

±2.06 kV Retard

±3.09 kV Reflect

Gain=2.3 kV

Output

Figure 5.10. A schematic diagram of a ToFSIMS instrument. Reproduced by permission from J.C. Vickerman, *Analyst*, **119**, 513–523 (1994)

spectra of a poly(methyl methacrylate) spin cast thin film. Figure 5.11 compares the spectra obtained using Quad-SIMS and ToFSIMS. The former required 10^{12} primary ions cm^{-2} whereas the latter required less than 10^{10} ions cm^{-2}. The yield of the negative ion at $m/z = 85$, which is characteristic of the methacrylate backbone, relative to the primary dose is 4×10^{-4} for ToF compared to 6×10^{-8} for the quad. Spectral acquisition is possible with primary doses in the 10^{9}–10^{12} ion region, which should be well within the static regime.

The ToF analyser has further benefits for the analysis of organic materials. The more complex the organic materials being studied, the greater the mass range required and the possibility of mass spectral overlap can cause serious problems for interpretation. Whereas the quadrupole analyser is limited in its mass range to about 1000 amu and unit mass resolution; ToF instruments currently can provide mass resolution, $m/\Delta m$, in the region of 5 000–10 000 with, in theory, a limitless mass range (usually in practice about 10 000 amu). Table 5.3 summarises the performance of the mass analysers.

The sensitivity advantages of ToFSIMS suggested that *scanning* or *imaging* ToFSIMS would in principle enable sub-μm molecular ion imaging. However, the critical parameter now becomes the secondary ion yield per pixel and in practice spatial resolution is limited by the number of molecules in a pixel area and the yields of molecular secondary ions from the material being studied, see Section 6. Such yields tend to be around 10^{-2} via the SIMS process. The only way to increase the yield and hence the ultimate spatial resolution is to post-ionise the vast number of neutral molecules in the sputtered plume (see Section 8).

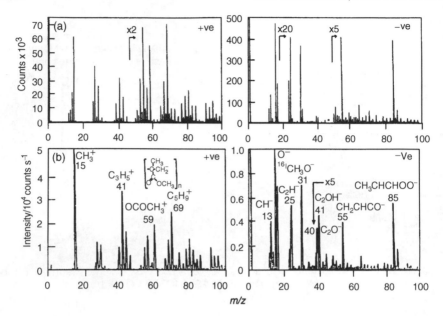

Figure 5.11. (a) Positive (i) and negative (ii) ToFSIMS spectra of PMMA, 30 keV Ga⁺, primary dose $= 10^{10}$ ions cm^{-2} (b) Positive (i) and negative (ii) quadrupole SIMS spectra of PMMA, 2 keV Ar°, primary dose $= 10^{12}$ atoms cm^{-2}. Reproduced by permission from A.J. Eccles and J.C. Vickerman, *J. Vac. Sci. Technol.*, **A7**, 234, (1989)

Table 5.3. Comparison of mass analysers for SIMS

Type	Resolution	Mass range	Transmission	Mass detection	Relative sensitivity
Quadrupole	$10^2 - 10^3$	$< 10^3$	0.01–0.1	Sequential	1
Magnetic sector	10^4	$> 10^4$	0.1–0.5	Sequential	10
Time-of-flight	$> 10^3$	$10^3 - 10^4$	0.5–1.0	Parallel	10^4

Whilst ToF instruments have significant advantages for surface mass spectral analysis, until very recently they were not the chosen instruments for depth profiling. Etching by a second ion source followed by static SIMS analysis is possible, but because the ToF instrument uses a pulsed analysis beam with a duty cycle of about 10^{-4} (i.e. the time the beam is on, divided by the time it is off) this is a time consuming process when analysing to a depth on the μm scale. Usually in a depth profile, perhaps six elements may be monitored, and there is no apparent advantage in collecting the whole spectrum. However, the developing requirement to analyse very shallow implants is making the ToFSIMS instrument attractive, see Section 7.1.2.

4 MECHANISM OF SECONDARY ION GENERATION

Ion formation in SIMS is a complex phenomenon. Simplistically the process can be divided into two components: the dynamical process by which atoms and multi-atomic clusters are desorbed; and the ionisation process in which a fraction of these sputtered particles become charged. Clearly, electronic factors are involved throughout the desorption event. Whilst a good deal of theoretical work has gone into understanding the process we are still a considerable way from a comprehensive theory which explains the experimental observations. Here we briefly outline a few of the main approaches to date. Since we are mainly concerned with surface analysis we will only consider the ideas which seek to explain the static SIMS process.

4.1 MODELS OF SPUTTERING

The simplest approach to sputtering of single component solids regards the atoms as hard spheres which obey Newtonian mechanics. Sigmund's linear cascade theory [14] has been the most successful model of the sputtering process so far. His model assumes that sputtering occurs by particle bombardment at small incident particle current and fluence. This excludes the situations where there is extensive heating and damage of the target and is close to the criteria for static SIMS. However, he also classifies sputtering events into knock-on sputtering and electronic sputtering. In his theory he disregards electronic sputtering. This approximation may well be valid for high incident primary particle energies, but in the low energy (few keV) region typically used in SSIMS, electronic interactions between incident particles and target atoms may not be negligible and the hard-sphere model may not be appropriate. The theory is developed on the basis of elastic collisions between point particles. Particular predictive success has attended the linear cascade ideas. In this process the incident particle transfers its energy to the target atoms and thereby initiates a series of collision *cascades* between the atoms of the solid within about 30 Å of the surface. Some of these collisions return to the surface and cause the emission of sputtered particles, see Figure 5.1. When applied to medium-to-high energy particle bombardment of single component materials, the data match the experimental results in terms of the dependence of yields on primary particle mass and energy rather well, see Figure 5.2. However, at lower energies collision energy may be exchanged over greater distances than envisaged in the point mass collision cross-section and in complex multi-component materials (e.g. polymers) the transport of energy will not be isotropic but highly directional.

To understand the sputtering process from complex materials, whilst Sigmund's theory gives some background ideas a different approach is required. In this regard various molecular dynamics simulations [15] have been very helpful in understanding the process occurring at the low primary flux densities encountered in SSIMS studies of inorganic materials. Much of the work in this area is due to Garrison and Winograd building on earlier models developed by Harrison [16]. Basically an ensemble of a few hundred to a few thousand atoms is selected to model a crystal with a surface plane with specified initial conditions of atomic mass, position and velocity. An atomic interaction potential function is devised to account for the bonding of the crystal. The sample is then bombarded by a number of primary particles of specified mass, velocity and angle of incidence. The classical (Hamiltonian) equations of motion are solved in a sequence of iterative steps and the motions of the target atoms are determined as a function of time from the initial impact. Early simulations could only treat small ensembles, but with the growth of computational power, larger ensembles of several thousand atoms can be treated. The early simulations calculated the dependence of sputter yield on the orientation of the exposed crystal plane, e.g. (100), (110) and (111) planes of copper and nickel [17]. The models did reproduce the observed relative dependence of yield and angular distribution on orientation, but the absolute calculated yields were too high. This was put down to the simplicity of the interaction potential used.

The interaction potential is crucial in influencing the accuracy of the simulation. It not only determines the motions of the target atoms within the collision cascade volume, it also influences the nature of any bonding interactions between emitted atoms as polyatomic species are emitted. If static SIMS is to be used to probe the chemical structure of a surface, the relationship between observed polyatomic ions and the surface chemical structure must be clearly defined. The simulations for copper and nickel surfaces suggested that there was a significant probability that the two atoms in an M_2 cluster, for example, might not arise from adjacent atoms in the surface. Clearly if this was reproduced in more complex materials, the relationship between a SSIMS spectrum and the surface chemical structure would be rather indirect and uncertain. However, this conclusion was based on the use of simple pair potentials to assess whether emerging atoms were likely to be bound in polyatomic clusters. A cluster was postulated if the potential energy calculated via the pair potential was greater than the kinetic energies of the constituent atoms. This was too simple and the predicted kinetic distributions of emitted species did not match experiment.

More recently Garrison and Winograd have used the embedded-atom potential method (EAM) to account for the many-body interactions during sputtering. It assumes that the total electron density in a metal can be approximated by a linear superposition of contributions from the individual atoms. The electron density in the vicinity of any atom is then the sum of the electron density contributed by the atom plus that from the surrounding atoms. The atom can be said to be *embedded* in this constant background electron density. Short-range repulsive effects between the atomic cores are also included using a Molière repulsive wall potential [18]. In studies of Rh atom desorption from Rh(111) the comparison between the theoretical and experimental energy distributions now proved to be very good [19]. The sputter formation of cluster polyatomic species from silver surfaces has also been tested using the EAM approach. The sputter yields of dimers relative to atoms together with energy distributions give good agreement between theory and experiment [20].

However in static SIMS analysis we are frequently more interested in the sputtering of organic films. These multi-element materials with highly directional properties are more challenging to model. As a first step Garrison and co-workers have modelled the sputtering of an organic layer on a Pt (111) surface [21]. The first study was of a p(2 × 2) ethylidyne, C_2H_3 at coverages of 0.25 and 0.5 ML. Subsequent studies dealt with adsorbed C_5H_9 [22] Many-body potential energy functions are required to allow for interactions among all of the three constituents of the target. As the collision cascade evolves it was necessary to be able to follow, via the integration of Hamilton's equations of motion, the reactions occurring among the substrate atoms, the substrate and adsorbate atoms and between the individual adsorbate atoms. The Pt{111} crystallite consisted of between 1500 and 2300 Pt atoms arranged in seven rows with C_2H_3 molecules placed in three-fold sites. The EAM approach was used to mimic the Pt crystallite. The C−C, C−H and H−H attractive interactions were well described by a reactive many-body potential energy function developed by Brenner [23]. The repulsive interactions were found to be improved by adding a Molière repulsive potential. The many-body interactions which could occur between Pt and C in the Pt−C complexes as sputtering proceeded were difficult to model with existing interaction potentials. A combination of the Brenner hydrocarbon potential and Pt−C and Pt−H Lennard−Jones pair potentials were used. Using this approach 'mass spectra' of the sputter yield following bombardment by 500 eV Ar were obtained. Although this was a representation of the frequency of yields of particles as a function of mass with no account taken of ionisation or stability. It is an encouraging start to the theory of surface mass spectra, Figure 5.12 (a). The most common mechanism of ejection observed was the fragmentation of the single C_2H_3 adsorbate. The main particles emitted were H and CH_3 with significant yields of Pt, and C_2H_3. Figure 5.12 (b) shows one of the more common mechanisms for the formation of CH_3. About 50 fs into the collision event, a second layer Pt atom collides with a first layer Pt atom causing it move outwards. As it tries to leave the surface in the 85 fs event it collides with the C_2H_3 adsorbate, which in the 125 fs event results in C−C bond rupture and the release of a CH_3 radical, by 200 fs the emission of a Pt atom and an intact C_2H_3 are also observed. This adsorbate was bound to the first layer Pt which was struck from below. The momentum of the first layer Pt is directed away from the adsorbate so the C_2H_3 *rolls off* the surface intact rather than fragmenting. Other fragments are formed by similar processes. If the emerging species has significant internal energy unimolecular fragmentations are observed. Other species such as H_2, CH_4 and HCCH were also observed although, with the exception of H_2

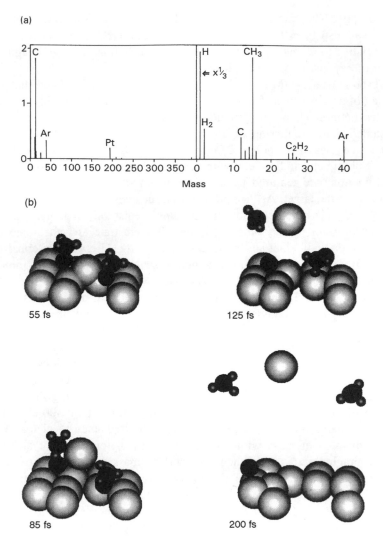

Figure 5.12. (a) Calculated mass distributions of sputtered particles from a C_2H_3 film adsorbed on Pr(111). (b) One of the more common mechanisms for the emission of CH_3 (see text for details). Reproduced by permission from [21], *Langmuir*, **11**, 1220. Copyright (1995) American chemical Society

in relatively minor proportions. The simulations suggest that these arise from reactions between emerging particles with either of the adsorbates still bound to the surface or with other emerging fragments.

These calculations are beginning to approach the types of chemistry found in the analysis of real organic layers. They yield results which begin to have the feel of real SSIMS spectra. However, the whole problem of ionisation has yet to be addressed.

4.2 IONISATION

The fraction of sputtered particles which are in the ionised state is in fact very small. In most cases over 99% of the sputtered yield is neutral. Whether a sputtered particle escapes from the surface as an ion depends on the relative probabilities of ionisation and de-excitation as it passes through the near surface region. Hence the high dependence of ion yield on the electronic properties of the matrix (the so called *matrix effect*). For metals, the rapid electronic transitions (10^{14}–10^{16} s^{-1}) make de-excitation a high probability during the 10^{-13} s required for a sputtered particle to traverse the near surface region. The probability, P_a, of escape as an ion from a metal can be approximated by

$$P_a \approx 2/\pi \ \exp[-\pi(\varepsilon_a - \varepsilon_F)/\hbar\gamma_N v_1]. \tag{6}$$

where ε_a and ε_F are the energies of the ionised state and the Fermi level, v_1 is the velocity of the emerging atom and γN^{-1} is the distance over which the level width decreases to 1/2.781 of the bulk value.

However, the need to develop an understanding of secondary ion emission from adsorbate and organic materials demands that we take account of the 'molecular' covalent type of bonding and consider that ionisation may take place at emission and also by subsequent fragmentation of vibrationally excited molecular units.

Two helpful qualitative models suggested to describe the process of secondary ion generation from molecular solids are briefly described here.

4.2.1 Nascent Ion Molecule Model

This model is due to Gerhard and Plog [24] and suggests that the rapid electronic transition rates which occur in the surface region will neutralise any ions before they can escape. Secondary ions are thought to result as a consequence of dissociation of sputtered neutral molecular species some distance from the surface. In the terminology of the model, ions are formed by the non-adiabatic dissociation of *nascent ion molecules* (neutral molecules). For inorganic oxides, most of the neutral molecules originate from direct emission of ion pairs such as MeO and keep their molecular character after leaving the surface. Only a few molecules have enough internal energy to dissociate into their constituents. The dissociation is considered to take place some distance from the surface where the electronic influence of the surface will be much smaller. The bond-breaking models which are used to explain emission from ionic materials consider the system solid-Me$^+$, whereas the nascent ion molecule considers the system Me$_x$O$_y^o$.

4.2.2 The Desorption Ionisation model

This is due to Cooks and Busch and introduces the concept that vibrational excitation may be important in understanding the emission of cluster or molecular ions from organic materials [25]. This model also emphasises that the processes of desorption and ionisation can be considered separately and however we understand the initial excitation process, the energy is transformed into thermal/vibrational motion as far as the molecules are concerned. A wide variety of ion emission processes is possible. Some pre-formed ions may be directly emitted. These are species which exist as ions within the material prior to bombardment and no ionisation step occurs. It is suggested that neutral molecules are desorbed in high yield, but to be detected must undergo an ionisation process such as cationisation. To generate other ions the model suggests that desorption is followed by two types of chemical reaction: (i) in the selvedge or top surface layers fast ion/molecule reactions — or electron ionisation can occur; (ii) in free vacuum, unimolecular dissociations may occur, governed by the internal energy of the parent ion giving rise to fragment ions.

According to these ideas the desorption event is of relatively low energy. The linear cascade ideas are not wholly appropriate when considering molecular solids. It is more helpful to think of energy being transferred to the vibrational modes of the molecule thus leading to fragmentation and ionisation. This is consistent with the observation that SSIMS is a relatively soft-ionisation phenomenon: for many materials, there is relatively little low mass fragmentation and large yields of molecular ions are observed.

5 STATIC SIMS — THE RELATIONSHIP BETWEEN SPECTRA AND SURFACE STRUCTURE

Whilst these qualitative models help us to understand what might happen during sputtering and ionisation of molecular surface species, in no sense do they provide any *a priori* grounds for expecting a direct relationship between SSIMS spectra and surface chemical structure, nor do they help to contribute towards predictive rules for the interpretation of spectra. Progress in the application of static SIMS as a method of surface chemical analysis has had to rely on a pragmatic experimental approach. The initial strategy has been to investigate surface chemistry which has been well characterised by other techniques and to assess the SSIMS data in the light of that knowledge. Confidence has then grown that SSIMS is indeed a surface mass spectrometry generating valuable chemical structure information and, as the experience of the technique has developed, so the rules of spectral interpretation are being defined. Recent research using SSIMS in adsorbate structure analysis and the characterisation of the surface structure of organic materials, demonstrates this capability. To have real confidence in the technique, an understanding of the fundamentals of the ion generation process is required.

Progress is being made in this direction using tandem MS/MS techniques to investigate the mechanism of ion formation from polymer surfaces [26].

5.1 SURFACE SCIENCE STUDIES OF THE ADSORBATE STATE

The surface scientist has a very wide range of sophisticated techniques available to study the state of single crystal surfaces and molecules adsorbed thereon. On less ideal surfaces, such as polycrystalline or supported catalysts, many of the techniques are very difficult to apply. Most catalytic processes involve the reaction or production of compounds with stoichiometries beyond the three or four atoms with which many surface science techniques are comfortable. Mass spectrometry has enormous power in structural characterisation and the application of that capacity to surface studies of catalysts, adsorbates and surface transformations could radically advance our understanding of surface processes.

Two groups which have made particularly significant contributions to the application of SSIMS to the study of adsorbates are the UMIST group in the UK and White's group at the University of Texas.

5.1.1 CO Adsorption on Metals

The early studies at UMIST demonstrated that SSIMS could distinguish molecular from dissociative adsorption of CO on metal surfaces. Dissociative adsorption, which was observed on tungsten at 300 K, was characterised by M_xC^+ or M_xO^+ ions, whereas molecular adsorption, observed on Cu, Pd, Ni and Ru in the temperature range 100–300 K, was distinguished by M_xCO^+ ions [27] Figure 5.13

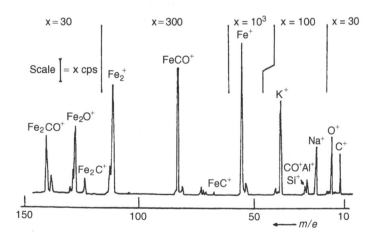

Figure 5.13. Static SIMS spectrum following the exposure of a clean iron foil to 10^8 torr of carbon monoxide. Reproduced by permission from [27], *J. Chem. Soc. Faraday I*, **71**, 40 (1976)

shows the spectrum observed for CO adsorption on iron at 300 K. CO is known to adsorb in both modes at this temperature and the spectrum shows both types of ions. There was no evidence from the data, when the low primary flux conditions were used, that the SSIMS process modified or destroyed the surface state. The data was in complete agreement with what was known from other techniques.

This conclusion was further supported when it was shown that the surface mass spectrum would characterise the surface chemical structure. SSIMS and HREELS investigations of CO adsorption on a number of metal single crystal surface planes, demonstrated that the relative intensities of the $M_xCO^+(x = 1-3)$ ions defined the adsorbate structure, whether linear, bridged or triply bridged to the metal surface atoms [28]. Table 5.4 lists the surfaces studied and the relative ion yields observed. It is significant that although the enthalpy of adsorption on Cu(100) is some 100 kJ mol^{-1} less than on Ru(0001), linearly adsorbed CO found predominantly on both surfaces yields the same relative ion distribution. Thus the 'fragmentation' patterns reflect the structure rather than adsorption strength. Figure 5.14 shows how the coordination of CO changes with coverage on a number of metals. Both HREELS and SSIMS show that at low coverage the coordination on Ni(111) is predominantly bridge; at a coverage of 0.5 coordination it becomes about 70% linear. The coordination changes on other single crystal surfaces are also shown and also, very significantly, from supported 6 nm Pd particles. This latter result demonstrates the applicability of the analysis to supported catalyst systems.

Very early in the application of SSIMS to studies of adsorbed layers on metals the fact of the matrix effect led the UMIST group to suggest the use of ion ratios to eliminate the influence of electronic effects which influenced the intensity of all ions [29]. As X was adsorbed on the surface of metal M electronic changes occurred (which could be monitored by measurements of the work function). Whereas MX_{ads}^+ increased as expected as the surface coverage increased, these electronic changes caused changes in the ionisation probability and thus the intensity of both M^+ and MX_{ads}^+ were also influenced as a consequence of this parameter. The ion ratio MX_{ads}^+/M^+ should therefore cancel out the

Table 5.4. Secondary ion cluster emission and CO adsorbate structure

Surface	SSIMS			CO structure IR/HREELS/LEED*
	MCO^+	M_2CO^+	M_3CO^+	
Cu(100)	0.9	0.1	—	Linear only
Ru(0001)	0.9	0.1	—	Linear only
Ni(100)	0.8	0.2	—	Linear + bridged
Pd(100)	0.3	0.6	0.1	Bridged only
Pd(111)	0.3	0.4	0.3	Triple bridged
Pt(100)	0.65	0.35	—	Linear + bridged

*LEED = Low energy electron diffraction.

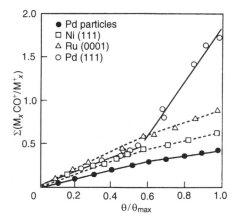

Figure 5.14. Variation of the ion fractions $MCO^+/\Sigma(M_xCO^+)$ with coverage for CO adsorption on several metal surfaces. Reproduced by permission of Elsevier Science – NL from [30]

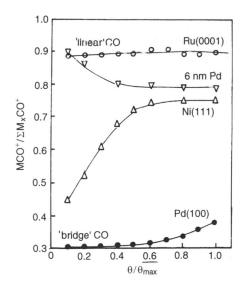

Figure 5.15. Variation of the sum of ion ratios, $\Sigma(M_xCO^+/M_x^+)$, as a function of CO coverage. Reproduced by permission of Elsevier Science – NL from [30]

electronic contribution. Figure 5.15 shows that the ion ratio is indeed directly proportional to CO surface coverage [30]. It was further demonstrated that using ion ratios $\Sigma(M_xCO^+/M_x^+)$ the relative surface concentrations of the different adsorbate states could be quantitatively monitored and these measurements used to determine the enthalpy of adsorption as a function of surface coverage [31].

5.1.2 Adsorption of Complex Hydrocarbons

Mass spectrometry is particularly helpful in the investigation of more complex adsorbates. Studies of the more complex hydrocarbon adsorbate states of ethene, propene and the butenes on Ru(0001) have indeed emphasised this capability [32,33,34]. In early work, vibrational spectroscopy has been used to cross-check the interpretation of the SSIMS data.

At first sight the SIMS mass spectral data look complex and difficult to interpret, but they are certainly no more difficult than vibrational spectra. Careful analysis of the data provides a great deal of useful information as to the surface science of quite complex molecules. Taking propene adsorption as an example, there are two mass spectral regions of particular interest: CH_x^+ ($x = 0$–3) and $RuC_yH_x^+$ with some weak signals in the $Ru_2C_yH_x$ region. In essence the $RuC_yH_x^+$ region is analogous to the parent ion in mass spectrometry. In this region, ions characteristic of each of the adsorbed species are formed by addition of the stoichiometric mass of the adsorbate to that of Ru. The spectra are complex solely because Ru has seven significant isotopes (m/z 96–104) and consequently the spectra from individual surface species interleave and overlap, Figure 5.16 [35]. It is possible to isolate signals at m/z values which are *uniquely* associated with each of the surface species formed as alkene adsorbs at 150 K and decomposes as the temperature is raised. Thus we can follow the loss of adsorbed propene from m/z 146; the decomposition to propylidyne from m/z 145 and the formation of ethylidyne from m/z 131. Normalising these signals to

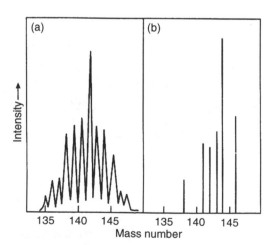

Figure 5.16. (a) Positive ion spectrum in m/z range 138–146 resulting from the adsorption of propene on a Ru(0001) surface at 150 K. (b) The expected SSIMS isotope pattern for molecular adsorption of propene on Ru(0001). Reproduced by permission from [35], *Proc. Roy. Soc. London A, Math. Phys. Sci.*, **330**, 147 (1990)

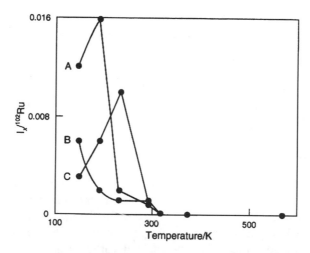

Figure 5.17. The variation in surface coverage as a function of surface temperature of (A) adsorbed propene from intensity of m/z 146; (B) adsorbed propylidyne from intensity of m/z 145 (C) adsorbed ethylidyne from intensity of m/z 131 after exposure of Ru(0001) to propene at 150 K. Reproduced by permission from [35], *Proc. Roy, Soc. London A, Math. Phys. Sci.*, **330**, 147 (1990)

the most abundant Ru m/z 102 isotope, enables us to follow the variation in relative surface coverage of these species with temperature, Figure 5.17. Cross-comparison of this SSIMS data with the vibrational data from HREELS and RAIRS provides close agreement as to the species present and their temperature ranges of stability.

Whilst these studies highlight the power of mass spectrometry, showing that it is able to be very detailed and precise in its analysis of adsorbed organic molecules, they also emphasise the importance of the synergy between SSIMS and vibrational spectroscopy: RAIRS does not detect di-σ adsorbed alkenes at 150 K but does detect the transitions to the alkylidenes; HREELS does detect di-σ adsorbed alkenes at 150 K and the transitions to the alkylidenes; SSIMS shows that the alkene adsorption at 150 K is accompanied by decomposition to alkylidenes, further transitions to alkylidenes and finally to CH_x. The interpretation of the SSIMS data was helped by the vibrational spectroscopy, but the SSIMS data, via the use of ion ratios, was semi-quantitiative as a function of temperature.

A similar approach has been used by White's group in a study of the influence of potassium on the adsorption of ethene on Pt(111) [36]. By a careful study of the positive and negative mass spectral fragmentation patterns in the C_2H_x and CH_x regions it was possible to clearly distinguish di-σ adsorbed ethene (fragments CH^+, CH_2^+ and $C_2H_2^+$) from π-adsorbed ethene (fragment $C_2H_2^-$) and then to show that potassium promoted the transition from di-σ adsorbed to π-adsorbed ethene and hence reduces the activity to dehydrogenation of ethene and changes its mechanism. The SSIMS data demonstrated that intermediates, ethylidene and ethylidyne are formed which decompose via C_2H to C_2 species which may then polymerise. As the potassium surface coverage increases, first the ethylidene and finally the C_2H are lost from the decomposition sequence. These studies also demonstrated that negative ion SSIMS is a very sensitive detector of surface carbon residues.

5.1.3 Surface Reactions

Whilst the above studies use the classic surface science approach, studying processes on single crystal surfaces, exactly similar data can also be accessed from small metal particles supported on oxide films which model supported catalysts. During CO oxidation on 6 nm Pd particles supported on alumina [37] the surface state of adsorbed CO, oxygen, and carbon has been followed at the same time as the gas phase processes. As the reaction temperature rises up to 350 K, although there is no evidence of gas phase CO_2, the surface coverage of CO (measured using $\Sigma Pd_xCO^+/\Sigma Pd_x^+$) falls. As this occurs, the coordination changes (measured from $PdCO^+/\Sigma Pd_xCO^+$) from a mix of linear and bridge CO, to predominantly bridge. Concurrently the surface carbon coverage ($\Sigma Pd_xC^+/\Sigma Pd_x^+$) rises. Clearly some dissociation occurs. At about 400 K as the production of gas phase CO_2 is rising, the surface coverage of carbon is seen to fall whilst the surface coverage of oxygen (measured by Pd_2O^+/Pd_2^+) begins to rise. In this region of activity the CO coverage continues to fall but the coordination of the CO changes progressively towards linear. Beyond 500 K the surface coverage of CO and carbon is very small, whilst the coverage of oxygen is high. The CO_2 production falls. The data suggests that besides the well known surface reaction between adsorbed CO and adsorbed atomic oxygen, there is also a route to CO_2 via adsorbed carbon presumably arising from CO dissociation. The study shows that the latter route becomes more significant on smaller Pd particles.

Another application of SSIMS in surface science is to combine it with temperature programmed desorption to give TPDSSIMS. This approach has been used in detailed studies of hydrogen adsorption on Ni(100)[38] acetylene and ethylene on Ni(111) and ethylene on Pt (111)[39,40].

5.2 SURFACE CHEMISTRY OF ORGANIC MATERIALS

Because of their complex chemistry, organic materials have not been amenable to study by surface science techniques. A number of groups have seen the potential power of a surface mass spectrometry to study such materials. The UMIST group with that of Briggs' group at the ICI Materials Research Centre were responsible for demonstrating the ability of SSIMS to fingerprint surface chemical structure by producing the first library of standard spectra in the Handbook for Static SIMS which was published by Briggs, Brown and Vickerman in 1989 [7]. This formed the prototype for the development of a much more extensive and ongoing Library of Static SIMS Spectra which enables the analyst to interpret unknown spectra by comparing them with the standard library spectra [41]. A large and increasing number of standard inorganic, and organic materials are featured. It would, however, be naive to suggest that spectral interpretation is always straightforward. The mechanism of sputtering and secondary ion emission from organic materials is still an active area of research. Thus we do not always know enough to interpret all the details of a spectrum nor, on occasion, to explain why features we expect to observe do not appear. It is clear, however, that much of the experience of conventional mass spectrometry is appropriate to spectral interpretation.

5.2.1 Static Conditions for Organic Analysis

Organic materials have never been easy for surface scientists to handle! Not only are they chemically complex, they are usually electrically insulating and they are also very sensitive to particle bombardment.

For inorganic materials it is assumed that 'static conditions' are about 1 nA cm^{-2}, with a total primary particle dose of about 10^{13} cm^{-2}. Briggs and co-workers have studied the generation of surface damage at PVA and PMMA surfaces under ion bombardment [42] using both XPS methods and SSIMS to identify changes in surface chemistry as the dose levels rise. At high dose levels, aromatic ions were generated which they judged were formed by the decomposition of polyene structures formed as a consequence of primary beam damage. This work established that a total dose of 10^{13} primary ions cm^{-2} was the *maximum* permissible before new spectral features become evident due to bombardment induced effects. As expected it was also demonstrated that total ion yield increased with particle mass (Xe > Ar = Ne ≫ He).

Studies on the sputter degradation of polymers have investigated the role of primary particle charge on the generation of surface damage in PET and PTFE [43]. The spectral time dependence was studied under xenon and argon ion and atom bombardment using equivalent particle flux densities. A comparison of the variation in the signal intensities of three structurally significant positive secondary ions from PET, m/z 104 and 149 and the pseudo-molecular ion m/z 193, as a function of primary particle dose for ions and atoms, showed that under ion bombardment there is a dramatic fall in the intensities of all the significant secondary ions, Figure 5.18. In contrast under atom bombardment little loss of intensity was observed by the 10^{13} total dose. The influence of particle mass is also significant. Assuming a simple exponential decay process where the sputter

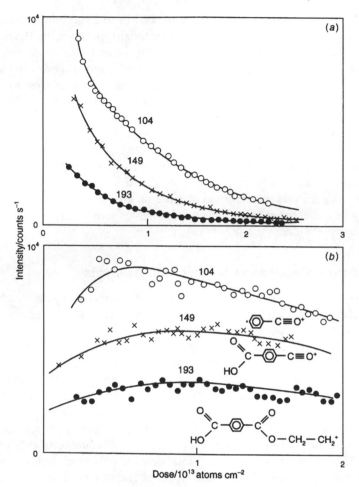

Figure 5.18. Variation of signal for some PET cluster ions with: (a) primary ion dose; and (b) with primary atom dose. Reproduced with permission from [43]. Copyright (1991) American Chemical Society

yield at time t is given by

$$S_t = S_0 \exp(-kt) \tag{5}$$

enabled the half-life for signal decay to be compared for bombardment by argon and xenon ions and atoms.

	Ar^o	Xe^o	Ar^+	Xe^+
Half-life/10^{12} particles cm^{-2}	50	26	9	6

Atoms are least damaging whether they are argon or xenon. However, the higher mass of xenon does generate more damage than argon. Similar data was obtained from a number of polymers including PVC, PMMA and PS [44]. The observation that neutral beams generate significantly less structural damage in organic materials can be understood along the following lines. It is generally believed that incident ions are neutralised prior to impact with a solid, thus the explanation for the increased damage rates under ion bombardment must lie with the occurrence of particle — surface electronic interactions (PSEIs). These interactions involve the transfer of an electron from the solid to the approaching ion via resonant or Auger processes. This results in charging and in the creation of electronically excited states in the surface region. If these states are antibonding, bond dissociation may occur, possibly accompanied by desorption. The effect of charge in the disruption of insulator structures is a recognised phenomenon. The presence of localised charges in a small region especially close to the surface creates large electrostatic forces that can lead to disruption of the surface by a Coulomb explosion mechanism. The potential energy of two electron charges separated by one atomic distance is 3.5 eV whilst a cluster of five ions has Coulomb energy of about 30 eV. These amounts of repulsive energy are significantly greater than the binding energy of atoms in a solid and hence the ejection of species from the surface can occur. The extensive surface damage and erosion which can result has been modelled by a molecular dynamics study [45].

We have seen that, to minimise surface damage during the analysis, static SIMS studies of organic materials should be performed using a ToF analyser. However, a ToFSIMS system requires the use of a pulsed ion beam. The above studies suggest that the static limit for organic materials under ion bombardment is well below 10^{13} ions cm^{-2}. It would seem to be safer to stay under 10^{12} ions cm^{-2}.

5.2.2 Spectral Interpretation

Three spectral regions can be identified which provide information on the surface state of materials. There is what might be termed the *sub-monomer* region, that is in the m/z region below the pseudo molecular ion $M^{+/-}$ or, if it is cationised, $(M+C)^{+/-}$. Then there is the *n-mer* region where ions and fragments are generated as a consequence of the bonding of two or three monomer units. Finally there is the *oligomer* spectral region usually generated via metal ion cationisation.

The sub-monomer region provides the fragmentation pattern spectrum which has been the main area of reference so far. Information on the basic chemical structure of the components of the material of the monomer(s) and any contamination, should be accessible. This data is obtained by analysis of the fragmentation pattern. If the material being analysed is completely unknown this can really only be accomplished by comparing the major peaks in the spectrum with a library of standard spectra. Whilst enormous libraries are now available in conventional mass spectrometry, they are only just appearing for SSIMS [41]. The methods of spectral interpretation are very similar to those used in conventional mass spectrometry although there are some significant differences. In positive ion SSIMS even electron quasi-molecular ions are predominantly found, whereas in electron impact mass spectrometry (EI-MS) odd electron molecular ions are formed where a non-bonding electron is lost. The difference can be easily understood from the different routes to ion formation, see Section 5.3.

The other major difference is that negative ion spectra are a very productive source of structural information in SSIMS, particularly for materials containing strongly electronegative heteroatoms, whereas negative ions are almost never found in EI-MS due to the inefficiency of electron attachment. The positive ion spectrum is frequently cluttered with other hydrocarbon originating ions which do not of themselves seem very diagnostic. The negative ion spectrum is free of these ions and hence when negative ions are formed they are frequently of significant diagnostic value. For example, the negative ion spectra of the methacrylate polymers are far more informative than the positive [46, 47]. The mechanism of ion formation seems to be much the same as that for positive ions, that is simple chain scission with unimolecular fragmentation processes. Since this region usually provides the most precise information on the chemical structure of the material, it is important that spectral assignment of the major peaks is accurate. Sometimes there are uncertainties where two or more possible ionic assignments have the same nominal mass. This problem may be overcome by derivitisation or chemical labelling, by the use of MS/MS CAD techniques or by accurate mass measurement. The latter approach is only possible using a high mass resolution, $m/\Delta m > 4000$, ToFSIMS instrument. Even this is not helpful if one wishes to identify the structure of a secondary ion species from an number of possibilities having the same stoichiometry. However, accurate mass measurement has been very helpful.

Recently it has been realised that, whilst there are peaks which are obviously diagnostic of particular surface chemistry, it is the *whole* spectrum which reflects the surface state. Some elements of the spectrum change more radically than others when the chemistry is modified, say in a plasma treatment. A weighting procedure is required to analyse the spectra.

Multivariate statistical methods are being explored to enable static SIMS spectra of organic materials to be used to define the surface chemistry in more detail [48].

There are ions formed from a combination of a small number of monomer units or monomer units plus fragments. The exploration of the potential information content of this type of so called *n-mer* ion is very much in its infancy. They may contain valuable micro-structural information of the same type as from the fragmentation of the monomer units. However, they can also potentially provide more macromolecular information about the polymer structure. In co-polymers the yield of *n*-mers composed of different combinations of the two components can indicate the extent, if any, of segregation [49]. Gardella *et al.* have suggested

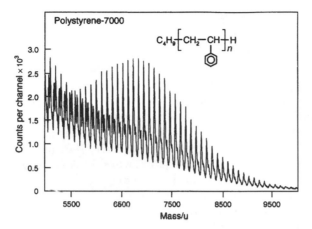

Figure 5.19. Positive secondary ion spectrum of polystyrene 7000 on silver. Reproduced with permission from [52]. Copyright (1991) American Chemical Society

that the statistical yield of n-mers of different lengths may point to folding or looping of a polymer chain at the surface [50]. There is also evidence that the cross-linking of polymers can be followed by changes in the yield of n-mers and even monomers. This would logically follow from the simple segmentation mechanism for sputtering.

In the high-mass region, the most interesting ions are generated by the use of cationisation. Benninghoven and Hercules and co-workers have demonstrated that laying down a thin layer of an organic compound or polymer on a silver substrate and using the high mass range capability of ToFSIMS, it is possible to generate spectra of oligomer distributions $(M_x + Ag)^+$ to $m/z > 8000$, Figure 5.19 [51, 52]. In principle these spectra permit an estimate of the average molecular weight distributions to be determined. In some cases, good agreement has been obtained with measurements from more conventional techniques, such as gel permeation chromatography. However, the response of the channel plate detector can fall off with ion mass and care has to be taken in interpretation. Sample preparation is also crucial. The best results have been obtained when the polymer is laid down by evaporation or spin-casting to give a very thin patchy covering on silver. This removes charging problems and aids cationisation. This is not a possible arrangement for the analysis of thick polymer films or devices.

5.3   EXAMPLES OF SURFACE CHARACTERISATION

The ability of static SIMS to characterise complex chemical systems has been most widely applied in the analysis of polymer surfaces. This work has been very recently reviewed [53].

5.3.1 Surface Analysis of an Adhesive System

A good illustration of the application of static SIMS to the surface analysis of a complex material is some recent work on adhesive systems. The surface and interfacial chemistry of adhesives and organic coatings can have a major effect on their properties and performance. Whilst XPS and AES have been used widely for adhesion studies, these techniques inherently lack the specificity to provide the level of molecular information that is essential for a full understanding of their interfacial chemistry. In the present study, four epoxides have been investigated using ToFSIMS [54]. They are Epikotes 828, 1001, 1007 and 1009. Epikote 828 is commonly used in adhesive formulations, whilst Epikotes 1001, 1007 and 1009 are used in lacquers, coatings and thermosetting agents. The Epikote resins are based on the diglycidyl ether of bisphenol-A epoxide structure, Figure 5.20.

Two sets of studies were carried out. First, silver cationisation was used to obtain detailed information about the oligomer distributions and composition of the polymers. In these experiments, thin layers of the Epikote were deposited on silver from butan-2-one solutions. In another set of experiments, to obtain data relevant to the study of the real adhesive and coating systems, thick films were laid down on aluminium foil.

Part of the silver cationised spectrum for the 1007 sample is shown in Figure 5.21. At intervals corresponding to the mass of the single monomer unit of the di-epoxide-terminated diglycidyl polyether, 284, ions are seen due to the oligomer + the two silver isotopes 107 and 109 for $n = 2$ to $n = 11$. Due to the wide mass range of the spectrum, the peak splitting due to the two isotopes cannot be seen. Each of the Ag cation peaks is accompanied by a signal at $m/z = 18$ higher. This corresponds to the presence of epoxide-*glycol* oligomers formed by hydrolysis of one of the two terminal epoxy groups. Signals at m/z 56 higher

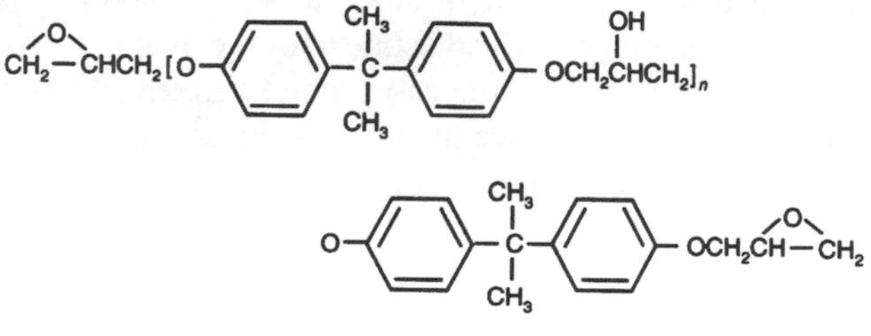

Resin	Mean molecular mass	*n* distribution
Epikote 828	450	mainly $n = 0$
Epikote 1001	880	centred $n = 3$
Epikote 1007	2870	centred $n = 10$
Epikote 1009	4000	centred $n = 14$

Figure 5.20. General structure of Epikote samples, listing their mean molecular masses and the central value of *n*. Reproduced with permission from J.C. Vickerman, *Analyst*, **119**, 513–523 (1994)

may correspond to the presence of *propanol* derivatives. In contrast to the 828 and 1001 Epikotes, in the 1007 and 1009 samples the epoxide-*glycol* terminated oligomer signals were higher relative to the normal di-epoxide terminated species. In addition, the 1007 and 1009 resins show other prominent Ag cationised species at m/z values 36 higher, 56 lower and 72 lower than the normal di-epoxide peaks. These additional signals correspond respectively to the presence of *diglycol-*, *phenol-* and *phenyl-* terminated oligomers. Thus it can be seen that there is a significant amount of information about the detailed chemical state of polymers. The preparation of the silver supported samples means that the analysis may not reflect the actual *surface* state of the resins, nevertheless the composition of the resin mixture is qualitatively indicated.

The positive ion spectra from the thick films generally did not give rise to signals above $n = 2$ oligomers. Fragments from the $n = 1$ and $n = 2$ oligomers were observed for Epikote 828. In the higher mass region of the negative ion spectra, fragments from $n = 1$ and $n = 2$ oligomers could be identified. However, the most useful signals were fragments of the monomer which could be attributed to terminal epoxide or bisphenol-A components. Figure 5.22 shows part of the positive ion spectra for the samples 1001, 1007 and 1009. The ions at m/z, 191, 252 and 269 can be assigned to fragments containing terminal epoxide groups:

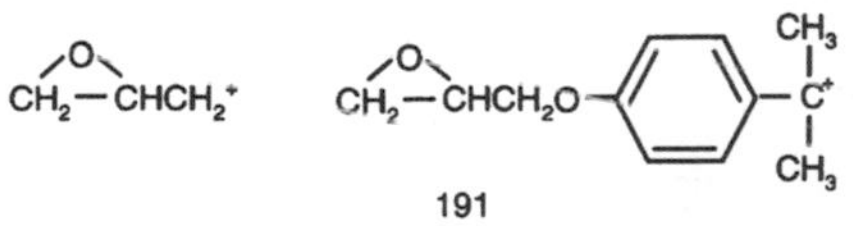

191

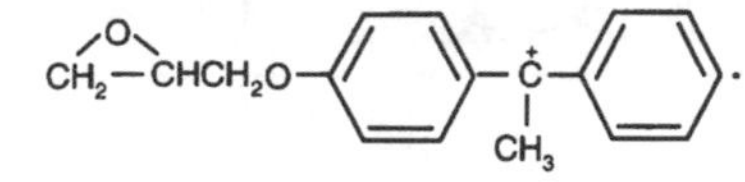

252

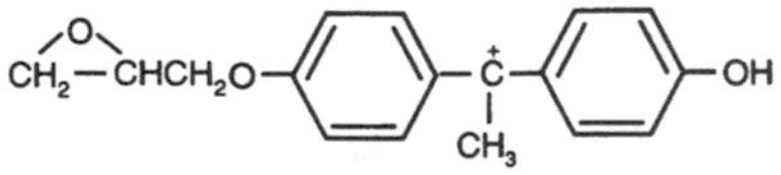

269

Scheme 5.1

In addition, positive ion signals specifically diagnostic of the bisphenol-A component were observed at m/z 135 and 213 corresponding to structures:

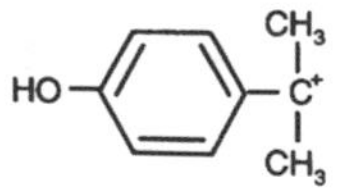

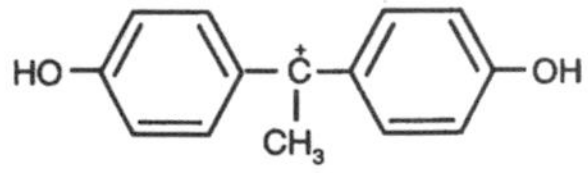

135   213

Scheme 5.2

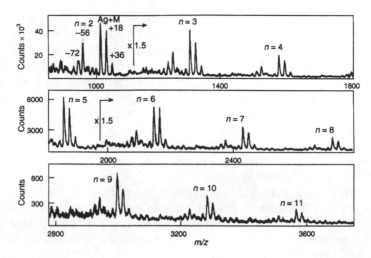

Figure 5.21. Spectrum of Epikote 1007 deposited on a silver substrate. Reproduced with permission from [54]. Copyright (1993) John Wiley & Sons Ltd

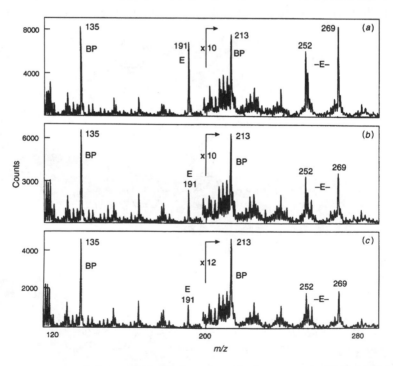

Figure 5.22. Part of the positive ToFSIMS profiles (m/z 115–290) recorded for Epikote 1001 (a), 1007 (b), 1009 (c). BP = bisphenol-A component; E = epoxide end group. Reproduced with permission from [54]. Copyright (1993) John Wiley & Sons Ltd

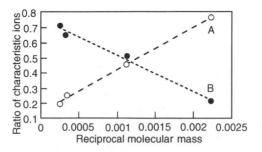

Figure 5.23. Plot of peak area ratios 191(epoxide)/[191(epoxide) + 135 (bis-phenol-A)] and 213(bisphenol-A)/[269(epoxide) + 213(bisphenol-A)] the Epikote resins. A, epoxide ratio; B, bisphenol-A ratio. Reproduced with permission from J.C. Vickerman, *Analyst*, **119**, 513–523 (1994)

In the negative ion spectra, similar assignments can be made.

Examination of these spectra makes it clear that changes in the molecular weight have a marked effect on the relative intensities of the signals specific to the terminal epoxide groups on the one hand, compared to those characteristic of the bisphenol-A groups on the other. As the molecular weight of the polymers increases, the ratio of epoxide end group to the bisphenol-A component would be expected to fall. A plot of the component peak ratios as a function of the reciprocal of the mean molecular weight shows a straight-line decrease for the epoxide ratio and an increase for the bisphenol-A, Figure 5.23. There is a clear quantitative relationship between the ToFSIMS spectral data and the composition of the Epikote resins.

This study demonstrates some of the information accessible from a static SIMS study of the surface of a complex organic system. The cationised spectra showed that a range of glycol and phenol terminated oligomers were present for the higher molecular weight resins. The array of secondary ions specifically diagnostic of terminal epoxide and bisphenol-A obtained from the spectra of the thicker films have been shown to be quantitatively related to the concentration of terminal epoxide and bisphenol-A groups. This latter observation should be useful in probing the effect of cross-linking in cured adhesives. The bisphenol-A part of the molecule is relatively unaffected by cross-linking. Thus most of the spectral features associated with this group should continue to be observed, whereas new features would be expected due to changes in the rest of the molecule. Preliminary studies of fully cross-linked adhesives have indeed indicated that the bisphenol-A fragments are observed in the spectra. Bisphenol-A fragments have also been observed in ToFSIMS spectra of cured epoxy paint systems [55].

5.3.2 Quantification of Surface Composition

Surface composition is of prime interest in the use of co-polymers. Surface segregation is of great interest whether it is a consequence of differences in surface free energies, contamination, surface treatment or technological use. Ion ratios have been used by two groups to monitor the surface composition of methacrylate co-polymers. Lub *et al* investigated the use of ion ratios to monitor the co-polymer system poly(methyl methacrylate)-poly(ethyl methacrylate) [56]. Two sets of negative ions were chosen to characterise the PMMA [m/z 31, CH_3O^- and 141, (M+41) where M is the methacrylate monomer] and PEMA [m/z 45, $C_2H_5O^-$ and 155, (M+41)] components, Figure 5.24. The m/z 85 ion (M- alkyl group) was used to characterise the methacrylate polymer backbone. Thus if m/z 45 and 155 are held to be unimportant (although they do appear in small yield) in the spectrum of PMMA and m/z 31 and 141 in the spectrum of PEMA and the presence of PMMA or PEMA does not influence the relative intensities of the ions from PEMA or PMMA respectively, then PMMA should be characterised by

$$I_{31}/I^m_{85} = \alpha \text{ and PEMA by } I_{45}/I^e_{85} = \beta \qquad (8)$$

where I^m_{85} and I^e_{85} are the intensities of the m/z 85 signal from PMMA and PEMA respectively. If the two ratios are plotted against each other at varying co-polymer composition, a linear relationship is obtained. It should be noted that there is a small yield of m/z 45 ions from PMMA. Thus when the ratio I_{45}/I_{31} is plotted against co-polymer composition, the line does not pass through zero, Figure 5.25. This was similarly found for the I_{155}/I_{141} ratio. Thus the composition of the co-polymer is related to the SSIMS ions

$$n_{EMA}/n_{MMA} = 1/\gamma(I_{45}/I_{31}) = 1/\delta(I_{155}/I_{141}). \qquad (7)$$

It is interesting that the coefficients give an estimate of the relative ion formation probabilities and the extent to which they may be influenced by ion fragmentation reactions. γ for the small ions is 0.59, whereas δ for the larger ions is 0.88.

Although the value of ion ratios in assessing composition is clear from this work, this study does show that care has to be used in using the m/z 85 ion as a standard peak. Analysis of the proportionality coefficients for the 50:50 copolymer shows that

$$I^e_{85}/I^m_{85} = \gamma\beta/\alpha \quad \text{(where } n_{EMA}/n_{MMA} = 1) \qquad (8)$$

which yields $I^e_{85}/I^m_{85} = 2.2$. Thus the formation of the methacrylate m/z 85 is more than twice as efficient from PEMA as from PMMA. It appears that the structure of the alkyl group influences the efficiency of formation of the 'side chain independent' methacrylate ion.

This study also highlighted the possibility of using *n-mer* ions to probe the statistical distribution of the component monomers. Dimer (2M-15) ions are formed reflecting the structures MMA-MMA, MMA-EMA and EMA-EMA having m/z 185, 199 and 213. If the distribution of the EMA and MMA

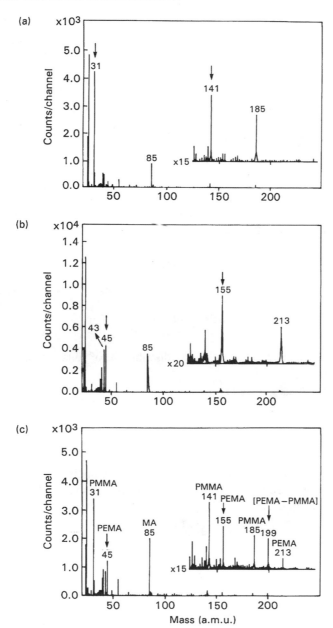

Figure 5.24. Negative ion spectra of PMMA (a), PEMA (b) and a copolymer of PMMA and PEMA (c). Reproduced with permission from [56], *J. Polym. Sci. B. Polym. Phys.*, **27**, 2071. Copyright (1989) John Wiley & Sons Ltd

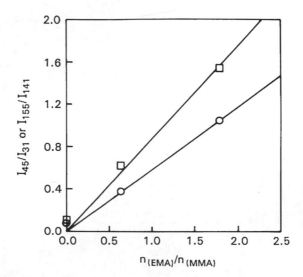

Figure 5.25. Plot of I_{45}/I_{31} and I_{155}/I_{141} against PMMA–PEMA copolymer composition. Reproduced with permission from [56], *J. Polym. Sci. B. Polym. Phys.*, **27**, 2071. Copyright (1989) John Wiley & Sons Ltd

monomers was statistical, the following distribution of ions would be expected:

$$I_{\text{MMA–MMA}} : I_{\text{MMA–EMA}} : I_{\text{EMA–EMA}} = n^2_{\text{MMA}} : 2n_{\text{MMA}}n_{\text{EMA}} : n^2_{\text{EMA}}. \qquad (9)$$

The distribution has been calculated by this formula and is shown in Table 5.5 for two different co-polymers. The relative intensities of the three ions are also presented. It can be seen that the surface composition agrees with the formula for the first polymer whereas for the second polymer the analysis value for m/z 199 is low and for m/z 185 is high. This may reflect a deviation from the random co-polymer structure. The possibility of valuable quantitative information is clear as we understand the SSIMS process more clearly.

Table 5.5. Comparison of calculated and measured SIMS intensity ratios for dimeric (2M-15) ions arising from MMA-MMA, MMA-EMA and EMA-EMA units from two samples of PMMA-PEMA co-polymers

Polymer composition	Method of determination	$I_{\text{MMA–MMA}}$ m/z 185	$I_{\text{MMA–EMA}}$ m/z 199	$I_{\text{EMA–EMA}}$ m/z 213
$n_{\text{MMA}} : n_{\text{EMA}}$ = 0.36 : 0.64	equation (9) from spectrum	0.13 0.16	0.46 0.44	0.41 0.40
$n_{\text{MMA}} : n_{\text{EMA}}$ = 0.61 : 0.39	equation (9) from spectrum	0.37 0.47	0.48 0.41	0.15 0.12

5.4 MS/MS STUDIES OF THE ION GENERATION PROCESS

Tandem mass spectrometry is extensively used in conventional mass spectrometry as an aid to spectrum interpretation and as a method for fundamental studies of fragment ion formation. It entails operating two mass spectrometers in series, usually separated by a collision cell. The basic idea is that an ion of interest is selected by tuning MS 1 to the appropriate m/z. The collision cell usually contains a low pressure of some inert gas species — argon or xenon. The ion collides with these gas species and fragments. The product ions (sometimes called daughter ions) are collected and analysed by MS 2. The method is useful if there is uncertainty as to the stoichiometry of a detected ion in a mass spectrum. Fragmenting the ion can usually clarify its composition.

The main application of the technique to data in SIMS has been to try to gain more understanding as to the secondary ion formation and fragmentation processes in the static SIMS of polymers. There has been a two-fold aim to these studies: first the development of generalised rules for the analysis of SSIMS data and secondly, at a deeper level, to see if it is possible to utilise gas phase fragmentation data to understand the processes which lead to secondary ion formation during sputtering.

The most detailed SIMS studies to date have utilised an experimental arrangement consisting of two high mass range quadrupole mass analysers, Q1 and Q3, separated by a quadrupole collision cell region, Q2. Collision energies are in the 5–20 eV range. In the MS/MS mode as well as the conventional daughter ion operation, it is possible to analyse for the neutral fragments lost from ions in the course of fragmentation. This is accomplished by scanning both Q1 and Q3 at the same time with Q3 offset by fixed mass difference corresponding to the expected neutral loss.

A number of polymer systems have been studied in some detail and the results reviewed [57]. These include PTFE low density poly(ethylene), LDPE,; poly(ethylene terephthalate), PET; polystyrene, PS; poly(4-hydroxystyrene), P4HS. A very significant general conclusion has been reached, namely that the processes which occur in the collision cell are very akin to those that occur at the surface as ions are emitted from the surface and fragment. The approach and conclusions are illustrated by reference to the studies on PET.

A diagnostic section of the positive ion spectrum of PET is shown in Figure 5.26(a). The majority of ions are odd mass, even electron ions. However, significant yields of radical cations are observed at m/z 104 and 148. The question which the study sought to answer was, 'How are the observed ions related to the surface structure of PET and can we discern the mechanism by which the ions are generated as the surface is sputtered?' The m/z 149 ion is significant. Using argon as the collision gas, a daughter ion spectrum of m/z 149 was generated which consisted of most of the major peaks seen in the normal SSIMS spectrum below m/z 149. For example, in the range of Figure 5.26(a) loss of 28 amu gave rise to the m/z 121 ion and the loss of 44 amu resulted in the m/z 105.

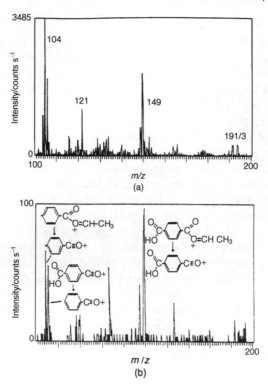

Figure 5.26. (a) A region of the positive ion SSIMS spectrum of PET. (b) Neutral loss spectrum for the loss of mass 44 from PET positive secondary ions. Reproduced with permission from G.J. Leggett and J.C. Vickerman, *J. Chem. Soc. Faraday Trans.*, **88**, 297 (1992)

In Figure 5.26(b) part of the 44 amu *neutral loss spectrum* is shown. It is particularly interesting to observe that the relative intensities of the ions at m/z 105, 149 and 193 are very similar to those in the SSIMS spectrum itself. This implies that the following fragmentation steps occur with equal probability:

$$HOOCPhCOO^+ \rightarrow PhCO^+ + CO_2 \quad \text{in the collision cell}$$

$$polymer \rightarrow PhCO^+ + CO_2 \quad \text{in SSIMS at the polymer surface}$$

and consequently that the probability of forming the m/z 105 ion from its gas phase progenitor is equal to the probability of forming the same ion (relative to the probability of forming any other ion, e.g. m/z 149) at the sample surface during sputtering. This, as we have said, suggests that there is a great similarity between the type of fragmentation processes which occur in the collision cell and during sputtering. The energies involved must be very similar, which suggests

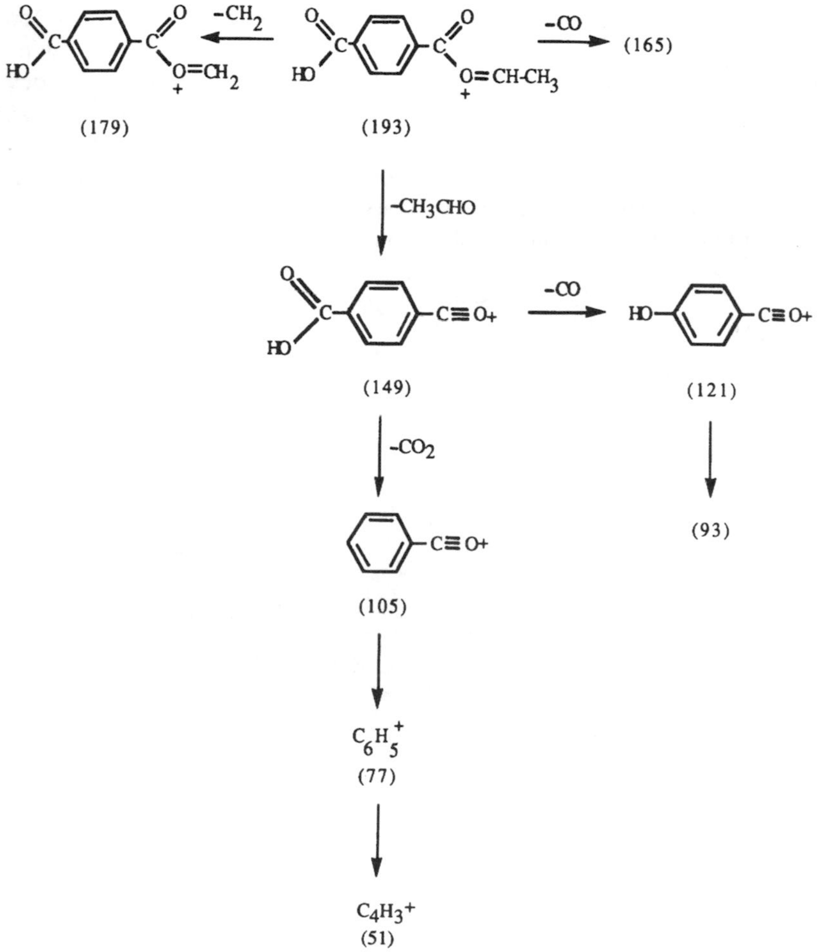

(179) $\xleftarrow{-CH_2}$ (193) $\xrightarrow[\;-CO\;]{}$ (165)

(193) $\xrightarrow{-CH_3CHO}$ (149)

(149) $-C\equiv O+$ $\xrightarrow{-CO}$ (121) HO—⟨ ⟩—$C\equiv O+$

(149) $\xrightarrow{-CO_2}$ (105)

(121) $\rightarrow$ (93)

(105) —$C\equiv O+$ $\downarrow$

$C_6H_5^+$
(77)

$\downarrow$

$C_4H_3^+$
(51)

Scheme 5.3

that the collisions which result in particle emission from the surface have energies in the 5–20 eV range.

By assembling all the daughter ion and neutral loss spectra it was possible to build up a single schema which represented the entire fragmentation sequence starting with the quasi-molecular ion $(M+H)^+$ at $m/z = 193$, Scheme 5.3. It can be seen that all the daughter ions observed in the fragmentation sequence could be envisaged as having arisen from the fragmentation of the largest ion observed in the spectrum. Furthermore, the ions most commonly observed as daughters were also the most intense peaks in the SSIMS spectrum (e.g. $m/z = 77$, 105 and 149. This observation accords with similar conclusions for PTFE and LDPE. This provided further evidence that the collision cell processes, model the processes which occur at the sample surface. Clearly many of the secondary ions observed in the SSIMS spectrum are the products of simple low-energy dynamical degradation processes: in other words the ion formation process is a soft one under low-flux static conditions. We can conclude that these simple low energy collisional processes lead, below a certain low dose threshold, to the generation of secondary ions without observable chemical degradation of the surface structure. The process may be considered to be physical degradation in which the polymer chains are simply segmented.

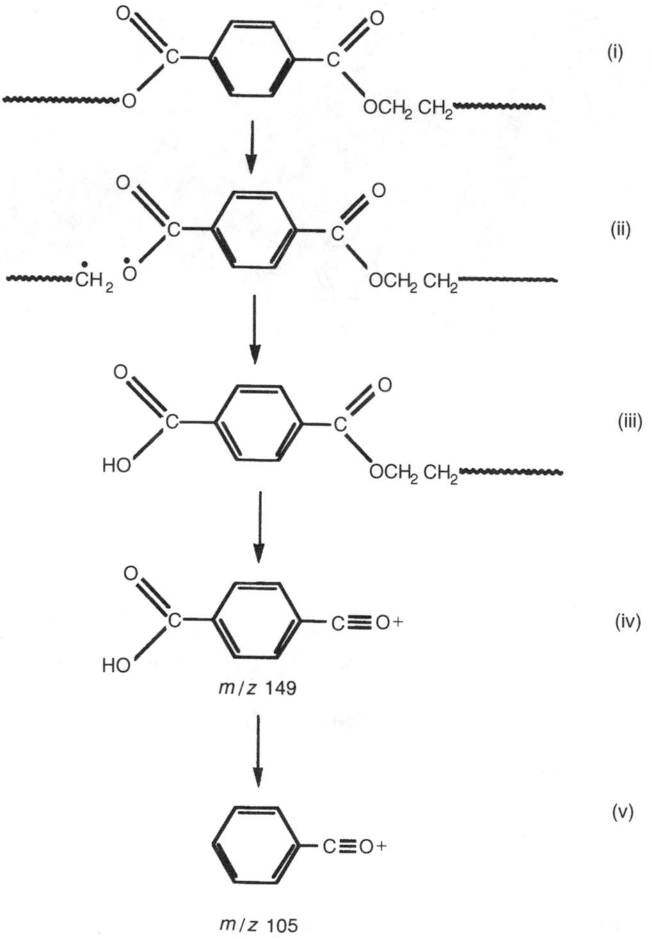

Scheme 5.4

On the basis of these ideas it is possible to propose a mechanism by which the secondary ions are formed from the polymer chain. Consider a PET polymer chain as shown in Scheme 5.4 If the chain is broken by a primary particle, impact (ii), capture of a hydrogen by the radical which is created, forms a surface bound species which is equivalent to the pseudo-molecular ion (iii). Separation of the species yields the m/z 193 ion, but fragmentation and loss of CH_3CHO yields the $m/z = 149$ ion (iv) which in turn can lose CO_2 to form the m/z 105 ion. Neither of the odd electrons, m/z 148 or 104 radical cations are formed in the collision process from the above ions. Thus it would appear that the even electron rule is obeyed (even electron ions will fragment to even electron ions, but odd electron ions may fragment to *either* even or odd electron ions).

The m/z 104 ion is, however, very intense in the SSIMS spectrum. It is a daughter ion of the m/z 148 ion. It can be envisaged that if the energy input was somewhat higher, before a hydrogen could be captured by the radical species in Scheme 5.3 (ii), CO_2 is eliminated and an m/z 148 radical cation separates from the surface. Fragmentation with the loss of CH_3CHO would yield the m/z 104 ion.

This study demonstrates that at low primary beam dose levels ($<5 \times 10^{12}$ cm^{-2}) there is a clear and simple relationship between the ions generated and the chemical structure. It has been shown that the mechanism of ion generation can be envisaged to follow from simple free-radical fragmentation steps.

The overall conclusions from these MS/MS studies in the static SIMS regime are:

(a)   that the collision cell processes very closely model those occurring at the surface during sputtering;

(b)   that this implies that in the main the mechanisms leading to ion formation are simple low-energy, non-adiabatic and unimolecular processes;

(c)   that it may be assumed that polymer chain cross-linking and other intermolecular processes may be neglected and the ions formed are characteristic of the virgin polymer;

(d)   that such rearrangements as do occur, effectively involve the formation of end-groups following chain scission, these end-groups are formed from the original polymer and bear a definite relationship to the virgin polymer structure;

(e)   that the rearrangements and ion formation processes are very similar to those observed in conventional mass spectrometry;

(f)   that the even electron rule is obeyed, apparent violations are possible due to the proximity of the surface, for example, as a consequence of the exchange of hydrogen atoms between a departing species and the surface.

6 SIMS IMAGING OR SCANNING SIMS

In principle, combining one of the mass analyser systems with a liquid metal ion source enables surface analysis with high spatial resolution to be carried out.

Indeed there is the potential for scanning electron microscopy-type images to be generated with the very considerable added facility of full chemical sensitivity. The highest spatial resolution is accessible when operating in microprobe mode.

As indicated earlier, liquid metal ion beam systems can be operated with beam diameters at the sample down to 50 nm, although the more usual range is 200 nm to 1 μm. The beam is digitally rastered across the surface of interest such that there are, say, 256 × 256 pixels in an image. At each pixel point it is possible either to collect ions of a single m/z, of a few specified m/z or a whole mass spectrum, dependent on the detail required and the sophistication of the analyser and data system. Elemental or chemical state images can then be generated of the areas of interest.

The image in Figure 5.27 illustrates the basic capability. It is an oxygen image of gate oxide electronic device structure. The clear geometry of the structure is evident. However, a blurring of the image is evident such that bright oxygen-rich areas span a number of the gate regions and form what appears to be a stain. The analytical system can focus in on the regions of interest and collect a spectrum to clarify the situation. The ion beam is first focused on the gate oxide region. A negative ion spectrum of this region, with strong SiO_2^- and SiO_3^- ions, is characteristic of silicon oxide. The ion beam is then focused on the polysilicon region. The resulting mass spectrum only contains a strong Si^- signal. The stain region is then examined and the spectrum is very similar to silicon oxide. Oxide has been formed in a region which should be polysilicon. The F^- peak in the spectrum gives a clue to the problem. The wafer on which the device had been grown would have been cleaned in HF and rinsed in water. If the rinsing was not fully effective, HF left on the wafer could cause oxidation of the silicon.

This study was carried out with a quadrupole mass analyser which was only capable of collecting a few ions at each pixel. Modern ToFSIMS systems are able to collect a whole spectrum at each pixel. In this case multiple analyses of oxide, polysilicon and stain regions would not have been necessary. Interrogation of the data system for the spectra in each region would have provided the information immediately.

Analysis of complex chemistry with good spatial resolution is possible. Figure 5.28 illustrates the analysis of a simple tri-peptide valine-tyrosine-valine (VYV) localised on a 40 μm polystyrene bead. It is estimated that there are less than 200 pmoles on the bead of which a very small fraction is on the surface. The mass spectrum fragmentation pattern from the bead is characteristic of the structure of the peptide. (The $(M + H)^+$ ion at m/z 381; the $y_2 + 2$ peptide fragmentation of the YV end of the molecule at m/z 281; the b_2 (VY) fragmentation at m/z 263 and the a_a (VY) fragmentation at 235 with immonium ions from valine and tyrosine at m/z 72 and 136 respectively.) The distribution of the peptide on the bead can be visualised from the image of the molecular ion plus some of the principle fragments. Care has to taken in interpreting such images from curved samples. Ion generation and collection will vary with

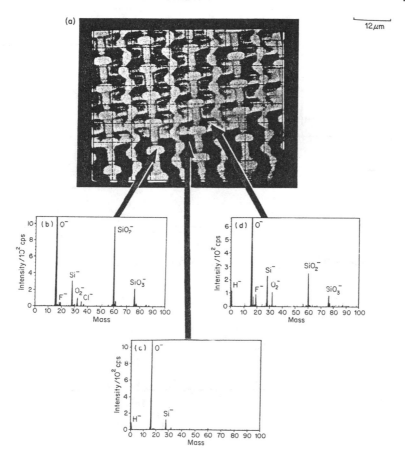

Figure 5.27. (a) An O$^-$ SIMS image of a memory device showing a stain region. (b) A SIMS pectrum of the gate oxide region of the device. (c) A SIMS spectrum of the polysilcon track area of the device. (d) A SIMS spectrum of a part of the stain area. Reproduced with permission from [8], *Handbook of Static SIMS* (John Wiley & Sons), p151 (1989)

angle, so signal intensity will be influenced to a significant degree by sample geometry [58].

While surface analysis at high spatial resolution is an attractive proposition, as the magnification increases it becomes more and more difficult to obtain images with adequate dynamic range and still maintain static conditions. SIMS is a destructive technique; as magnification increases, the number of atoms or molecules in a pixel area decreases, see Table 5.6. If we assume a sputter yield of 1, an ionisation probability of 10^{-3}, instrumental transmission of 10^{-1} and also assume that for a major component, at least 10 secondary ions are required

(a)

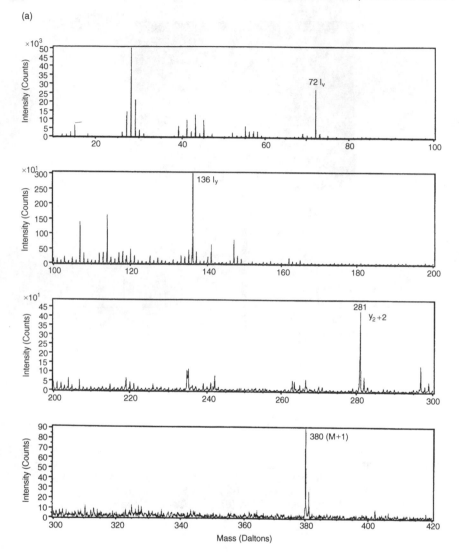

Figure 5.28. (a) ToFSIMS spectrum of tripeptide valine-tyrosine-valine supported on a 40 μm polystyrene bead, see text for interpretation

per pixel, then static SIMS analysis below 1 μm^2 is not really possible. It is clear that to have any real hope of chemical state analysis at high spatial resolution, secondary ion yields have to be improved dramatically. This is only possible if the ionisation probability can be raised. Since about 99% of the particles sputtered from a surface are neutral, if some effective method can be devised to ionise most

(b)

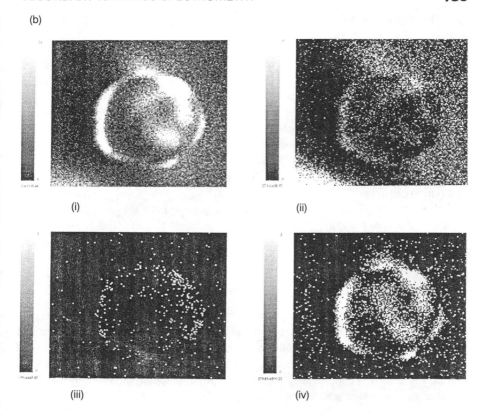

Figure 5.28. (b) Images of tri-peptide on bead: (i) total ion image; (ii) Si$^+$ image from supporting silicon wafer; (iii) image of (M+I)$^+$ ion; (iv) image of fragments made up of a composite of the M+H ion and fragments in range $m/z > 280$

of these neutrals there is the possibility of a startling improvement in sensitivity. This will be considered in the final section of this chapter.

Table 5.6. Estimation of the number of molecules and atoms per pixel area

Imaged area (μm)	Pixel size	Pixel area	Molecules per pixel	Atoms per pixel
100	10 μm $\times$ 10 μm	10^{-6} cm^2	4×10^8	2.5×10^9
10	1 μm $\times$ 1 μm	10^{-8} cm^2	4×10^6	2.5×10^7
5	500 nm $\times$ 500 nm	2.5×10^{-9} cm^2	1×10^6	6.25×10^6
1	100 nm $\times$ 100 nm	1×10^{-10} cm^2	40 000	2.5×10^5
0.2	200 Å $\times$ 200 Å	4×10^{-12} cm^2	1600	10 000

7   DEPTH PROFILE ANALYSIS BY DYNAMIC SIMS

The earliest application of SIMS was not as a surface analysis technique but as a means of detecting trace deposits of elements with very high sensitivity. To do this the primary ion flux is increased so that many layers are removed rapidly, consequently the secondary ion flux is increased such that elemental concentrations down to sub-ppm levels can be detected. The technique has been refined such that today very accurate concentration profiles can be monitored as a function of depth from the surface of the material. Although this capability finds important application in many areas of materials science, it is in the dopant analysis of semiconductor materials where dynamic SIMS has made its most dramatic contribution.

Indeed dynamic SIMS is the only technique capable of providing analysis down to the levels required. In this section we will illustrate the application of dynamic SIMS with reference to depth profile dopant analysis in semiconductors. A useful reference book is that by Wilson, Stevie and Mcgee [59].

7.1   THE DYNAMIC SIMS EXPERIMENT

To derive quantitative concentration data from the sputtered secondary ions is quite complex. The various parameters will be considered in this section, however, a basic underlying requirement is *steady state* sputtering. The various elements in a material will sputter at different rates. Initially, when sputtering starts, the surface concentration of high sputter rate elements will decrease relative to the lower sputter rate elements. Thus the actual yield of sputtered material of the easily sputtered elements will then fall and the yield of the less easily sputtered elements will rise. When steady state is reached the yields of sputtered atoms plus ions will reflect the relative concentrations of the components.

In carrying out a depth profile analysis there are two main parameters for which the instrumentation and experimental procedure have to be optimised. First, the *dynamic range* of concentration sensitivity for the element to be analysed and second, the *depth resolution*.

Usually a dopant element, for example boron in silicon, is incorporated either by ion implantation by bombarding a silicon wafer with high energy boron ions or during an epitaxial growth process. In either case there will be a variation of boron concentration with depth from the surface of the silicon wafer. Thus we want our experiment to be able to monitor the concentration of the dopant of interest from the maximum level which occurs down to the minimum (dynamic range). We also want to measure as accurately as possible the depth at which a particular dopant concentration occurs (depth resolution). Success in both areas is a function both of the material being analysed and quality of the equipment and its operation.

7.1.1 Depth Profiling — the Ion Beam

A focused ion beam is rastered over a defined rectangular area on the surface of the wafer to be analysed. The aim is to sputter etch a precise crater down several micron below the surface, see Figure 5.29. The x/y position of the beam on the surface will be controlled by digital electronics. As the beam steps across the surface, secondary ions are generated and may be collected by the mass analyser. Usually the beam energy will be in the range 1–20 keV, the beam flux will be in the μA cm^{-2} region; the beam will be focused dependent on beam flux in the μm–tens of μm diameter range (the smaller the ion beam flux, the smaller the minimum beam diameter possible); the beam will be rastered over an area ranging from 10 μm × 10 μm–100s μm × 100s μm. Where analysis of wafer product is required, larger area analysis, say 500 μm × 500 μm, is possible. However, where analysis of individual devices is required, very small area capability is required. Specialist procedures can analyse sub-micron areas. The angle of incidence of the primary beam on the surface depends on the instrumentation used. 90° incidence usually provides the highest sensitivity. However, where very shallow profiles are required with high depth resolution, low angles of incidence are used.

It will be recalled that secondary ion yield is very sensitive to the electronic state of the material being analysed. Very high positive ion yields are obtained from oxide matrices. It is found that the use of oxygen ion primary beams, O^- or O_2^+, at the usual operational flux density, generate an oxide layer which gives high yields of electropositive ions. Ions which have low positive ionisation probabilities usually have high negative ion yields (e.g. oxygen, halogens, arsenic, etc) when Cs^+ is the primary beam species. Thus the most widely used primary beams for depth profiling are derived from oxygen and caesium sources. Where

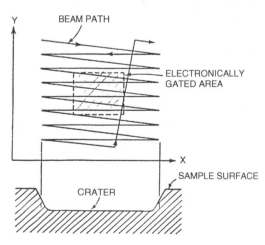

Figure 5.29. A schematic representation of the method for sputter etching a crater by rastering an ion beam of a material surface

very high spatial resolution is required, liquid metal ion sources have to be used, but unless oxygen flooding is used, the ion yields will be reduced.

7.1.2 Depth Profiling — the Mass Analyser

All three types of mass analyser are used in dynamic SIMS. However, to date the most widely used is the magnetic sector. The high extraction field (≈ 4 keV) gives good secondary ion collection efficiency and the potentially high mass resolution enables mass interferences to be tuned out. The Cameca range of analysers have been the instrument of choice in this area of analysis. Although their overall transmission and mass resolution is inferior, quadrupole analysers have some advantages. They operate with a low extraction field, usually in the range 10–300 or so eV. This means that low energy primary beams can be used and low angles of incidence are possible. The trajectory of the primary beam is much less influenced than it is by the high extraction field in the magnetic sector instruments. Low primary beam energies and low angles of incidence are advantageous when shallow profiles are required.

Recently ToFMS analysers have been used for depth profiling. It will be recalled that the move to ToFMS in static SIMS was to use most efficiently all the generated ions in order to maximise the information-to-damage ratio. As the requirement for shallow depth profiles has developed, it has been realised that there could be advantages in combining the high efficiency static analysis of ToFSIMS with low energy primary beam depth profiling [60]. Two primary beams are used. To depth profile, an oxygen or caesium beam is used with the ion extraction potentials off. After etching to a required depth, the etching beam is switched off, the analysis beam, usually a highly focused liquid metal beam, is pulsed on to analyse a small area at the bottom of the crater. A detailed profile is obtained by rapidly switching between profiling and analysis. The pulsed mode of operation makes analysis somewhat slower than using the other analysers, but higher sensitivity with nm depth resolution is obtained.

7.2 QUANTITATIVE ANALYSIS

The use of SIMS for quantitative analysis is complicated by the dependence of ion yields on the precise electronic state of the matrix. The successful application of dynamic SIMS to the analysis of small concentrations of dopants relies on the fact that the concentrations are *very small* such that their variation can be considered to have no real effect on the matrix and thus on the ion yield. The maximum concentration measured would be of the order of 10^{20} atoms cm^{-3} or less than 10^{-2} atom %. Variations in concentration above this level will generate a significant matrix effect and quantitative analysis would not be possible. Even though dopant concentrations are generally well below these levels, where a profile passes from one matrix to another, e.g. from a silicon oxide layer to a poly-silicon layer, secondary ion yields change.

This has to be factored into the data interpretation. In all cases, reliable quantitative analysis is only really possible if closely matched standards are available. The standards have to be the same as the analysed materials in terms of the chemistry, the host matrix and the identity and concentration region of the dopant. The exact concentration of the dopant in the standard has to be known, to provide a reference point for the instrumental variables. Quantification to an accuracy better than 5% is possible if the ion-implanted standards are known with sufficient accuracy. Thus, either immediately before, or after a depth profile analysis, the experiment has to be referenced by analysing a standard.

7.2.1 Sensitivity

SIMS can detect all elements of the periodic table. The sensitivity of the analysis is dependent on the secondary ion yield and the transmission of the analyser. We have seen that secondary ion yield varies with element across the periodic table by about four orders of magnitude, the yield is also dependent on host matrix. Thus, under oxygen beam bombardment, the yield of Li^+ per atom sputtered is between 3×10^{-3} (from a germanium matrix) to 8×10^{-3} (from silicon) whilst the yield of Cl^+ is between 3×10^{-6} (from germanium) to 1×10^{-7} (from silicon) [61]. The yield for chlorine is very much higher using a caesium primary beam. Figure 5.30 shows the ratio of ion yield for negative ions under Cs^+ bombardment compared to positive ion yield under O^- bombardment as a function of atomic number. Elements featured above 0 in the log M^-/M^+ axis

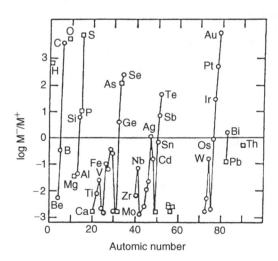

Figure 5.30. The ratio of negative ion yield (M^-) under Cs^+ bombardment to positive ion yield (M^+) under O^- bombardment as function of atomic number. Reproduced from data in H.A Storms, K.F. Brown, and J.D. Stein, *Anal. Chem.*, **49**, 2023. Copyright (1977) American Chemical Society

show the highest yield as negative ions using Cs^+ bombardment, whilst those below 0 are usually most sensitively analysed using positive ions under oxygen bombardment.

The relationship between dopant concentration and secondary ion count rate is as follows: Suppose the material to be analysed contains a dopant X present in the solid at atom fraction $n(X)$ well below the 10^{20} atoms cm^{-3} level. Then since N_s total target atoms are sputtered per second from a crater area of A cm^2,

$$N_s(X) = N_s(\text{total})n(X) \text{ atoms (X)s}^{-1} \quad (10)$$

atoms of type X are sputtered per second. The analyser is tuned to collect X^+ ions. The *useful yield* of X^+ ions will be less than the actual number of atoms sputtered because the ionisation probability, $\alpha(X^+)$, of X is much lower than 1 and the instrumental transmission, $\eta(X^+)$, of the ions emitted from the surface to the detector is much less than 100%. The useful yield τ_u is the ratio of ions detected to atoms sputtered within the analytical collection area (gated area — see below).

$$\tau_u(X^+) = \alpha(X^+)\eta(X^+) \text{ secondary ions per atom.} \quad (11)$$

Thus the number of ions of type X^+ detected per second is:

$$N_s(X^+) = N_s \text{ (total) } n(X)\tau_u(X^+)(a/A) \text{ ions s}^{-1} \quad (12)$$

where a is the analytical area which is usually less than A the crater area. Now $n(X)$ can be expressed as the ratio of the concentration of atoms X, $\rho(X)$, to the total concentration of atoms ρ of the solid, thus

$$\rho(X) = [\rho/\{N_s(\text{total}) \tau_u(X^+)(a/A)\}]N_s(X^+) \text{ atoms cm}^{-3}. \quad (13)$$

If the primary beam current is I_p of ions of type $A_n{}^+$ (eg $n = 2$ for $O_2{}^+$) then the number of target atoms sputtered per second will be

$$N_s = y_x I_p n/e \text{ atoms s}^{-1} \quad (14)$$

where y_x is the partial sputter yield of the matrix and e is the charge on the electron. Combining equations (13) and (14).

$$\rho(X) = [\rho \, e/\{y_x I_p n \, \tau_u(X^+)(a/A)\}]N_s(X^+) \text{ atoms cm}^{-3}. \quad (15)$$

The sputter rate z is the number of atoms sputtered per second divided by the total number of atoms in the crater area, A:

$$z = [y_x I_p n/(\rho e A)] \text{ cm s}^{-1}. \quad (16)$$

So combining equations (15) and (16)

$$\rho(X) = [1/\{z \, a \, \tau_u(X^+)\}]N_s(X^+) \text{ atoms cm}^{-3}. \quad (17)$$

The sputter rate is measured independently, usually at the end of the depth profile by measuring the crater depth using a Talystep profilimeter and dividing the depth by the sputter etching time.

We can now see how sensitivity is influenced by the experimental parameters. High sensitivity, namely the ability to detect low concentrations, requires a high sputter rate (high primary beam energy, high primary flux, low angle of incidence), large analytical area and high useful yield. Say we require 100 secondary ions per second to give good analytical statistics. A useful yield, τ_u of 10^{-3} is rather good, a sputter rate, z, of 10 nm s^{-1} is high and an analytical area of 1000 μm × 1000 μm is large. Under these conditions a concentration of 10^{13} atoms cm^{-3} could be detected, which would be very good. However, if we wished to analyse a very small device structure, say 10 μm × 10 μm where a sputter rate of 1 nm s^{-1} night be obtained, the minimum concentration detectable would be 10^{18} atoms cm^{-3}.

7.2.2 Dynamic Range

The aim is to be able to monitor concentration as a function of depth over the full range of its variation. The degree to which this is possible is very much dependent on the experimental procedure. Assuming that the material to be analysed was geometrically flat at the start and the ion beam has etched a well defined crater, only the ions emitted from the crater bottom should be characteristic of the dopant concentration at that depth. However, as the etch process proceeds, secondary ions are being generated from the crater edges as well. It would be possible for the ion collection system to detect these ions. The detected signal could thus be a complex convolution of the concentration at many depths. To eliminate the collection of ions other than from the crater bottom, two instrumental procedures are used. First, a lens is incorporated in the collection optics which cuts down the field of view such that only the crater bottom is 'seen'. This is the so-called *optical gate*. This helps to reduce the collection of ions from the crater edges. To improve matters further an *electronic gate* is also added. This permits the detection system to be switched off other than when the primary beam is located within a small defined area in the centre of the crater. Thus ion detection only occurs in the *gated area*. The combination of the optical and electronic gating is very effective in rejecting ions arising from the crater edges. The detected signal is thus characteristic only of the crater bottom. Figure 5.31(a) illustrates the improvement in the analytical dynamic range resulting from these ion collection arrangements. Another source of signal arising outside the analytical area is if there is a neutral component in the ion beam. Of course this is unaffected by the beam steering and scanning electrodes. This problem is usually overcome by placing bend in the low pressure end of the beam line. The ion beam can be steered around the bend, the neutral beam cannot. Incorporating all

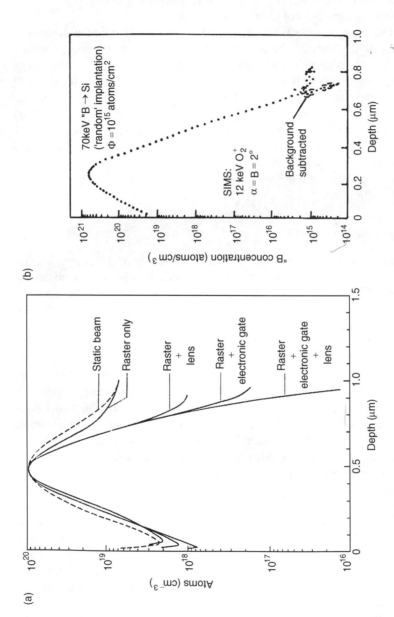

Figure 5.31. (a) A boron implant in silicon, profiled with increasingly sophisticated primary ion beam techniques and secondary ion optics. Reproduced by permission from [7], *Secondary Ion Mass Spectrometry – Principles and Applications*, Oxford Science Publications(1989) (b) An example of a high dynamic range analysis of boron in silicon when all instrumental parameters are optimum

these modifications, modern SIMS instruments are capable of at least six orders dynamic range, figure 5.31(b).

The analysis dynamic range of some elements, e.g. hydrogen, carbon or oxygen can be affected by residual gases in the SIMS vacuum system. These can cause a dramatic increase in the background counts for these elements such that a signal arising from the dopant in the sample cannot be detected to fall to the levels required. The problem can be alleviated by improving the vacuum to UHV quality using cryopumping around the sample.

Using optimum analytical arrangements, concentrations down to the 10^{13} atoms cm^{-3} can be attained, see Table 5.7.

7.3 DEPTH RESOLUTION

We require to define the depth relative to measured concentration as accurately as possible. If, as shown in Figure 5.32, the true concentration in a material changed abruptly, the actual measured change would deviate from it to a greater or lesser

Table 5.7. Typical detection levels for SIMS dopant analysis in silicon and gallium arsenide

SIMS — detection limits in silicon			
Element	Analytical ion	Atoms cm^{-3}	ppma
As	$(^{28}Si^{75}As)^-$	1.10^{16}	0.2
Au	$(^{197}Au)^-$	5.10^{15}	0.1
B	$(^{11}B)^+$	1.10^{14}	0.002
Cu	$(^{63}Cu)^+$	3.10^{16}	0.6
F	$(^{19}F)^-$	4.10^{16}	0.8
N	$(^{28}Si^{14}N)^-$	1.10^{18}	20
P	$(^{31}P)^+$	1.10^{16}	0.2
Sb	$(^{28}Si^{123}Sb)^-$	5.10^{15}	0.1

Detection limits in a depth profiling mode for elements in GaAs			
Element	Detected ion	Atoms cm^{-3}	ppma
Be	$^9Be^+$	8×10^{13}	0.002
B	$^{11}B^+$	5×10^{14}	0.01
Al	$^{27}Al^+$	4×10^{14}	0.01
Si	$^{28}Si^+$	4×10^{14}	0.01
Cr	$^{52}Cr^+$	4×10^{13}	0.001
Mn	$^{55}Mn^+$	4×10^{13}	0.001
Fe	$^{56}Fe^+$	7×10^{13}	0.002
Cu	^{63}Cu	1×10^{15}	0.02
Zn	$^{64}Zn^+$	5×10^{15}	0.1
Ge	$^{74}Ge^+$	1×10^{16}	0.2
Sn	$^{120}Sn^+$	5×10^{15}	0.1

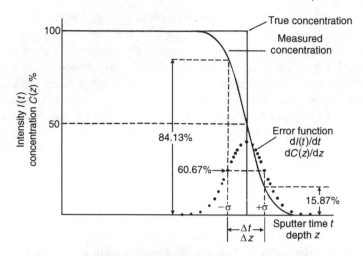

Figure 5.32. The definition of depth resolution

extent. The depth resolution is defined from the measured intensity/sputter time curve or the concentration/depth curve. It is either Δt or Δz which is the depth or time difference corresponding to the 84.13% and 15.47% levels on the concentration or intensity scale. It measures the degree to which the experiment is able to measure an abrupt interface. The parameters which affect the depth resolution are (a) instrumental effects, particularly the quality of the ion beam; (b) the surface topography of the material to be analysed; (c) radiation induced effects as a consequence of the ion bombardment process.

7.3.1 Instrumental Effects

The most important instrumental effect on depth resolution is the uniformity of the ion beam over the analysed area. The beam intensity profile does not have a flat top to it, it is usually approximately Gaussian, thus as it scans across a surface it will sputter more in the centre section of the beam compared to the edges. The beam edges will overlap at each pixel point, see Figure 5.33. Dependent on the ratio of the width of the beam profile, d, to the adjacent pixel distance, D, microroughness can be developed in the crater bottom. It is clear that this will significantly degrade the depth resolution of the profile. A squarer beam profile can be obtained by passing it through an aperture which cuts off the edges. Then by optimising d/Δ this effect can be minimised.

7.3.2 Surface Topography

If the surface of the material to be analysed is not flat to start with, it will not be possible to etch a well defined crater and good depth resolution will be

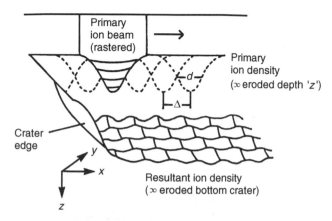

Figure 5.33. A schematic representation of the development of microroughness in a crater bottom dependent on the overlapping ration d/Δ assuming the ion beam has a gaussian profile

impossible to attain. The rougher the sample surface at the start the worse things will be. Sputter etching will never improve matters. Some materials develop rough structures as etching proceeds. Spectacular cones and columns can be formed as a consequence of various types of preferential etching. This may be chemically determined or a consequence of the angle of the ion beam to the material surface. Obviously the development of such structures has a terminal effect on depth resolution! One effective way around this in many cases is to rotate the sample whilst it is being depth profiled [62].

7.3.3 Radiation Induced Effects

When a high energy ion collides with the atoms of the material in the sputtering process, some of the atoms at the bottom of the crater can be moved deeper into the material, whilst some recoil upwards. Thus bombardment-induced mixing occurs. This effect occurs over a depth similar to the range of the primary ions. For a 2 keV beam this would be around 7 nm. Thus if we were analysing for a sharp dopant layer buried some distance below the surface in a material, these mixing effects would mean that we would start to detect dopant ions some distance before the true depth of the layer (due to recoil mixing) and after we had passed through the true depth we would continue to detect dopants (due to random atomic mixing). Thus the depth resolution will be degraded. These effects are greater at high primary beam energy and normal incidence. Figure 5.34 shows a profile of antimony on silicon. The profile is significantly broader using 12 keV O_2^+ than using 1.5 keV ions. By extrapolation the 'true' zero energy profile has been estimated. Thus mixing can be minimised using low beam energies and low angles of incidence.

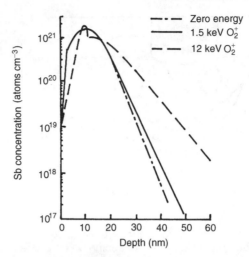

Figure 5.34. Depth profiles if ^{121}Sb in silicon using 1.5 keV and 12 keV O_2^+. The 'zero energy' profile is obtained by extrapolation

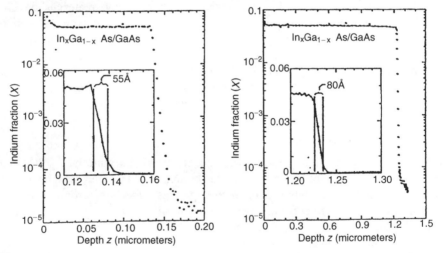

Figure 5.35. Depth profiles using 5 keV Ar^+ of $In_{0.05}$ $Ga_{0.95}$ As grown on GaAs to depths of 0.13 and 1.3 μm

The degradation of depth resolution due to these mixing effects becomes greater the deeper the profile proceeds. Figure 5.35 shows two profiles of InGaAs on GaAs. One layer was 0.13 μm, the other 1.3 μm. The measured interface width is greatly increased in the latter case.

Finally two profiles are illustrated one which highlights the problems which can arise, and one which shows the power of the technique. In figure 5.36 (i) is

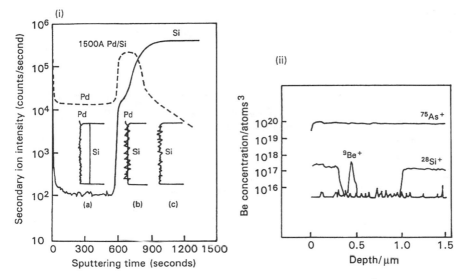

Figure 5.36. (i) A 5 keV Ar⁺ multi-element depth profile of 1500 Å Pd on silicon. Region (a) totally within the Pd layer. Region (b) surface intersecting both Pd and Si. Region (c) fully within the Si substrate. (ii) A depth profile of a silicon doped GaAs material doped with a Be spike

shown a depth profile of a 150 nm layer of Pd on silicon. The primary beam was 5 keV argon but a jet of oxygen was directed at the surface to raise the ion yield. As the profile progresses through the Pd layer, roughening occurs due to the generation of cone structures, see (b). This has two effects: first, it degrades the interface width because the Pd signal will not disappear until all the cones have been removed; second at the interface exposed Si holds oxygen in the region of the Pd (more effectively than oxygen is held on the Pd surface) and increases the ionisation probability causing an increase in the Pd⁺ yield. Finally, the slow decrease of Pd⁺ yield can be attributed to atomic mixing of Pd into the silicon lattice.

Figure 5.36 (ii), on the other hand, shows a good profile through a GaAs structure doped with Si to give p-type conductivity and a sharp Be marker spike.

8 SPUTTERED NEUTRAL MASS SPECTROMETRY

Over 95% of the particles emitted during sputtering are neutral. It is clear therefore that the SIMS experiment ignores the vast majority of the information generated. If this could be accessed there is the possibility of increased sensitivity and further types of information regarding the surface of materials. There is, however, a further potential attraction from accessing the neutral component.

Quantification by SIMS is bedevilled by the matrix effect. This arises because the particle emission and ionisation processes occur 'simultaneously'. If we can decouple ionisation from emission, ionisation occurring after the neutrals have moved away from the surface, the ion yield would be independent of the matrix and quantification would be easier.

Thus the principle of SNMS is to suppress the sputtered secondary ion component and post-ionise the sputtered neutrals some distance above the solid surface. The aim is easier quantification, higher sensitivity and augmented information.

Experimentally post-ionisation can be accomplished by either electron or photon bombardment. The possibility of a type of 'chemical ionisation' is also being explored in which caesium bombardment leads to the formation of MCs^+ ions which also appear to be less influenced by the matrix effect [63]. Only photon and electron post-ionisation will be described here. In each category there are a number of variants. We will consider each briefly.

8.1 ELECTRON POST-IONISATION

Electron bombardment ionisation is of course the basic method of ionisation in mass spectrometry. Simple bombardment by an electron beam generated from a hot filament is effective but not very efficient. Ionisation probabilities, α^0, in the range 10^{-4} to 10^{-3} are usual. Other than for elements which have very low secondary ion yields, eg As or Au, ionisation probabilities in this range would not give any increase in sensitivity over against SIMS. Ionisation probabilities can be increased to the range 10^{-3} to 10^{-2} if electron plasmas are used to increase the electron density in the ionising region. Again this does not give a significant increase in sensitivity compared to SIMS for most elements, but the easier quantification can be good reason for exploiting the technique.

8.1.1 Electron Beam Post-Ionisation

Figure 5.37 shows a typical arrangement for an electron beam post-ioniser [64]. The positive secondary ions emitted are suppressed by biasing the sample and applying a significant potential to the front deflector plate. Since the neutrals cannot be attracted into the ioniser, the geometry of the ioniser optics relative to the direction of the sputtered flux is crucial in determining the collection efficiency. The specular reflection direction of the primary beam is the direction of optimum yield.

To maximise ionisation, the electrons are confined within the ionising region by a magnetic field which causes the electrons to execute spiral motion around the field lines. This increases the probability of electron collision with the neutrals. Once formed the ions are extracted from the ioniser through a deflection energy analyser and injected into the mass analyser. The concentration of sputtered

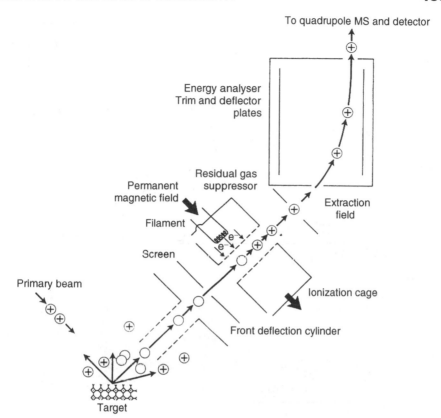

Figure 5.37. Schematic diagram of an electron bombardment ioniser combined with a parallel plate energy analyser. Reproduced by permission from [7], *Secondary Ion Mass Spectrometry – Principles and Applications*, Oxford Science Publications (1989)

neutrals will be small compared to residual gas molecules from the vacuum system. They will also be ionised, but to prevent interferences in the mass spectrum they have to be suppressed. Figure 5.38 shows the energy distributions of residual gas molecules compared to those for sputtered neutrals [65]. Fortunately the former have low energies and narrow energy distributions. It is possible to tune the energy analyser to filter them out before they enter the mass spectrometer. Using such a system, one can compare the yield of Ga and As from a GaAs surface using SIMS and SNMS. The surface concentration of Ga and As are roughly similar. Using SIMS the As^+/Ga^+ ratio is 5×10^{-4} ie the As yield is very low. Using e-beam SNMS the observed As^+/Ga^+ is 1.53. The post-ionisation probabilities are very similar for the two elements and the actual raw data much more closely reflects the actual concentrations. To obtain accurate

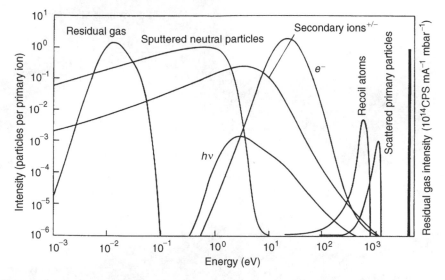

Figure 5.38. A schematic diagram of the energy distribution of residual gas ions, secondary ions, sputtered neutrals, electrons and other particles. Reproduced by permission from [65]

Table 5.8. Relative sensitivity factors for an electron beam SNMS instrument determined by the analysis of a number of NIST standard samples

Element	NIST source	RSF	St Deviation
Al	1256a	1.5399	0.060
C	C1152	0.3796	0.117
Co	C1154	1.1070	0.032
Co	1243	0.9784	0.022
Cr	*	0.8788	0.042
Cu	1103	0.5470	0.017
Mn	*	0.8628	0.024
N	TiN	0.0917	0.010
Ni	*	0.7901	0.020
P	C1154	0.8297	0.030
Pb	1103	0.1435	0.004
Si	*	1.4217	0.045
Sn	1103	1.7420	0.059
Ti	1243	1.3349	0.063
V	C1154	1.0443	0.102
Zn	1103	0.3863	0.010

All are measured relative to Fe. *these RSFs are mean values from several samples.

surface concentrations, the observed ion yields have to be calibrated to give instrumental sensitivity factors for each element. Table 5.8 shows such a calibration table of sensitivity factors relative to iron using a NIST steel standard. The great advantage of SNMS is evident. The elemental sensitivity factors are now *instrumental* parameters, they are matrix independent. It can be seen from the table that the sensitivities for the elements only vary within a factor of about 5. This compares with over four orders of magnitude for SIMS. However, whilst the sensitivity is uniform and almost matrix independent (there is a small matrix effect because sputter yields vary with matrix by a factor of about 3–5) the sensitivity is low. The ionisation probability is less than 10^{-3}, whilst in SIMS it varies between 10^{-1} and 10^{-5}. e-beam SNMS is therefore only really useful for elemental analysis under dynamic conditions. It cannot be used for static analysis or for molecular materials. Whilst dynamic SIMS is used for depth profiling in the dilute dopant region, SNMS can be used in the high concentration region.

8.1.2 SNMS Basic Equation for Elemental Analysis

The basic equation for SNMS yield from element X is given by

$$I(X^0) = I_p \ y_X \ \alpha_X^0 \ \eta_X(1 - \alpha_X^+ - \alpha_X^-) \tag{18}$$

where y_X is the sputter yield for element X; $\alpha_X{}^0$ is the post-ionisation probability for element X; η_X is the transmission for element X; α_X^+ and α_X^- are the secondary ion probability factors for positive and negative ions, these are usually very much less than unity. When $\alpha_X^{\pm} \ll 1$ then

$$I(X^0) = I_p y_X D_X{}^0. \tag{19}$$

If most of the sputtered species are atoms and there is negligible yield of molecular species, e.g. X_2, the partial sputter yield of X is given by

$$y_X = c_X Y_{tot}. \tag{20}$$

Under steady state conditions

$$Y_{tot} = \sum y_X \tag{21}$$

Thus $$\frac{c_i}{c_j} = \frac{I(X_i^0)D_j^0}{I(X_j^0)D_i^0}. \tag{22}$$

If the relative sensitivity factors D_j^0/D_i^0 are known, the concentrations in the sample can be determined.

8.1.3 Electron Plasma SNMS

The low ionisation probability can be improved by a factor of 10^2 by increasing the electron density. A low pressure inert gas is used to reduce the inter-electron

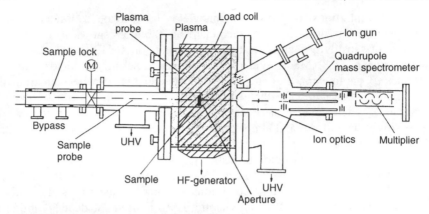

Figure 5.39. A schematic diagram of a commercial combined SNMS–SIMS instrument. With the plasma gas removed, SIMS is possible using the ion gun. Reproduced by permission from [7], *Secondary Ion Mass Spectrometry – Principles and Applications*, Oxford Science Publications (1989)

repulsions. Figure 5.39 shows one arrangement [66]. The plasma is excited by electron cyclotron wave resonance usually, in argon gas, at about 10^{-4} mbar pressure. The gas is contained in a ceramic tube surrounded by a weak magnetic field. An rf discharge at 27 MHz and 150 W excites the gas and generates a high density of electrons–about 10^{10} cm^{-3}. The ionisation volume is large so the neutrals have a higher probability of ionisation. Conversely, since the distance to be travelled to the analyser is high, there is a high probability of ion loss by scattering.

The sample to be analysed is held on a probe which is located in the ionisation region. One interesting feature of the design is that the Ar$^+$ ions which are also generated in the plasma can be used as the source of the primary ions. By biasing the sample to attract the ions to the surface, they will sputter the sample. The bias may be quite low so that the primary beam energy may be only a few hundred eV. This gives the technique the potential for depth profiling with very low bombardment induced mixing and hence high depth resolution.

Other types of discharge can be used. A low pressure dc discharge has been incorporated in the device shown in Figure 5.37 to attain even higher yields [67]. However, although overall some 10–100 times more sensitivity than the simple electron beam SNMS can be obtained, the device is still not of high enough sensitivity for static type surface analysis. Its main application is in elemental depth profiling of steels and other multi-component materials in the high concentration region.

8.2   PHOTON INDUCED POST-IONISATION

The most elegant and most efficient method of post-ionisation is to use pulsed high-energy laser photons to ionise the neutrals as they leave the surface. In this

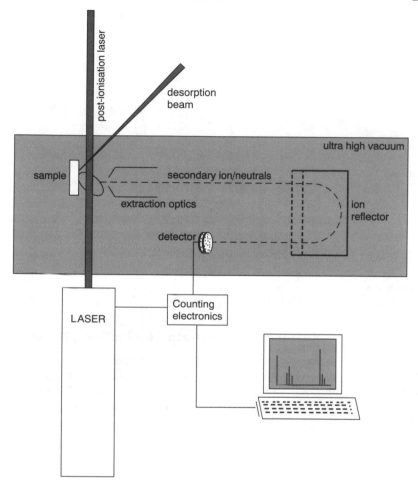

Figure 5.40. A schematic diagram of a ToFSIMS system with laser postionisation. The potentials on the various electrodes are set so that only post-ionised ions are transmitted to the detector. Reproduced with permission from N. Lockyer, PhD Thesis UMIST 1996

mode of operation a ToFMS must be used for mass analysis. Figure 5.40 shows the basic arrangement. There are two basic mechanisms by which the photons can ionise the neutrals. *Resonant* multi-photon ionisation (REMPI) in which one or more photon wavelengths are chosen to match the energy differences between electronic levels lying between the ground state and the vacuum level of the element or species which is to be detected. An electron is thus stimulated to climb the energy level ladder to ionisation. *Non-resonant* MPI (NRMPI) uses a very high power laser, usually in the UV, to stimulate electrons from the ground state to ionisation via *virtual* energy levels.

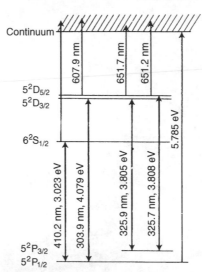

Figure 5.41. Partial electron structure of indium. Reproduced by permission from [68]. Copyright (1984) American Chemical Society

8.2.1 SNMS by Resonant Multi-photon Ionisation (REMPI)

This approach tunes the laser photons to the energy levels of the atom or species of interest. Thus if we want to analyse for indium, we consult the energy level diagram, e.g. Figure 5.41, and we find that we could ionise indium using two photons of 410.2 nm or one of 303.9 nm followed by one of 607.9 nm. Clearly the former arrangement would be more convenient since only one laser wavelength would be required. This example highlights the fact that REMPI is element or species specific. To ensure that the correct photons are available one has to know beforehand which elements are to be analysed. Some elements require three, four or even five photons. REMPI-SNMS has the great advantage that it is extremely efficient. The ionisation probability for neutrals in the ionisation volume can be close to 100%. To attain these sensitivity levels, the overlap between the laser photon field and the sputtered plume of neutrals is crucial. The laser is pulsed about 1 μs after the ion beam pulse to give the neutrals time to fly from the surface into the region of the laser beam. The factors which will affect the overlap between the laser field and the sputtered plume will be the primary pulse width; the energy of the sputtered neutrals; and the neutral spatial distribution; the laser photon timing; and the laser beam diameter. If all these factors are tuned in well, very high sensitivities are attainable down to sub-ppb levels [68] Figure 5.42 shows an analysis of ^{56}Fe in silicon. There is a difficult mass interference between ^{56}Fe and ^{56}Si$_2$ making the use of SIMS difficult. Using REMPI selective ionisation of ^{56}Fe is possible enabling a depth profile down to 2 ppb to be carried out [69].

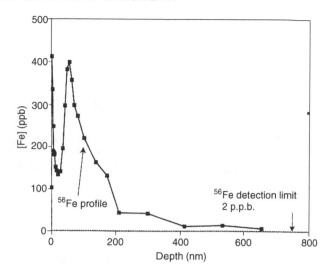

Figure 5.42. A depth profile of ^{56}Fe-implanted silicon performed by REMPI-SNMS of sputtered Fe atoms. Reproduced by permission of Elsevier Science – NL from [69]

The high sensitivities attainable mean that static conditions of ion bombardment can be used. The analysis of complex organic and inorganic materials is also possible. However, whilst the resonant requirements are not so stringent for many organic species because of the many vibrational levels associated with each electronic state, nevertheless some knowledge of the electronic absorption characteristics of the species to be ionised are required. Unknown analysis requires a less specific ionisation process.

8.2.2 SNMS Using Non-resonant Multi-photon Ionisation

Non-resonant multi-photon ionisation is not species specific. Ionisation occurs via virtual energy levels. Electron lifetimes in virtual levels are very short so it is essential that the rate of the photon arrival is high enough to elevate the electron to the vacuum level before it returns to the ground state. Thus, the process is not so efficient and high power densities are required: 10^9–10^{10} W cm^{-2} compared to $\approx 10^7$ W cm^{-2} for REMPI. Usually UV photons are used, e.g. 193 nm or 248 nm from ArF or KrF eximers or 266 nm from a frequency quadrupled Nd-YAG. The technique is very effective for elemental analysis of unknowns using a ToFMS analyser. Sensitivities are not quite as high as with REMPI but very close. Figure 5.43 shows the spectrum of a standard steel using NRMPI [70]. From such a material, tables of relative sensitivity factors (RSFs) can be derived to use the instrument for unknown analysis. As in the case of electron beam SNMS the RSFs vary by less than an order of magnitude across the periodic table. Sensitivities in the low ppm to ppb region can be attained. The technique

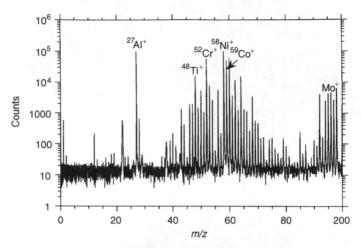

Figure 5.43. NRMPI-SNMS spectrum of NIST standard SRM 1243 steel. Reproduced by permission from [70]. Copyright (1995) John Wiley & Sons Ltd

offers easier quantification, rather uniform sensitivity for all elements and hence higher sensitivities than SIMS for many elements.

Static analysis is clearly obtainable with NRMPI-SNMS. This offers the possibility of higher sensitivity for surface chemical structure analysis of organic and inorganic materials. Although a number of studies of organic compounds, desorbed from glass substrates using a CO_2 laser followed by ionisation by UV-MPI, have been successful in generating molecular ions, early studies using the post-ionisation of sputtered PMMA were not encouraging [71]. The spectrum showed only carbon fragments, see Figure 5.44(a). If the power level is reduced there is no evidence of ion production. The observation was attributed to excitation of emitted clusters by a number of photons whose energy was not quite enough to ionise the cluster, but which increases the vibrational energy so that the molecule falls apart before it can absorb another photon to be ionised. There are two solutions to this problem. The first is to generate VUV photons which have sufficient energy to ionise most molecules and clusters with one photon. The other is to use femto-second photons so that the photon energy can be input more rapidly than the molecular vibration. Ionisation should then occur before

Figure 5.44. (a) Laser post-ionisation spectrum of PMMA obtained with pulsed Ar^+ sputtering with multi-photon ionisation (258 nm at 1×10^7 W cm^{-2}). Spectra were obtained with 1000 pulses. (b) Schematic diagram of the photon induced post-ionisation arrangement using VUV (118 nm) radiation for ionisation. (c) Spectrum of PMMA obtained with pulsed Ar^+ sputtering sith single photon ionisation (118 nm, 3×10^3 W cm^{-2}). Spectra were obtained with 1000 pulses. Reproduced by permission from [71]. Copyright (1988) American Chemical Society

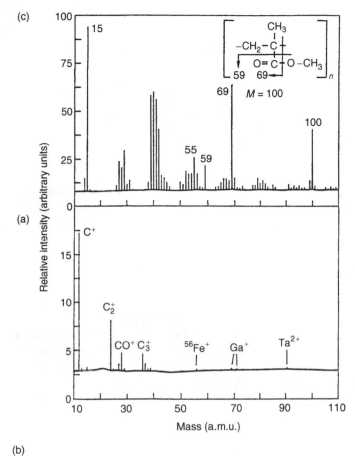

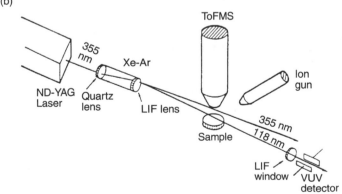

the molecule has a chance to fragment. Becker *et al.* demonstrated that the first approach is viable. They generated 118 nm radiation by frequency tripling the 355 nm Nd-YAG radiation in a xenon–argon gas mixture, Figure 5.44(b). The tripling process is of low efficiency ($\approx 10^{-4}$), but with a 20 mJ input pulse, 10 ns pulses of 118 nm radiation containing 1.3×10^{12} photons were obtained. This was sufficient to produce a good spectrum of PMMA, see Figure 5.44(c). The spectrum was somewhat different from the SIMS spectrum. Whilst m/z 59 and 69 are strong SIMS peaks the m/z 100 peak due to the monomer ion does not appear in SIMS. The sensitivity is at least on a par with SIMS and there is augmentation of the data. The technique has been commercialised under the acronym SALI — surface analysis by laser ionisation! As with REMPI, sensitivity is very much dependent on the efficiency of the interaction of the laser beam with the sputtered plume. Furthermore, the lasers used to date have very low repetition rates, so data acquisition can be slow. The technique has great potential if the instrumentation can be optimised.

One of the attractions of photon post-ionisation is to greatly increase sensitivity in molecular species analysis so that high resolution analysis in the imaging mode becomes possible under a close to static regime. Because the ion yield in SIMS is so low, in theory there is the possibility of an increase by at least 100, perhaps 1000. Since in imaging, the experiment has to interrogate say $(256)^2$ points on the surface there is also the requirement of high laser repetition rate if the acquisition of an image is to occur in a reasonable time. The new femto-second lasers based on Ti-sapphire technology have the advantage of kHz rep rates, so taken with their capability to rapidly inject energy into molecular species, they may be the laser of choice for photon induced static SNMS. Early data is very promising. It is clear that compared to ns photons, fs photons do result in much higher ionisation efficiency and less fragmentation [72, 73]. In the example shown in Figure 5.45 benzo[a]pyrene has been sputtered and ionised in SIMS and SNMS. Only in the SNMS modes is the molecular ion at m/z 252 clearly evident. However, there is significantly more fragmentation when the ns radiation is used as compared to the fs. The yield is also significantly higher for the fs case.

8.3 SUMMARY

To date, the main interest in SNMS has been in elemental quantification and it is does seem that matrix effects are greatly reduced. They are not wholly eliminated, however, because sputter yield is sensitive to surface bond strengths and to the angular distribution of emission. There is therefore some variation from matrix to matrix, but probably only by a factor of two or three. Accurate quantification can be disturbed by two other parameters. First, if secondary ion emission is high, as it can be for alkali metals, the neutral yield will be significantly reduced and will affect the post-ionised yield. Second, a significant yield of atom clusters (as can happen with metals such as silver) may also distort the elemental yield. Despite

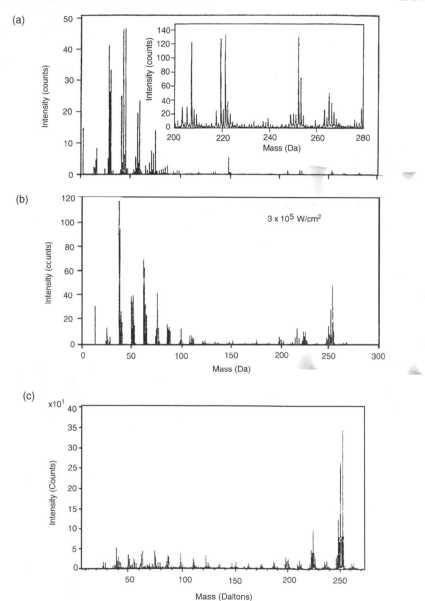

Figure 5.45. A comparison of spectra of benzo[a]pyrene (a toxic polyaromatic hydrocarbon) sputtered by a 25 keV gallium beam from a silicon surface. (a) A SIMS spectrum; (b) SNMS by 280 nm *ns* photons; (c) SNMS by 266 nm, 250 fs photons $(1.5 \times 10^{12} \text{ W cm}^{-2})$. Reproduced by permission from [72]

Table 5.9. A comparison of SNMS methods with SIMS

Technique	Ionisation	Ionisation efficiency	Neutrals in ionisation region	Mass spectrometry transmission	Relative sensitivity
e⁻ bombardment	General	10^{-4}–10^{-3}	10^{-2}	10^{-4}–10^{-3}	10^{-9}
e⁻ discharge	General	10^{-3}–10^{-2}	$\leqslant 1$	10^{-6}–10^{-4}	10^{-8}–10^{-7}
REMPI	Specific	≈ 1	$< 10^{-1}$	<0.5 (ToF)	10^{-2*}
NREMPI-UV	General	10^{-1}– ≈ 1 (fs)	10^{-2}	<0.5 (ToF)	$>10^{-3*}$
VUV single photon	General	$\approx 10^{-1}$	10^{-2}	<0.5 (ToF)	10^{-4}–10^{-3*}
quadSIMS	General	10^{-6}–$10^{-2\dagger}$ (element dep)	—	10^{-4}–10^{-3}	10^{-10}–10^{-5}
ToFSIMS	General	10^{-6}–$10^{-2\dagger}$ (element dep)	—	<0.5	10^{-7}–10^{-3*}

†ionisation probabilities in the region of 10^{-1} are observed for alkali metals and other elements under oxygen bombardment.
*Note that this comparison of relative sensitivities is only valid for single element detection. It ignores the parallel detection advantage of the ToF analyser (see Section 3.2.3). Because the ToF analyser collects the whole spectrum per unit input of primary ions, the relative sensitivity of ToF based instruments where a multi-component spectrum is required may be about 10^4 higher than a quadrupole based instrument

these provisos electron beam SNMS is being exploited by a number of industrial concerns world-wide for elemental analysis. Elemental analysis by laser-SNMS is much less widely used as yet. The cost and complexity of the equipment is inhibiting its application other than in one or two specialised contract laboratories.

The analysis of complex chemistry by laser-SNMS is still very much in the research and development laboratory. The potential is enormous, but much has to be done to realise it. Table 5.9 summarises the important parameters in the main methods of SNMS and compares them with quad and ToF based SIMS. The highest secondary ion yield in SIMS is quoted as 10^{-1}. This level is only encountered with alkali metals or using oxygen primary beams for sputtering. More usually the yields lie below 10^{-2}. For molecular species the neutral yields are certainly greater than 95%. The comparison in Table 5.9 is not intended to be accurate, but gives an impression in 'order of magnitude terms' how the SIMS related techniques compare.

REFERENCES

[1] J.J. Thomson, *Phil. Mag.*, **20**, 252 (1910).
[2] R.F.K. Herzog and F.P Viebock, *Phys. Rev.* **76**, 855L (1949).
[3] H.J. Liebl and R.F.K. Herzog, *J. Appl. Phys.*, **34**, 2893 (1963).
[4] R. Castaing and G. Slodzian, *J. Microscopic*, **1**, 395 (1962).
[5] A. Benninghoven, *Z. Physik*, **230**, 403 (1970).

[6]   A. Müller, A. and A. Benninghoven, *Surf. Sci.* **39**, 427; *ibid* **41**, 493 (1973).
[7]   J.C. Vickerman, A. Brown and N.M. Reed, (Eds) *Secondary Ion Mass Spectrometry, Principles and Applications.* Oxford University Press, (1989).
[8]   D. Briggs, A. Brown and J.C. Vickerman, *Handbook of Static SIMS*, John Wiley & Sons Ltd, Chichester and New York, (1989).
[9]   N.M. Reed, P. Humphrey and J.C. Vickerman, in *Proc. of 7th Int. Cong. SIMS.*, (Eds A. Benninghoven, C.A. Evans, K.D. McKeegan, H.A. Storms and H.W. Werner) 809, J. Wiley & Sons Ltd, Chichester, (1990);
      A. Brown and J.C. Vickerman, *Surf. Interface Anal.*, **8.**, 75 (1986).
[10] A. Brown, J.A. van den Berg and J.C. Vickerman, *Spectrochim. Acta B*, 1985, **40**, 871
[11] K. Wittmaack, *Rev. Sci. Instrum.*, **47**, 157, (1976).
[12] B Sakakini, A.J Swift, J.C Vickerman, C Harendt and K Christmann, *J Chem. Soc, Farad. Trans I*, **83**, 1975 (1987).
[13] K. Tang, R. Bevis, W. Ens, F. Lafortune, B. Schueler and K.G. Standing, *Int. J. Mass Spec. Ion Proc.*, **85**, 43, (1988). E. Niehuis, T. Heller, F. Feld and A. Benninghoven, *J. Vac Sci. Technol.*, **A5**, 1243 (1987). A.J. Eccles and J.C. Vickerman, *J. Vac. Sci. Technol.*, **A7**, 234 (1989).
[14] P. Sigmund in *Sputtering by Particle Bombardment Springer Series Topics in Applied Physics*, R. Behrisch (ed.) Springer, Berlin vol **47**, p9 (1981).
[15] N. Winograd, *Prog. Solid State Chem.*, **13**, 285 (1981).
[16] D.E. Harrison, J.P. Johnson and N.S. Levey, *Appl. Phys. lett.* **8**, 33 (1966).
[17] D.E. Harrison, P.W. Kelly, B.J. Garrison and N. Winograd, *Surf. Sci.* **76**, 311 (1978).
      N. Winograd, B.J. Garrison and D.E. Harrison, *Phys. Rev. Lett.* **41**, 1120 (1978).
[18] B.J. Garrison, N. Winograd, D.M. Deavan, C.T. Reimann, D.Y. Lo, T.A. Tombrello, D.E. Harrison and M.H. Shapiro, *Phys. Rev.* **B37**, 7197 (1988).
[19] N. Winograd in Fundamental Processes in Sputtering of Atoms and Molecules (ed. P Sigmund) *Mat. Fys. Medd. Dan. Vid. Selsk.*, **43**, 223 (1993).
[20] A. Wucher and B.J. Garrison, *Surf. Sci.* **260**, 257 (1992).
      A. Wucher and B.J. Garrison, *Phys. Rev.* **B46**, 4855 (1992).
[21] R.S. Taylor and B.J. Garrison, *Langmuir* **11**, 1220 (1995).
[22] R.S. Taylor, C.L. Brummel, N. Winograd, B.J. Garrison and J.C. Vickerman, *Chem. Phys. lett.*, **233**, 575 (1995).
[23] D.W. Brenner, *Phys Rev. B* **42**, 9458 (1990).
[24] W. Gerhard and C. Plog, *Z. Phys. B, Cond. Matt.*, **54**, 59 and 71 (1989).
[25] R.G. Cooks and K.L. Busch, *Int. J. Mass Spectrom. Ion Phys.*, **53**, 111 (1983).
[26] G.J Leggett and J.C Vickerman, *Int. J. Mass Spec. Ion Proc.*, **122**, 281 (1992).
[27] M. Barber, J.C. Vickerman and J. Wolstenholme, *J. Chem Soc. Faraday I*, **71**, 40 (1976), *Surf. Sci.*, **68**, 130 (1977).
[28] A. Brown and J.C. Vickerman, *Surf. Sci.* **117**, 154 (1982), *ibid.* **124**, 267 (1983), *ibid.* **151**, 319 (1985).
[29] R.S. Bordoli, J.C. Vickerman and J. Wolstenholme, *Surf. Sci.*, **85**, 244 (1979).
[30] J.C. Vickerman, *Surf. Sci.* **189/190**, 7 (1987).
[31] A. Brown and J.C. Vickerman, *Vacuum*, **31**, 429 (1981).
[32] M.A. Chesters, A.B. Horn, L.M. Ilarco, I.A. Ransley, B.H. Sakakini and J.C. Vickerman, *J. Electron Spec. & Rel. Phenomen.*, **54/55**, 677 (1990).
[33] I.A. Ransley, B.H. Sakakini, J.C. Vickerman and M.A. Chesters, *Surf. Sci.*, **271**, 227–236 (1992).
[34] B.H. Sakakini, C. Harendt and J.C. Vickerman, *Spectrochim. Acta*, **43A**, 1613 (1987).

[35] A.J. Paul and J.C. Vickerman, *Proc. Roy. Soc. London A, Math. Phys. Sci.*, **330**, 147 (1990).

[36] X-L. Zhou, X-Y Zhu and J.M. White, *Surf. Sci.*, **193**, 387 (1988).

[37] V. Matolin, E. Gillet, N.M. Reed and J.C. Vickerman, *JCS Faraday Trans.*, **86**, 2749 (1990).

[38] X-Y Zhu and J.M. White, *J. Phys. Chem.*, **92**, 3970 (1988).

[39] X-Y Zhu, M.E. Castro, S. Akhter and J.M. White, *J. Vac Sci. Technol.*, **A7**, 1991 (1989).

[40] X-Y Zhu and J.M. White, *Surf., Sci.*, **214**, 240 (1989).

[41] J.C. Vickerman, D. Briggs and A. Henderson, *The Static SIMS Library* John Wiley & Sons Ltd, (1996).

[42] D. Briggs, and M.J. Hearn, *Vacuum* **36**, 1005 (1986).

[43] G.J. Leggett and J.C. Vickerman, *Anal. Chem.*, **63**, 561 (1991).

[44] G.J. Leggett and J.C. Vickerman, *Appl. Surf. Sci.*, **55**, 105 (1992).

[45] E. Yue Wu, R.J. Friauf and T.P. Armstrong, *Surf Sci.*, **249**, 350 (1991).

[46] A. Brown and J.C. Vickerman, *Surf. Intface Anal.*, **8**, 75 (1986).

[47] D. Briggs and M.A. Hearn, *Surf. Intface Anal.* **11**, 198 (1988).

[48] A. Chilkoti, B.D. Ratner and D, Briggs in *Proc. 8th Int. Congr. SIMS* (John Wiley & Sons Ltd, Chichester, 1992)

[49] J. Lub, FCBM van Vroonhoven, D. van Leyen and A. Benninghoven, *J. Polym. Sci. A, Polym. Chem.* **27**, 4035 (1989).

[50] K.J. Hook, T.J. Hook, J.H. Wandass and J.A. Gardella Jnr, *Appl. Surf. Sci.*, **44**, 29 (1990)

[51] D. van Leyen, M. Deimel, B. Hagenhoff and A. Benninghoven, *Proc. of 7th Int. Cong. SIMS.*, (Eds) A. Benninghoven, C.A. Evans, K.D. McKeegan, H.A. Storms, and H.W. Werner, 757, John Wiley & Sons Ltd, Chichester, (1990).

[52] I.V. Blestos, D.M. Hercules, D. van Leyen, B. Hagenhoff, E. Niehius and A. Benninghoven, *Anal. Chem.*, **63** 1953 (1991).

[53] N.M. Reed and J.C. Vickerman, in *Surface Characterization of Advanced Polymers* (eds) L Sabatini and P G Zambonin, VCH, Weinheim, (1993).

[54] J.A. Treverton, A.J. Paul and J.C. Vickerman, *Surf. Interface Anal.*, **20**, 449 (1993).

[55] W.J. van Ooij, A. Sabata and A.D. Appelhans, *Surf. Interface Anal.*, **17**, 403 (1991).

[56] J. Lub, F.C.B.M. van Vroonhaven, D. van Leyen, and A Benninghoven, *J. Polym. Sci. B Polym. Phys.*, **27**, 2071 (1989).

[57] G.J. Leggett and J.C. Vickerman, *Int. J. Mass Spectro. Ion Phys.*, **122**, 281 (1992). G.J. Leggett and J.C. Vickerman, *Ann Rep. Roy Soc. Chem.*, C, 77 (1991).

[58] C. Brummel, J.C. Vickerman, S. Benkovic and N. Winograd, *Anal. Chem.*,

[59] R.G. Wilson, F.A. Stevie and C.W. Magee, *Secondary Ion Mass Spectrometry — A Practical Handbook for Depth Profiling and Bulk Impurity Analysis*, John Wiley & Sons (1989)

[60] E. Niehius, H-G, Cramer and M. Terhorst, *Proc. 10th Int. Conf. SIMS*, (Eds) A Benninghoven *et al.* SIMS X, Munster, (1995), John Wiley & Sons Ltd (1996)

[61] D.P. Leta and G.H. Morrison, *Anal. Chem.* **52**, 514 (1980)

[62] M.R. Houlton, O.D. Dosser, M.T. Emeny, A. Chew and D.E. Sykes, *Proc. 8th Int Conf. SIMS (Amsterdam, 1991)* 343 John Wiley & Sons (1992)

[63] H. Gnaser, *Surf. Interface Anal.*, **24**, 483 (1996).

[64] R. Wilson, J. van den Berg and J.C. Vickerman, *Surf. Interface Sci.*, **14**, 393 (1989)

[65] D. Lipisky, R. Jede, O. Ganshow, and A. Benninghoven, *J. Vac. Sci. Technol.*, **A3**, 2007 (1985).

[66] H. Oechsner, W. Ruhe and E. Stumpe, *Surf. Sci.*, **85**, 289 (1979).

[67] S. Smith, R. Wilson and J.C. Vickerman, *Inst Phys. Conf Ser.*, **130**, 411 (1993).

[68]   F.M. Kimock, J.P. Baxter, D.L. Pappas, P.H. Kobrin and N. Winograd, *Anal Chem.*, **56**, 2782 (1984).
[69]   C.E. Young, M.J. Pellin, W.F. Calaway, B. Jorgenson, E.L. Schweitzer, and D.M. Gruen, *Nucl. Instrum. Methods Phys. Res.*, **B27**, 119 (1987).
[70]   E. Scrivenor, R. Wilson and J.C. Vickerman, *Surf. Interface Anal.*, **23**, 623 (1995).
[71]   U. Schühle, J.B. Pallix and C.H. Becker, *J. Am. Chem. Soc.*, **110**, 2323 (1988).
[72]   C.L. Brummel, K.F. Willey, J.C. Vickerman and N. Winograd, *Int. J. Mass Spectrom. Ion Phys.*, **143**, 257 (1995).
[73]   N. Lockyer and J.C. Vickerman, *Proc. 10th Int. Conf. SSIMS*, (Eds) A. Benninghoven *et al.* Münster, 1995, John Wiley & Sons (1996).

PROBLEMS

1.   Define the terms 'sputtering' and 'surface sensitivity' and 'surface damage' as they apply to SIMS.

2.   The secondary ion signal intensity from element m in a SIMS experiment is given by

$$I_{ms} = I_p \theta_m y_m \alpha^+ \eta$$

Identify and explain the importance of each parameter.

Describe and explain the variation of the Al^+ signal as a function of depth, as a depth profile analysis is performed through an aluminium oxide film on aluminium. What are the implications of this observation for *quantitative* elemental analysis using SIMS?

3.   Explain why in secondary ion mass spectrometry (SIMS) the yield of secondary ions arising from different chemical species in a surface is not directly proportional to the chemical composition of the surface.

4.   Explain why the *static* SIMS primary particle bombardment dose limit has been set at 10^{13} ions cm^{-2}. Outline why this may be too high for the analysis of organic surfaces.

5.   An X-ray photoelectron spectrum of poly(ethylene terephthalate) shows three peaks in the C 1s region with binding energies of about 284, 287 and 290 eV. In the static SIMS positive ion spectrum there are three significant peaks at m/z 193, 149 and 104. Provide an interpretation for this spectral data and discuss the relative advantages and disadvantages of XPS and SSIMS for the surface analysis of polymers.

6.   Surface analysis at high spatial resolution is becoming increasingly important in studies of corrosion, carbon and glass fibre based composite manufacture and electronic device architecture. Outline the problems associated with the analysis of a component or contaminant at 0.01 at% in a feature 5 μm in diameter on each of the above types of material. How would the analytical problems be eased or exacerbated if the feature to be analysed was 20 μm or 100 nm in diameter?

7. There is a requirement to analyse a 1 μm × 1 μm area of a material by static
    SIMS. Carry out a calculation to determine whether this is possible, assuming
    the sputter yield of all the components is 1, the ionisation probability is
    10^{-3}, the transmission of the mass analyser is 0.1 and a yield of at least 100
    secondary ions are needed across the spectrum for analysis. How could the
    calculated yield be increased whilst keeping within the static limit?

8. If the analysis of Fe in silicon shown in Figure 5.42 were to be carried
    out by SIMS rather than REMPI-SNMS what mass resolution would be
    required? If the transmission of the mass analyser was 0.05; the sputter
    yield of both Fe and Si was 3; the ionisation probability of Fe was 10^{-3};
    the minimum acceptable count level was 5 counts, the analysis area was
    500 μm × 500 μm the primary beam current was 1 μA cm^{-2}. What would
    the minimum detectable Fe concentration be? How quickly would the depth
    profile take to be completed to the depth of the minimum detectable concen-
    tration of Fe; to 600 nm?

9. In Section 8.2.2, in producing VUV radiation, Becker *et al.* generated 10 ns
    pulses containing 1.3×10^{12} photons of 118 nm radiation from 20 mJ input
    pulses of 355 nm radiation. What was the efficiency of the conversion?

CHAPTER 6

Low-energy Ion Scattering and Rutherford Backscattering

EDMUND TAGLAUER

Max-Planck-Institut für Plasmaphysik, Garching bei München, Germany

1 INTRODUCTION

Energetic ions are unique probes for surface analysis. The impact of an ion with a few electron volts of kinetic energy on a solid surface causes a series of collisional processes and electronic excitations; analysis of the energy spectra of backscattered ions shows that they provide detailed information about the atomic masses on the surface and about their geometric arrangement. Since the time for the scattering processes is very short compared to thermal vibrations or the lifetime of a collision cascade, this information can be generally considered as being representative of the instantaneous condition of the surface, unperturbed by the ion beam (this holds for an individual scattering process and does not exclude surface modifications due to high fluence bombardment).

There are two distinct parameter regimes (see Table 6.1) for ion scattering analysis of surfaces and near-surface layers, and accordingly two different techniques.

In low-energy ion scattering (LEIS) or ion scattering spectroscopy (ISS) [1], primary ion energies of 0.5–5 keV are used with noble gas ions (He^+, Ne^+, Ar^+) and also alkali ions (Li^+, Na^+, K^+). With this method, information is obtained from the topmost atomic layer, under certain circumstances also from the second or third layer [2].

In Rutherford Backscattering (RBS) [3] the primary ion energy ranges from about 100 keV (for H^+) to several MeV (for He^+ and heavier ions). The ion–target atom interaction can be described using the Coulomb potential from which the Rutherford scattering cross-section is derived, which allows absolute quantification of the results. Information in principle arises from a thickness of the order of 100 nm (10^{-5} cm), but surface analysis is also possible by using channelling/blocking techniques. Scattering of H^+ with energies around 100 keV is sometimes referred to as MEIS (Medium Energy Ion Scattering), probably because it only needs a smaller type of accelerator, but physically it is within the RBS regime.

Surface Analysis — The Principal Techniques. Edited by John C. Vickerman
© 1997 John Wiley & Sons Ltd

Table 6.1. Typical physical parameters in ion scattering

Method	Ions	Energy (eV)	de Broglie wavelength λ (Å)	Distance of closest approach r_0 (Å)
ISS	He^+, Ne^+ Li^+, Na^+	10^3	10^{-2}	0.5
RBS	H^+, D^+ He^+	10^6	10^{-4}	0.01

The physical principles are the same for both techniques (ISS and RBS): an ion beam is directed onto a solid surface, a part of the primary projectiles is backscattered from the sample and the energy distribution of these ions is measured (see Figure 6.1). Since the ion–target atom interaction can be described by two-body collisions, the energy spectra can be easily converted into mass spectra. The difference between ISS and RBS arises from the difference in the cross-sections and the influence of electronic excitations and charge exchange processes, which result in different information depths. In both cases, structural information is obtained from crystalline samples by varying the angles between beam and sample. Deduction of structural information from the data is straight-forward, since both techniques are 'real space' methods which are based on fairly simple concepts.

In general, models are required for structure analysis. These models are usually based on results from diffraction techniques (X-ray crystallography or low-energy

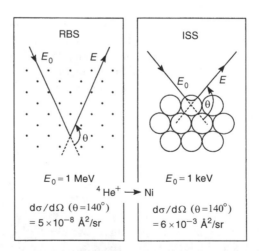

Figure 6.1. Schematic illustration of Rutherford Backscattering Spectroscopy (RBS) and Ion Scattering Spectroscopy (ISS)

electron diffraction), which provide the symmetry of the unit cell but not, directly, the real atomic positions. A large number of such studies is published in the literature, demonstrating the useful and unique contributions of ion-scattering techniques to surface analysis.

In the following section the physical basis of ion scattering is explained in detail, for low and high energy techniques together, since they are based on the same principles. In the subsequent sections, experimental instrumentation, physical characteristics and typical results are described for RBS and ISS individually. The descriptions and the examples are selected with the intention to demonstrate and explain the potential and typical achievements of the methods. It is not useful in the present context to review the vast amount of published results for which the reader has to refer to the relevant literature.

2 PHYSICAL BASIS

2.1 THE SCATTERING PROCESS

An important feature of analysis techniques based on energetic ions is that the scattering process can be considered as one or a sequence of classical two-body collisions. A simple estimate of the situation is possible using the parameters given in Table 6.1. Quantum effects are negligible for scattering angles larger than Bohr's critical angle θ_c [4], which is determined by the ratio of the de Broglie wavelength λ and the distance of closest approach r_0 (see also Figure 6.2), $\theta_c \approx \lambda/r_0$. This value is indeed very small compared to all practically used scattering angles. Also diffraction effects from periodic crystal lattices are negligible since

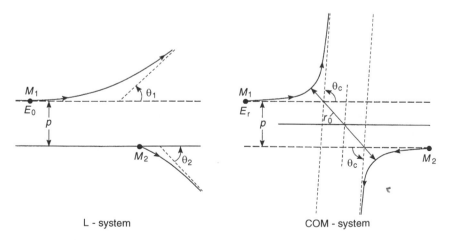

Figure 6.2. Trajectories for the elastic collision between two masses M_1 and M_2 in the laboratory system (left) and the centre-of-mass system (right), p is the impact parameter, r_0 the distance of closest approach

$\lambda \ll d$; typical lattice constants d are of the order of a few Angstroms. Let us finally consider in brief the role of thermal vibrations of lattice atoms on the scattering process. Phonon energies are of the order of 0.03 eV and thus very small compared to the ion energies, i.e. phonon interaction cannot be detected in the ion energy spectra. Another way is to look at the collision times, which are about 10^{-15} s, or less for ISS and even shorter for RBS energies, whereas thermal vibration periods are about $10^{-12}-10^{-13}$ seconds. Therefore, the energetic ions virtually 'see' a snap-shot of a rigid lattice with atoms thermally distributed around their ideal lattice positions. In suitable experiments ion scattering can well be sensitive to interatomic distances with an accuracy of 0.1 Å or less and therefore thermal displacements can be detected by ion scattering, as will be shown in Sections 3 and 4. It clearly follows from this discussion that the interaction between an ion and a target atom can be treated as a classical two-body collision and this is dealt with in the following.

2.2 COLLISION KINEMATICS

We consider the collision between two masses M_1 and M_2 which interact through a centrosymmetric potential $V(r)$. Figure 6.2 shows the trajectories in the laboratory and in the centre of mass system (COM). The projectile mass M_1 has the initial energy E_0 and the target mass M_2 is initially at rest. From the conservation of energy and momentum the particle energies can be calculated [5] as a function of the scattering angle θ_1 and θ_2 in the laboratory system (see Figure 6.2). For the projectile we write

$$E_1/E_0 = K, \tag{1}$$

where K is the so-called kinematic factor

$$K = \left(\frac{\cos\theta_1 \pm (A^2 - \sin^2\theta_1)^{1/2}}{1+A} \right)^2 \tag{2}$$

K only depends on the mass ratio $A = M_2/M_1$ and the scattering angle. The positive sign holds for $A > 1$ and both signs for $A < 1$. In this latter case, i.e. heavy projectile on a lighter target atom, the scattering angle is limited to $\theta_1 < \arcsin A$, while there are two scattering angles possible in that region. The function $K(\theta_1)$ is plotted in Figure 6.3. We also note that the kinematic factor, being based on the conservation laws, does not depend on the shape of the potential function.

Equation (2) becomes particularly simple for $\theta_1 = 90°$:

$$K(90°) = \frac{A-1}{A+1} = \frac{M_2 - M_1}{M_2 + M_1} \tag{2a}$$

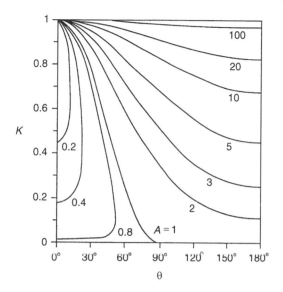

Figure 6.3. The kinematic factor K as a function of the laboratory scattering angle (see equation (2)). The parameter is the mass ratio $A = M_2/M_1$

and for $\theta_1 = 180°$:

$$K(180°) = \left(\frac{A-1}{A+1}\right)^2 . \tag{2b}$$

The corresponding expression for the recoiling target atom is

$$\frac{E_2}{E_0} = \frac{4A}{(1+A)^2} \cos^2 \theta_2 \tag{3}$$

The applicability of equation (2) and (3) for the identification of the scattering masses from ion energy spectra has been demonstrated in many cases [2, 3] and is extensively used in Sections 3 and 4. A schematic representation of energy spectra for scattering ^{4}He at an angle of 140° from a sample containing ^{108}Ag, ^{28}Si, and ^{16}O is shown in Figure 6.4. The arrows indicate the peak positions according to equation (2). This figure shows the principal similarity of both techniques, ISS and RBS, and the specific differences which are discussed in the respective sections.

Figure 6.4 also demonstrates that scattered-ion energy spectra are transformed into mass spectra by virtue of equation (2). Consequently, the mass resolution also can be calculated from this equation:

$$\frac{M_2}{\Delta M_2} = \frac{E}{\Delta E} \frac{A + \sin^2 \theta_1 - \cos \theta_1 (A^2 - \sin^2 \theta_1)^{1/2}}{A^2 - \sin^2 \theta_1 + \cos \theta_1 (A^2 - \sin^2 \theta_1)^{1/2}} \tag{4}$$

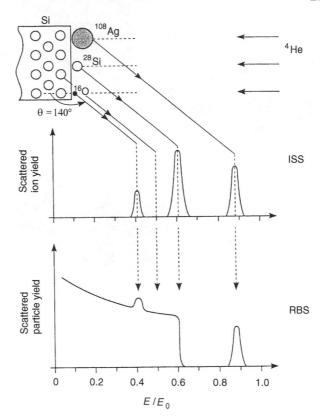

Figure 6.4. Schematic representation of energy spectra of He$^+$ ions scattered at an angle of 140° from a Si substance with Ag, Si and O on the surface. Upper spectrum; ISS ($E_0 \approx 1$ keV), lower spectrum: RBS ($E_0 \approx 1$ MeV)

which for the special case of $\theta_1 = 90°$ becomes

$$\frac{M_2}{\Delta M_2} = \frac{E}{\Delta E}\frac{2A}{A^2 - 1}.$$

It can be deduced from equation (4) and its representation in Figure 6.5 (assuming a constant relative energy resolution of the detector of $E/\Delta E = 100$), that mass resolution is best for large scattering angles and about equal ion and target atom masses. So the primary projectile mass has to be selected accordingly if mass resolution is important.

2.3 INTERACTION POTENTIALS AND CROSS-SECTIONS

The collision kinematics treated in the previous section yields the positions of the peaks in the energy spectra. The peak intensities, i.e. the probability for

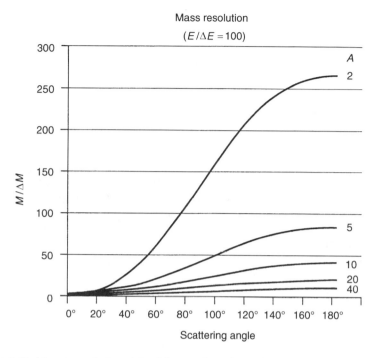

Figure 6.5. Mass resolution $M/\Delta M$ as a function of the scattering angle for a given analyser resolution $E/\Delta E$. The parameter is the mass ratio $A - M_2/M_1$ (see equation (4))

scattering into a certain angular and energy interval, are usually given by a scattering cross-section which is determined by the interaction potential. This is briefly explained in the following.

We first need a relation between the scattering angle θ (in the COM system) and the impact parameter p (see Figure 6.2). It can be obtained by considering the conservation of angular momentum, which yields the so-called scattering integral:

$$\theta = \pi - 2 \int_{r_0}^{\infty} \frac{p \, dr}{r^2 \left(1 - \frac{p^2}{r^2} + \frac{V(r)}{E_r}\right)^{1/2}} \tag{5}$$

where $E_r = E_0 M_2/(M_1 + M_2)$ is the relative energy in the COM-system. Equation (5) gives the connection between the impact parameter p and the scattering angle θ (see Figure 6.2) which is required for calculating the differential scattering cross-section $d\sigma = 2\pi p \, dp$ and therefore also $d\sigma/d\Omega$ for scattering into a solid angle of $d\Omega$. The angles in the COM and laboratory system are connected through the mass ratio

$$\tan\theta_1 = \sin\theta_c/[(M_1/M_2) + \cos\theta_c] \tag{6}$$

$$\theta_2 = 1/2(\pi - \theta_c).$$

An analytical solution of equation (5) is only possible for certain simple potential functions $V(r)$, e.g. the important Coulomb potential

$$V(r) = \frac{1}{4\pi\varepsilon_0}\frac{Z_1 Z_2 e^2}{r} \tag{7}$$

Z_1 and Z_2 are the nuclear charges of projectile and target atom, respectively, e is the unit of electrical charge (in SI units). The Coulomb potential describes correctly the interaction in the RBS regime, i.e. the solution of equation (5) with the potential of equation (6) yields the well known Rutherford scattering cross-section [6]:

$$(d\sigma/d\Omega)_c = \left(\frac{Z_1 Z_2 e^2}{4E_r \sin^2\theta_c/2}\right)^2 \tag{8}$$

in the COM-system, which also holds in the laboratory system if $M_1 \ll M_2$.

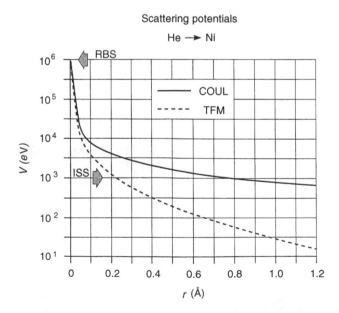

Figure 6.6. Interaction potentials as a function of atomic distance for He–Ni scattering: solid line, Coulomb potential (equation (7)); dashed line, screened potential in the Thomas–Fermi–Molière approximation (equations (10) and (12). The operation regimes for RBS and ISS are indicated

The general formula in the laboratory system is [7]:

$$\frac{d\sigma}{d\Omega} = \left(\frac{Z_1 Z_2 e^2}{2E_r \sin^2 \theta}\right)^2 \cdot \frac{[(1 - \sin^2 \theta/A^2)^{1/2} + \cos \theta]^2}{(1 - \sin^2 \theta/A^2)^{1/2}} \tag{9}$$

As an example, the Coulomb scattering potential He–Ni is shown in Figure 6.6 and the Rutherford scattering cross-section as a function of Z_2 in Figure 6.7 (for $E_0 = 1$ MeV).

For the lower energies used in ISS, the screening of the nuclear charges by the electron cloud has to be taken into account and therefore frequently screened Coulomb potentials of the form

$$V(r) = \frac{Z_1 Z_2 e^2}{r} \Phi \left(\frac{r}{a}\right) \tag{10}$$

are used where a is the screening parameter in the screening function Φ and can be given according to Firsov [8] by

$$a_F = \frac{0.8854 \, a_0}{(Z_1^{1/2} + Z_2^{1/2})^{2/3}} \tag{11}$$

in which a_0 is the Bohr radius of 0.529 Å and the numerical factor is $(9\pi^2/128)^{1/3}$. A similar expression has been developed by Lindhard et al. [9].

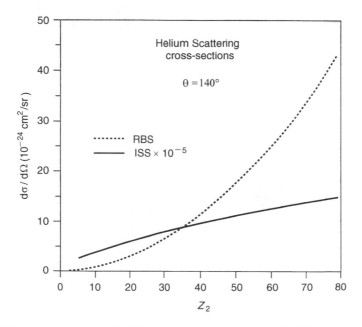

Figure 6.7. Cross-sections for He scattering at an angle of 140° as a function of target nuclear charge Z_2. Solid line, ISS ($E_0 = 1$ KeV); dashed line, RBS ($E_0 = 1$ MeV)

For many cases, a value of the order of $0.8\, a_F$ has been found to give best agreement with experimental results.

Several analytical expressions are given in the literature for the screening function $\Phi(r)$. The Molière approximation to the Thomas–Fermi function has become most widely used in ISS. It is given by a sum of three exponentials:

$$\Phi(x) = 0.35\ e^{-0.3x} + 0.55\ e^{-1.2x} + 0.10\ e^{-6x} \qquad (12)$$

with $x = r/a$.

For cross-section calculations, usually no distinction is made between charged and neutral projectiles, i.e. He^0 and He^+ or Ne^0 and Ne^+, because there is sufficient overlap in the electron clouds [8]. The interaction potential for He–Ni and the scattering cross-section for 1 keV He are also shown in Figures 6.6 and 6.7, respectively. From the different regimes in the internuclear distance, the relevance of the electronic screening becomes obvious.

Figure 6.7 shows that the scattering cross-section for ISS is several orders of magnitude larger than the RBS cross-section. While the latter increases with Z_2^2, the dependence on Z_2 is much weaker in the lower energy regime used for ISS.

2.4 SHADOW CONE

A very useful concept in structural surface analysis is the so called shadow cone [10, 11]. It is formed by the distribution of ion trajectories downstream of a scattering target atom, see Figure 6.8. The flux of primary ions, described by

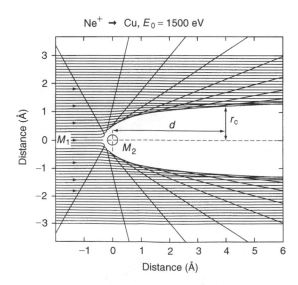

Figure 6.8. Trajectories of a parallel beam of projectiles of mass M_1 forming a shadow cone behind the scattering atom of mass M_2

a beam of parallel trajectories, is deflected by the scattering atom such that a trajectory-free region is formed behind the scatterer. The envelope of trajectories forming this region is called the shadow cone. Obviously, scattering from another atom is not possible if it is located inside the shadow cone, but deviations from a static lattice position can lead to a temperature dependent scattering intensity and therefore to a determination of vibrational amplitudes.

In order to obtain structural information, i.e. to locate atomic positions, it is necessary to know the radius of the shadow cone as a function of distance d from the scattering atom and the intensity distribution across the shadow cone. These quantities can be gained analytically for the case of a Coulomb interaction potential and by using the momentum approximation (i.e. small scattering angles for which $\tan\theta \approx \theta$ and only momentum change but negligible energy loss in the scattering process). In that case the scattering angle θ is related to the impact parameter p according to

$$\theta = \frac{Z_1 Z_2 e^2}{E_0 p}. \tag{13}$$

Hence the scattering angle is inversely proportional to the impact parameter, in this case.

Since we consider small-angle scattering, the distance R_s between the ion trajectory and a second atom located at a distance d from the scatter can be written as

$$R_s = p + \theta d. \tag{14}$$

The function $R_s(p)$ is plotted in Figure 6.9. It represents a situation corresponding to classical rainbow scattering since two primary impact parameters p can lead to the same radius (or secondary impact parameter) R_s. The function $R_s(p)$ has a minimum, the corresponding radius is the Coulomb shadow cone radius R_c given by

$$R_c = 2(Z_1 Z_2 e^2 d/E_0)^{1/2}. \tag{15}$$

In this approximation, the shadow cone has a square-root shape, R_c varies as $d^{1/2}$. From this treatment it can also be deduced that shadow cone formation leads to a flux peaking at the edge of the cone. The flux distribution at the position R_s is

$$f(R_s)\, 2\pi R_s\, dR_s = f(p)\, 2\pi p\, dp \tag{16}$$

and if we normalize the primary flux $f(p)$ to one, we obtain

$$f(R_s) = \frac{p}{R_s} \left| \frac{dR_s}{dp} \right|^{-1}. \tag{16a}$$

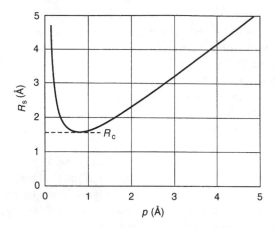

Figure 6.9. Distance R_s to the second atom as a function of the impact parameter to the first atom for 500 eV He scattered at a pair of Mo atoms with an interatomic distance of 3.15 Å. R_c is the corresponding shadow cone radius

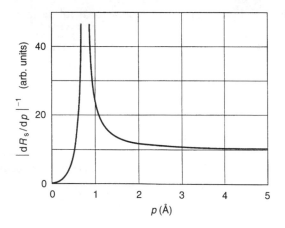

Figure 6.10. Intensity function across the shadow cone (see text)

This equation can be solved analytically by inserting equation (14) giving the result:

$$f(R_s) = 0 \quad \text{for } R_s < R_c \quad \text{(inside the shadow cone)}$$

$$f(R_s) = \tfrac{1}{2}(1 - R_c^2/R_s^2)^{-1/2} + (1 - R_c^2/R_s^2)^{1/2} \quad \text{for } R_s > R_c \quad (17)$$

(outside the shadow cone).

Equation (17) shows that we obtain a flux peak at the edge of the shadow cone represented by a square-root singularity. The distribution obtained from numerical differentiation of equation (14) is plotted in Figure 6.10.

For the low-energy case in which a screened Coulomb potential has be used, an analoguous expression for the shadow cone radius was calculated by Oen [12]:

$$R_s = R_c(1 + 0.12\alpha + 0.01\alpha^2) \tag{18}$$

with $\alpha = 2R_c/a$ being between 0 and 4.5 and a similar expression for larger values of α. Values calculated using equation (18) agree quite well with experiment if a screening length a between 0.8 a_F and 0.9 a_F is used [2] (see equation (11)).

3 RUTHERFORD BACKSCATTERING

3.1 ENERGY LOSS

An energetic ion penetrating into a solid loses its energy by a variety of collisional processes. At large impact parameters (of the order of lattice parameters, i.e. ≈ 1 Angstrom) it transfers energy to valence electrons, about 10 eV per collision with virtually no deflection. The cross-sections for such processes are high, about 10^{-16} cm^2. At smaller impact parameters, excitation of inner shell electrons can occur, which subsequently leads to de-excitation by X-ray emission, a process which is the basis for proton-induced X-ray analysis (PIXE) of surface layers. Only a small fraction of the primary ions come close enough to a target nucleus (impact parameters of the order of 10^{-12} cm) to undergo an elastic nuclear collision which is described by the kinematics given in the previous section. If such an ion is backscattered, its final energy is determined by the 'elastic' nuclear collision in a certain depth of the sample and the additional 'inelastic' energy loss to electrons on its way in and out of the target. The penetration of ions in matter and the corresponding energy loss processes have been subject of a number of fundamental papers in this field [4, 10, 13]. Here, we are interested in the energy loss per unit length $-dE/dx$ given in eV/Å and commonly called stopping power. The stopping cross-section S [eV/(atoms/cm^2)] relates this quantity to the atomic density N and is therefore more specific for a certain atomic species:

$$S = -\frac{dE}{dx}\frac{1}{N}. \tag{19}$$

Extensive tables of stopping power data have been collected by Ziegler *et al.* [14] and they have become indispensable for practical analysis of RBS data. An example is given in Figure 6.11 which shows the stopping power for helium in nickel over a wide range of primary energies. Nuclear stopping (included in the dashed curve) is only significant at the lower energy part. The stopping power curve exhibits a broad maximum around 1 MeV and that is the operational regime of RBS: here, the stopping power does not depend very much on the ion energy and hence can be assumed to be constant as a sufficient approximation in many cases; the stopping power there has its maximum value and therefore

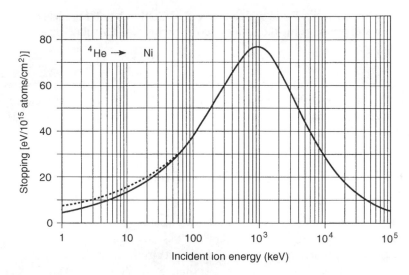

Figure 6.11. Stopping power for He ions in Ni as a function of incident energy. The solid line corresponds to electronic stopping, the dashed line also includes nuclear stopping (from [14])

RBS its best depth resolution and in this energy range the nuclear interaction is exactly given by the Coulomb potential, giving RBS the advantage of an absolute analytical method. The shape of the stopping power curves is very similar for various ions and materials but the position of the maximum of course depends on the considered species. The high-energy part of the stopping power curve is theoretically well described by the famous Bethe–Bloch formula [14].

An RBS energy spectrum as, for instance, schematically shown in Figure 6.4 is not only determined by the two-body collision kinematics but also by the broad distribution of ions backscattered from deeper layers. One generally observes a surface 'edge' in the spectrum for each atomic species which is present in the target material, and an increasing scattered ion yield towards lower energies. A heavy adsorbate on the surface (Ag in the example) gives rise to an isolated peak at higher energy, obviously an analytically favourable situation. The peak of a lighter constituent (oxygen) generally sits on a broad background and is sometimes difficult to detect without special measures.

The energy spectra in RBS are therefore transformed into mass spectra via equation (2) and into a depth distribution through the stopping power. The basic equation which relates the final energy E_1 of an ion to the scattering depth t in the approximation of a constant energy loss (dE/dx) (E_0) on the way in and (dE/dx) (E_1) on the way out, is for normal incidence:

$$E_1(t) = K \left[E_0 - t \left(\frac{dE}{dx} \right) (E_0) \right] - \frac{t}{|\cos \Phi|} \left(\frac{dE}{dx} \right) (E_1) \qquad (20)$$

A more complete analysis is based on the same concept but applies appropriate numerical techniques.

In the continuous distribution of the scattered ion spectrum, an energy interval ΔE_1 therefore corresponds to a layer of thickness Δt in the depth of the sample. In other words, an energy resolution ΔE_1 given by the apparatus, results in a depth resolution Δt that can be derived from equation (20):

$$\Delta t = \Delta E_1 / \left[K \left(\frac{dE}{dx} \right) (E_0) + \frac{1}{|\cos\theta|} \left(\frac{dE}{dx} \right) (E_1) \right]. \tag{21}$$

It can be seen that the depth resolution strongly depends on the stopping power and it is best for most elements in the energy range of 1–2 MeV (for He) where the stopping power has its maximum. For the same reason, better depth resolution can be expected for heavier materials, higher Z elements having larger stopping power. If we consider a typical energy resolution of 15 keV for a solid state detector and backscattering geometry ($\theta \approx 180°$), we get a depth resolution of about 220 Å for He in nickel. This can be improved by setting the detector to a grazing exit angle and thus increasing the path length of the scattered particles in the material. Taking, in the example, a scattering angle of $\theta = 95°$ improves the depth resolution to about 40 Å.

In compound material the stopping is commonly calculated as the sum of the weighted elemental stopping cross-sections, this is called Bragg's rule [15]. If we consider, e.g., a compound of two constituents A and B with the relative abundances m and n, respectively ($m + n = 1$), then Bragg's rule yields

$$S(AmBn) = mS(A) + nS(B) \tag{22}$$

and the specific energy loss is

$$\frac{dE}{dx}(AmBn) = N(AmBn)S(AmBn) \tag{22a}$$

where $N(AmBn)$ is the atomic density of the compound material. It turns out that this simple rule implies an uncertainty of less than 10% in most practical cases [16].

Depth resolution can be optimized by detector resolution and grazing exit angles only for scattering from near-surface layers. In larger depths, energy straggling occurs due to the statistical nature of the energy-loss process, i.e., if a number of particles has penetrated to a certain depth in the sample, their energies have a distribution of a certain width. The variance of this (Gaussian) distribution was calculated by Bohr [17] to be for normal incidence:

$$\Omega_B{}^2 = 4\pi Z_1{}^2 e^4 N Z_2 t (1 + 1/|\cos\theta|). \tag{23}$$

We see that the mean-square value of the straggling in Bohr's treatment increases linearly with the nuclear charge Z_2 of the target material and with depth t and that it is independent of the ion energy. Bohr's calculation is only an approximation, improved values for straggling were obtained by Chu *et al.* [18]. For He scattering from Ni at a depth of 1000 Å and a scattering angle of 95°, equation (23) yields $\Omega_B = 17$ keV. This is the value of the standard deviation; for comparison with detector resolution we have to take the full width at half maximum (FWHM), that is Ω_B has to be multiplied by a factor of $2\sqrt{(2\ln 2)} = 2.335$ which gives us an energy width of 40 keV, i.e. much larger than a typical detector resolution of 15 keV. Obviously in such a case the resolution of the system is determined by energy straggling.

We now give an estimate for the shape of the continuous energy spectrum determined by scattering from a thick target. The scattering yield from a slab of width Δt at a depth t can be written as

$$Y(t)\Delta t = \frac{d\sigma}{d\Omega} N \Delta\Omega Q \Delta t \tag{24}$$

where $\Delta\Omega$ is the solid angle subtended by the detector (of 100% efficiency) and Q denotes the number of primary ions. We know from equation (8) that the cross-section depends on the energy E_t in the depth t like

$$\frac{d\sigma}{d\Omega} \approx E_t^{-2}.$$

E_t can be estimated by assuming a constant ratio a of the energy loss on the inward and outward ion path

$$a = \frac{\Delta E_{out}}{\Delta E_{in}} = \frac{KE_t - E_1}{E_0 - E_t}. \tag{25}$$

For light projectiles we can assume $K \approx 1$ and then also $a \approx 1$ which yields

$$E_t = \tfrac{1}{2}(E_0 + E_1). \tag{26}$$

Combining equations (24) and (26), we see that the scattering yield $Y(E_1)$ in which E_1 varies with depth is

$$Y(E_1) \sim (E_0 + E_1)^{-2}. \tag{27}$$

This is the form indicated in Figure 6.4 schematically and shown in the experimental results in the following sections.

3.2 APPARATUS

The principal components in RBS are those of a typical scattering experiment: (i) a source which provides energetic primary ions; (ii) a sample holder that allows to position the target with the necessary degrees of freedom and the required precision and; (iii) a detection system for measuring the energy distribution of the scattered particles. In most cases, particularly in fundamental surface research, manipulator and detector are mounted in an ultra-high vacuum (UHV) chamber.

The most widely used accelerator type for the energy region of interest for RBS is the Van de Graaf type in which a high voltage is built up by a fast moving belt that carries electrons from a charging screen to a high-voltage terminal. This charging belt and the ion source at the terminal are housed in a gas-filled tank. Voltages up to about 2 MV, which are most useful for surface work, are readily accomplished, but much higher values (up to 30 MV) are achieved in large machines. The ion beam from the tank is directed through an evacuated beam line to the target chamber. For this purpose, a system of switching magnets, focusing quadrupole magnets, collimators etc. has to be set up. For surface analysis work a beam current of about 100 nA on a beam spot of about 1 mm in diameter is typical and sufficient to give counting rates of 10 kHz or more. The high voltage requires tank dimensions of the order of metres and also the length of the beam lines is usually several metres. Therefore, major construction and financial investments are necessary for an accelerator and this limits the proliferation of instruments for that powerful method.

The UHV chamber and the manipulator are nowadays very much of a standard type used in surface science research, e.g., also in low-energy ion scattering (see Section 4). The manipulator usually has two rotational degrees of freedom, one around the main axis (defining the angle of incidence) and one around an axis perpendicular to the sample surface (defining the azimuthal position of the scattering plane). For channelling measurements, a precision for setting these angles of much better than 1°, about 0.1°, is usually required.

The manipulator can also provide heating (by electron bombardment) and cooling (liquid nitrogen) facilities if necessary. It is also very useful that it contains a calibrated scattering standard with a known concentration of a heavy element on a light support (e.g. Au on Si). It is then easy to determine the absolute aerial atomic density on a sample by comparison, without exact knowledge of the solid angle accepted by the detector.

An extremely useful instrument in the development of the RBS analysis technique turned out to be the silicon solid state particle detector. This is a fairly simple device which allows to detect particles and their kinetic energies, i.e., to record an energy spectrum without sweeping an energy window. Energetic particles which penetrate through the gold surface barrier, deposit their kinetic energy in the silicon by creating electron–hole pairs, losing 3.6 eV/pair. The reverse bias voltage separates the charges and creates a corresponding voltage pulse $\Delta V = Ee/3.6C$ (C being the capacity of the device), i.e., the signal is proportional to the energy of the incoming particle. The pulse height distribution, after appropriate amplification recorded by a multichannel analyzer or a computer, therefore represents the energy spectrum of the scattered particles. In this high energy range, ions and neutrals are equally well detected with high probability. Internal fluctuations in the solid state detector, limit the energy resolution to about 10 keV, in practical cases often 15 keV. This is then also a limitation of the mass resolution (see Section 2) and the depth resolution of the system.

Better resolution (at the cost of more complicated equipment and sequential recording) can be obtained by magnetic or electrostatic energy analyzers which have a constant relative energy resolution $\Delta E/E$. Typical values are of the order of 5×10^{-3}. For an electrostatic analyzer used with an accelerator in the 50 keV–400 keV range, a depth resolution of about 10 Å has been reported [19].

3.3 BEAM EFFECTS

An energetic primary ion beam impinging on a sample, modifies its surface by removing atoms from this surface. This effect, called sputtering [20] and for removal of adsorbed layers occasionally ion impact desorption, limits the sensitivity of ion-scattering methods. The questions of interest are: how many atoms are removed (sputtered) from the surface for obtaining a given scattered ion yield Q_D in the detector and what is the minimum atomic areal density which can be detected if we accept a fraction q of the layer to be sputtered away?

The scattered ion yield can be expressed as

$$Q_D = N \frac{d\sigma}{d\Omega} \Omega Q, \qquad (28)$$

where Q is the number of primary ions, N the surface density of atoms of the considered species (atoms/cm^2), $d\sigma/d\Omega$ the differential Rutherford scattering cross-section (cm^2/sr), which in the following will be written as σ for simplicity, and Ω is the solid angle subtended by the detector (of unit efficiency). It follows from equation (28) that we need an amount of primary ions $Q = Q_D/(N\sigma\Omega)$ in order to obtain a given signal Q_D. The number of atoms sputtered by these ions is QY, where Y (atoms/ion) is the sputtering yield that depends on the ion mass and energy and on the target material. Considering very thin layers (monolayers and below) it is convenient to define a total sputtering cross-section σ_s (cm^2/ion) [21] since the sputtering yield (the probability that a surface atom of the considered species is sputtered) depends on the coverage, in contrast to bulk elemental sputtering, σ_s can be connected to the sputtering yield via the monolayer density N_{ML}:

$$Y = \sigma_s N_{ML}. \qquad (29)$$

For a given Q_D the number of atoms removed by the beam with area a (cm^2) is then determined by the ratio of the sputtering and scattering cross-sections:

$$\Delta N = \frac{Q_D}{a\Omega} \frac{\sigma_s}{\sigma}. \qquad (30)$$

If we then require that not more than a fraction q (e.g. 1%) of the initial coverage should be removed during the measurement, $\Delta N < qN$, we get an estimate for the minimum detectable coverage

$$N_{\min} = \frac{Q_D}{a\Omega} \frac{\sigma_s}{q\sigma} \tag{31}$$

and also a condition for the maximum primary ion fluence by using equation (28):

$$\frac{Q}{a} < \frac{q}{\sigma_s} \tag{32}$$

Obviously, we get a good sensitivity for heavy elements since the Rutherford scattering cross-section σ varies with Z^2. Let us consider an example by using typical numbers for an Au layer on a light substrate [22]. We assume a minimum of 100 ion counts to give a statistically significant signal, i.e. $Q_D = 100$; a beam spot size of 10^{-2} cm^2, a solid angle of 4×10^{-2} sr (i.e. 1 cm^2 at a distance of 5 cm), σ_s is about 10^{-18} cm^2 [21]. From these numbers we obtain a detection limit of 2.5×10^{12} Au atoms/cm^2 or about 1/400 of a monolayer for this special case, 1% therefore would be removed during the measurement which requires 10^{14} primary ions or slightly more than 10 µC. These estimates are not too far from experimental experience. An important general conclusion which can also be drawn from these considerations is that RBS can quite generally be regarded as a virtually non-destructive method for the analysis of surfaces and near-surface layers, another essential feature which makes this method so attractive.

3.4 QUANTITATIVE LAYER ANALYSIS

In Figure 6.4 we showed schematically the features of an RBS energy spectrum and, with the relations derived in the previous sections, we are able to understand in principle the quantitative interpretation of an actual RBS spectrum. As an example, Figure 6.12 shows an energy spectrum taken from a model catalyst consisting of an Al_2O_3 layer on Al as support material and a Rh metal overlayer as the active component [23]. This is the situation of a heavy adsorbate on a light substrate and the amount of Rh can easily be calculated using equations (24) or (28) to calculate N. In practice the solid angle $\Delta\Omega$ and the detection efficiency are taken into account by using a calibrated standard as mentioned above. For the present example we obtain an Rh coverage of 10^{15} Rh/cm^2 which is about one monolayer on the nominal surface and demonstrates nicely the high sensitivity of RBS in such a case. In the continuous part we first see the Al edge from the Al_2O_3 layer and a shoulder from the Al at the Al/Al_2O_3 interface. The oxygen (from the alumina layer) sits on the continuous part of the spectrum which is related to scattering from Al in the depth.

The decomposition of the spectrum of such a layered structure is demonstrated in more detail in the example given in Figure 6.13. It shows a spectrum taken

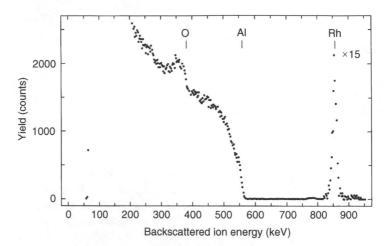

Figure 6.12. RBS spectrum for scattering of 1 MeV He$^+$ ions from Al$_2$O$_3$ with about a monolayer coverage of Rh (from [23])

with 2.5 MeV He^{2+} from a superconducting high-T_c film of YBa$_2$Cu$_3$O$_7$ on (100) SrTiO$_3$ [24]. The upper part shows the actual spectrum (noisy trace) and a numerical fit (solid line). The solid arrow lines indicate the edges corresponding to scattering from the respective elements at the sample surface, (cf. equation (1)) and the dashed arrow lines correspond to scattering from the substrate surface, i.e. they include the additional energy loss of the ions in the film (cf. equation (20)). This is further demonstrated in the lower part of Figure 6.13 where the spectral contributions of the film components are separated and the scattered ion distribution for each of the elements in the film (Ba, Y, and Cu) shows a leading edge for scattering from the surface and a trailing edge corresponding to the film-substrate interface.

The stoichiometry of the film can now be calculated from the height H_i of the spectra of the various elements i. Let us for simplicity consider normal incidence and a scattering angle close to 180° and assume the validity of the constant dE/dx approximation. Then a detector with an energy resolution ε counts backscattered particles from a layer with thickness Δt and from equation (21) we find

$$\varepsilon = [E]n\,\Delta t = [S]\Delta t \tag{33}$$

$[S]$ is the so-called energy loss factor,

$$[S] = K\left(\frac{\mathrm{d}E}{\mathrm{d}x}\right)_{\mathrm{in}} + \left(\frac{\mathrm{d}E}{\mathrm{d}x}\right)_{\mathrm{out}} \tag{34a}$$

and $[E]$ is the stopping cross-section factor,

$$[S] = [E]n, \tag{34b}$$

both expressions in our approximation; n is the number of atoms per unit volume.

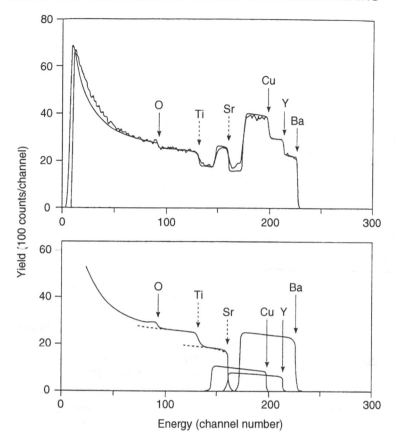

Figure 6.13. RBS spectrum for scattering of 2.5 MeV He^{2+} from a superconducting YBa$_2$Cu$_3$O$_7$ film. Upper part: experimental spectrum (noisy line) and
numerical simulation (smooth line) (redrawn from [24]). Lower part: decomposition
into the scattering contributions from the various constituents. Solid arrows correspond to the surface of the film, dashed arrows to the interface with the SrTiO$_3$
substrate

The number of counts from species i in a layer Δt at the surface is then

$$H_i = \frac{d\sigma_i}{d\sigma}(E_0)\Delta\Omega Q\frac{\varepsilon}{[E]_i} \tag{35}$$

and the ratio for two species i and k is

$$\frac{H_i}{H_k} = \frac{n_i\sigma_i[E]_k}{n_k\sigma_k[E]_i} \tag{36}$$

from which the stoichiometric ratios $n_i/n_k/\ldots$ can be determined using the experimentally gained peak-height ratios. If we assume $[E]_n = [E]_i$ we obtain:

$$\frac{n_i}{n_k} = \frac{H_i}{H_k} \left(\frac{Z_k}{Z_i}\right)^2$$

as a crude estimate of the stochiometry.

Detailed numerical analysis, as e.g., shown in Figure 6.13 have to take the correct angular situation, the dependence of $\mathrm{d}E/\mathrm{d}x$ on stoichiometry (Bragg's rule) and energy into account, but are based on the same principles.

3.5 STRUCTURE ANALYSIS

Analysis of crystalline surface structures can be achieved with RBS by exploiting collective scattering phenomena similar to the flux modifications discussed in connection with the shadow cone (Section 2.4). Structure analysis is based on the channelling effect [25–27] and the (in this respect) inverse process of blocking. Using these techniques, surface layer analysis can be performed, although in principles the depth resolution in RBS is of the order of 30 Å–100 Å, as pointed out in the previous section.

Channelling occurs if a collimated ion beam impinges on a monocrystalline target along a low-index direction (i.e. close to a high-symmetry axis). In that case most primary particles have large impact parameters with the atoms of the first layer, i.e. they suffer only small angle deflections. This can be continued in the following deeper layers and the ions are then steered in the channels (axial channelling) or between crystal planes (planar channelling). The situation is schematically shown in Figure 6.14. It is obvious that the backscattering yield is reduced in such a case compared to the 'random' situation. In the idealized limit (rigid lattice, no beam divergence) only scattering from the top atomic layer would be possible. If the detector also defines a scattered ion direction along a high-symmetry axis, scattering from atoms in regular lattice positions is blocked. This double-alignment technique has been proven to be particularly useful for surface analysis [27, 28].

The analytical principle is then, that deviations from the idealized lattice structure change the backscattered flux intensity and angular distribution and so can be used for investigating effects such as thermal vibrations [29], lattice relaxation [30], lattice reconstruction [31], interstitial atom positions [32] adsorbate locations [33] and surface disorder [34].

The role of thermal vibrations has been studied in a number of investigations on silicon and metal single crystals [29]. It could be shown that the intensity of the surface peak scales with the ratio ρ/R_c where ρ is the root-mean-square amplitude of the thermal vibrations and R_c the shadow cone radius at the second atom. The surface peak intensity (in atoms/row) varies from 1 for $\rho/R_c < 0.3$ up to values of 4 atoms/row for $\rho/R_c > 1.5$.

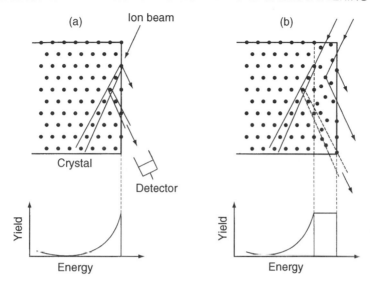

Figure 6.14. Schematic of the channeling-blocking technique: scattering geometry (ion incidence and detection along high-symmetry directions) and energy spectrum for: (a) a well ordered crystal; and (b) a crystal with a disordered overlayer (after [34])

The channelling and blocking effects cause minima in the angular scan curves of the scattered ion yield. This is schematically shown in Figure 6.15 for a double alignment situation [28]. The figure also shows how the angular shift is related to the lattice spacing between the top surface layer and the bulk. Using this technique, surface relaxation and reconstruction has been identified on a number of metal (Ag, Ni, Pt, W) and Si single crystal surfaces.

An illustrative example for adsorbate position determination by channelling techniques, is a study of deuterium adsorption on Pd (100) [33]. Here transmission channelling through a thin (3000 Å) Pd crystal was used. The principle for site location and the experimental results are shown in Figure 6.16 giving data for 2 MeV He$^+$ scattering from Pd and D elastic recoil detection. Deuterium is obviously in a four-fold hollow site position, with different vertical displacements for two adsorbate phases, $\Delta z = 0.3$ Å for $p(1 \times 1)$ and $\Delta z = 0.45$ Å for the $c(2 \times 2)$ phase. This kind of application of the channelling technique is, in principle, very similar to the method used for analysing the positions of solute atoms in a bulk crystal, e.g. for discriminating between substitutional, interstitial or random solute atom positions [32].

Another very successful application of the channelling–blocking technique relates to the thermodynamics of ordered surfaces, in particular to the presently very actively studied field of surface melting. These studies again nicely demonstrate the power of the technique and an example is illustrated in Figure 6.17

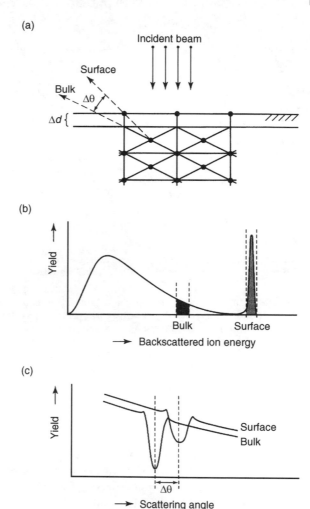

Figure 6.15. Double alignment experiment for measuring surface relaxation: (a) Scattering geometry showing the different blocking directions for surface and bulk scattering, (b) corresponding energy spectra taken with an electrostatic analyser, (c) angular intensity distributions showing the shift in the blocking minimum due to surface relaxation (from [28])

[34]. On a well ordered Pb (110) surface at 295 K the backscattered ion yield in double alignment geometry only shows the surface peak. By increasing the crystal temperature, the development of a disordered surface layer on an ordered bulk crystal can be seen. A typical 'random' spectrum is obtained only after heating to the bulk melting temperature of 600.7 K.

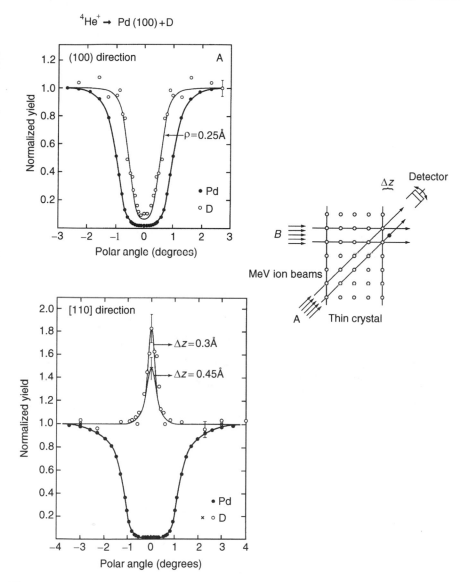

Figure 6.16. Transmission channelling experiment for determining the position of deuterium adsorbed on Pd(100) using 1.9 MeV ^{4}He$^+$. Pd scattering and D recoil intensities both show a minimum for incidence along the [100] direction (upper left panel), whereas in the [110] direction the D recoil intensity has maxima (lower left panel), demonstrating the D position in a four-fold hollow site, 0.3 Å (for p (1 × 1), open circles) and 0.45 Å (for C (1 × 1), crosses) above the surface (from [33])

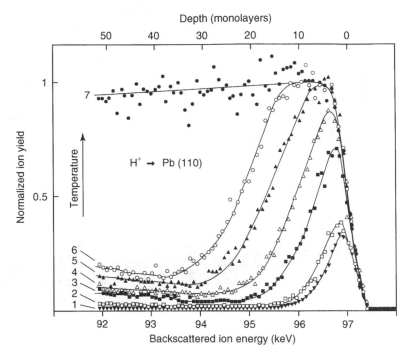

Figure 6.17. RBS spectra for 97.5 MeV protons scattered from a Pb(110) surface in the scattering geometry shown in Figure 6.14. The temperature variation (1:295 K, 2:452 K, 3:581 K, 4:597 K, 5:599.7 K, 6:600.5 K, 7:600.8 K) demonstrates the development of a disordered layer by surface melting. Spectrum 7 (above the bulk melting temperature of 600.7 K) corresponds to scattering from a 'random' solid (after [34])

3.6 THE VALUE OF RBS AND COMPARISON TO RELATED TECHNIQUES

The outstanding strength of RBS is that it provides absolute quantitative analysis of elemental compositions. Surface coverages can be measured easily with an accuracy of 5% or better. For surface layers and thin films it provides quantitative depth profiles for a thickness of up to about 1 μm with depth resolution of the order of 30–100 Angstroms. In many practical cases, the high-energy ion beam used for the analysis, causes relatively little damage to the sample. This is particularly so for metals and most semiconductors which are virtually unaffected by the energy transfer through electronic energy loss processes. Only the small nuclear energy loss (see Figure 6.11) contributes to permanent damage. The situation is different for insulating material such as oxides, alkali halides, polymers etc. in which considerable radiation damage can be caused by electronic energy loss processes. Moreover, charging of the sample often disturbs the analysis in these cases.

Depending on the type of analysis, RBS requires high vacuum or ultra-high vacuum (for surface analysis) in the target chamber. In special cases analysis under atmospheric pressure has also been reported [35].

Using the channelling and blocking techniques, RBS can provide information with monolayer resolution and thus it can be very successfully used as a near-surface structural tool.

The high-energy ion beam used in RBS also provides the possibility to apply related analytical techniques in the same apparatus. Forward recoil detection or elastic recoil detection (ERD) is useful for light sample constituents such as hydrogen isotopes. It is a natural complement to the scattering process, the kinematics is similar (Section 2.2) and an example is given in the previous section.

Another related technique is the analysis of proton induced X-ray emission (PIXE) [36]. It is also a quantitative method, the element identification is often more unambiguous than in the case of RBS which sometimes suffers from limited mass resolution. Compared to electron induced X-ray analysis (EIX), protons

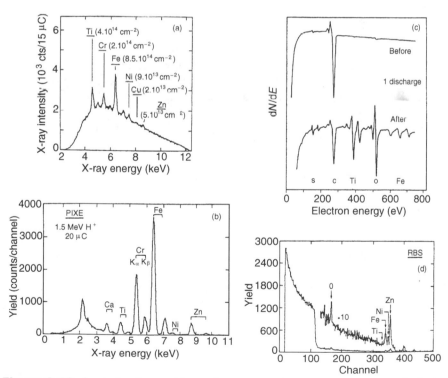

Figure 6.18. Analysis of a graphite surface exposed to one fusion plasma discharge with different techniques: (a) electron induced X-rays ($E_0 = 15$ keV, 175 μm Mylar filter); (b) proton induced X-rays; (c) AES ($E_0 = 3$ keV, 50 μA); (d) RBS with 2 MeV ^{4}He$^+$ (from [37])

have the advantage of producing much less background due to bremsstrahlung. Some of these features can be seen in the example shown in Figure 6.18 [37] which shows the analysis of the identical surface area by RBS with 2 MeV He^+, PIXE with 1.5 MeV H^+, EIX with 15 keV electrons and Auger Electron Analysis (AES) [38] with 3 keV electrons. The sample is a graphite foil whose surface was exposed to the flux of particles coming from one discharge in the fusion device ASDEX [37] to the vessel wall. It can be seen that the metal components are better separated by their core level electron energies showing up in the X-ray spectra, whereas light constituents (oxygen) are only accessible to RBS and AES. In this study RBS was again most useful due to its capability of absolute quantification and was used to calibrate the other methods.

4   LOW-ENERGY ION SCATTERING

4.1   NEUTRALISATION

A fundamental property of low-energy ion scattering (ISS) that distinguishes this technique from RBS shows up in Figure 6.4: while ISS spectra exhibit a peak for each atomic species on the sample surface, RBS generally yields an edge in the spectrum, followed by a broad distribution towards lower energies.

This is a consequence of the fact that in ISS only particles backscattered from the top surface layer have a significant chance to survive the scattering process as ions and can thus be detected in an apparatus using an electrostatic analyser (see Section 4.2).

It is due to this selective property of the neutralization effect that ISS can be used as an extremely surface sensitive method in the sense that the scattering signal only arises from the topmost atomic layer. This is most pronounced for noble gas ion scattering (He^+, Ne^+, Ar^+) in combination with an electrostatic analyser (modifications which occur by using alkali ions or neutral particle detection are discussed below). For these noble gas ions, the probability P for surviving as ions is of the order of about 5% for scattering from the first layer and at least an order of magnitude lower for scattering from deeper layers [2]. From this it immediately follows that the scattered ion yield is not only determined by the cross-section (cf. equation (28) for the RBS case) but also to a large extent by the neutralization effect, expressed as the ion survival probability P.

The ion current I_i^+ arising from scattering from species i with a surface density N_i therefore can be written as;

$$I_i^+ = I_0^+ T N_i \frac{d\sigma_i}{d\Omega} \Delta\Omega P_i, \tag{37}$$

where I_0^+ is the primary ion current and T is a factor taking the transmission of the apparatus and the detector sensitivity into account. The cross-section $d\sigma/d\Omega$ can be calculated with sufficient accuracy as described in Section 2.3. The ion survival probability P, however, is generally not well known. As a rough estimate for scattering from metal surfaces, values of about 10% for 1 keV He^+ and 5% for 1 keV Ne^+ can be taken. For quantitative composition analysis, calibration with elemental standards has been used successfully, particularly for metal alloys [39, 40]. But it has also been observed [41, 42] that the survival probability can be trajectory dependent, i.e., it not only depends on the specific target atom i but also on the electron density encountered by the projectile on its way into and out of the target. This behaviour was studied in some detail e.g. for oxygen adsorbed on Ni surfaces [41, 43].

The theoretical description of the various neutralization processes is based on Hagstrum's work [44] and is schematically shown in Figure 6.19. For noble gas ions with large ionization potentials (between 16 eV and 24 eV), Auger neutralization (AN) is the dominant process. The ion survival probability can be described within a model in which the electron transition rate depends exponentially on the distance of the ion from the surface. This results in an expression for the ion survival probability which depends on the ion velocity perpendicular

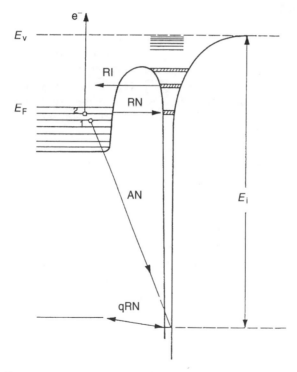

Figure 6.19. Energy levels for an ion (atom) close to a surface: E_V; vacuum level; E_F, Fermi energy; RI and RN, resonance ionization and neutralization, respectively: AN; Auger neutralization; qRN, quasi resonant neutralization; E_i; ionisation energy of the atom

to the surface, v_p:

$$P = e^{-v_0/v_p} \qquad \cdot \tag{38}$$

The parameter v_0 depends on the ion–target combination but is very generally of the order of 10^7 cm/s. Basically an expression like equation (38) has to be taken into account for the incoming and the outgoing trajectory, but experimental results can often be described with sufficient accuracy by only considering the final velocity perpendicular to the surface. In addition to these neutralization processes due to charge exchange with the surface, a contribution from the close encounter during the large angle collision has been postulated [42, 45] (see also [46] for a recent discussion of this matter and the relevant literature). These observations indicate that the final charged state is primarily determined on the outgoing trajectory. The incoming ion is effectively neutralized and re-ionization can take place during the violent collision with a target atom. The related energy loss, corresponding to the ionization energy of the projectile, was in fact observed for ions above a certain threshold of kinetic energy [47, 48, 49].

For alkali ions, resonance neutralization (RN) is most important. Their ionization potentials are close to the values of work functions of many materials, particularly metals, i.e. of the order of a few eV. Depending on the energetic position of the valence level involved, the ion yield can decrease or increase with increasing kinetic energy of the projectile. This behaviour has been reviewed and theoretically discussed in [50] and [51]. Generally, ion yields are very much larger for alkali ions than for noble gas ions, of the order of 50% to 100%. If the work function of a surface is reduced, e.g. by alkali ion adsorption, resonant charge exchange with excitation levels of noble gas ions also becomes important and a dependence on the distance of closest approach is observed [46].

A special case of so called quasi resonant (qRN) charge exchange occurs if the ionized energy level of the projectile lies in energetically close vicinity to a target atom core level, as e.g., in the case of He 1s and Pb 5d levels. Then a Landau–Zener type of charge exchange occurs which results in an oscillatory dependence of the ion yield on the projectile velocity [52, 49].

From the analytic point of view the discussion given above can be summarized as follows: in general the ion escape probability for noble gas ions cannot be given *a priori* and therefore it poses a problem on quantitative analysis. However, if proper calibration can be provided, quantitative analysis is possible and a linear dependence on N_i, as suggested by equation (37) has been established in a number of cases (see also Section 4.3). One way to circumvent these problems is to use alkali ions for which the values of P are close to one if the target work function is not too small. The other possibility consists of the detection of neutral (or ion plus neutral) scattered particles. This can indeed be done with substantial advantage, as discussed in the following sections. In both cases, however, the beneficial part of the neutralization effect, i.e. the exclusive surface sensitivity, is largely lost.

4.2 APPARATUS

Because of its extreme surface sensitivity, ISS requires vacuum conditions which allow the sample surface to be cleaned or prepared and maintained in a defined state for a sufficiently long period. Therefore the scattering chamber must be a UHV system with a base pressure for reactive gases (H_2, CO, H_2O) below 10^{-9} mbar. If, for example, a surface with a contamination below 10^{-2} monolayers should stay clean for an hour, a pressure below 10^{-11} mbar is required for gases with a sticking coefficient around one. For reactive gases such values are not uncommon on metal surfaces [53]. In many systems the partial pressure of noble gases from the ion source reaches pressures between 10^{-7} and 10^{-6} mbar in the scattering chamber. These values are tolerable since the sticking probability of thermal noble gas atoms is virtually zero on all surfaces at room temperature [54].

A typical ion-scattering apparatus using an electrostatic energy analyser is schematically shown in Figure 6.20. The essential components are the ion source, the target manipulator and the analyser and detector system. Electron impact ion sources are most convenient for noble gas ions. For energies around 1 keV they easily provide a constant ion current of 10–30 nA which is sufficient for surface analysis and causes limited surface erosion due to sputtering. For a beam spot diameter of 1–2 mm a total fluence of the order of 10^{13} He^+/cm^2 is required to record an energy spectrum over the whole range of secondary energies. With a typical He^+ sputtering yield of about 10^{-1} atoms/ion [55] only about 10^{-3} of a monolayer is removed from the surface for one spectrum. For light adsorbates, however, much higher yields are possible [56] and the adsorbed layer must be restored after short bombardment intervals. Lower ion fluences are necessary for alkali ions and for neutral particle detection, as discussed below. If, on the other hand, erosion of the surface by ion bombardment is desirable for cleaning purposes or in order to obtain near-surface depth profiles, then higher current densities can generally be applied [56]. Mass separation is not absolutely necessary for electron impact noble gas ion sources, but with plasma ion sources and solid dispenser sources it is a necessity, since they produce a variety of ionic species and they also emit reactive neutral gas particles. Sources for alkali ions are commercially available. They contain appropriate minerals that release Li^+, Na^+ or K^+ ions at elevated temperatures.

For sample holding, various commercially available UHV-target manipulators exist and for special requirements, appropriately designed manipulators have been developed [57]. For structure investigations, generally two axes of rotation are necessary, one in the sample surface defining the incidence and exit angles, and one perpendicular to the surface for azimuthal variations. For the angular settings, generally an accuracy of $0.5°$–$1°$ is sufficient. For composition analysis a fixed scattering geometry, preferably with large incidence and exit angles, can be used. The manipulators also provide facilities for sample heating by electron bombardment and in some instances also cooling by liquid nitrogen. The electrical and mechanical leads necessary for temperature control and measurement,

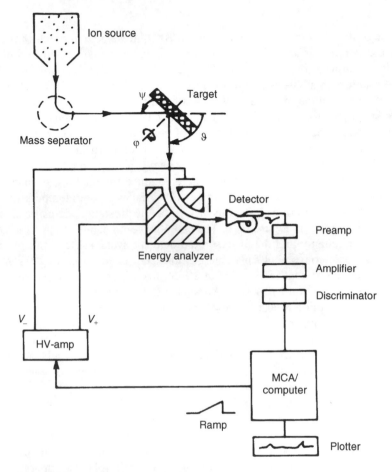

Figure 6.20. ISS arrangement with a 90° spherical sector electrostatic analyzer [58]

together with the required rotational degrees of freedom call for intricate and precise constructions.

For the energy analysis of the scattered ions, electrostatic sector fields are very convenient and most commonly applied. Their relative energy resolution $\Delta E/E$ is given by the ratio of the aperture width to the radius of the central trajectory, i.e., $\Delta E/E = s/r$. For ISS purposes a resolution of 1%–2% is sufficient. It follows from this relation that the energy window of an electrostatic analyser increases in proportion to the energy detected and this has to be corrected for absolute measurements. The possibility to operate with constant pass energy is not often reported to be used with ISS. Spherical sector analysers (with a sector angle of 90°, see Figure 6.20) can be mounted on a UHV manipulator system, such that the scattering angle is variable from 0° to large angles of 160° or more. This is very useful for the experimental determination of the direction, the energy, and the width in angle and energy of the incident beam. Variation of the scattering angle is often useful for peak assignment and large scattering angles have become important for structure analysis (see Section 4.4). For the detection of charged particles, usually channeltron electron multipliers are used in the counting mode.

Cylindrical mirror analysers (CMA) have the advantage of large acceptance angles and consequently higher scattering signals (up to a factor of 30 compared to spherical sectors [58]). This high intensity and the relatively good mass resolution due to the large scattering angle (137°) make CMAs very useful for standard surface composition analysis. Since the scattering geometry is fixed, they are generally not suited for structure determination. For very high detection efficiency, special systems were developed with a spiral shaped position-sensitive detector for simultaneous angular and energy distribution recording [59]. An energy dispersive toroidal prism has also been successfully used to measure energy and angular distributions simultaneously in a multichannel mode [60].

A very successful alternative means for measuring the energy distributions of scattered or recoiling particles is the time-of-flight (TOF) method [61–67]. Since here the energy, or rather the velocity, is determined by the flight time, this method is applicable equally to charged and neutral particles. Typical elements of a TOF systems are shown in Figure 6.21 [66]. After mass separation, the primary ion beam is pulsed by a square-wave voltage applied to two orthogonal

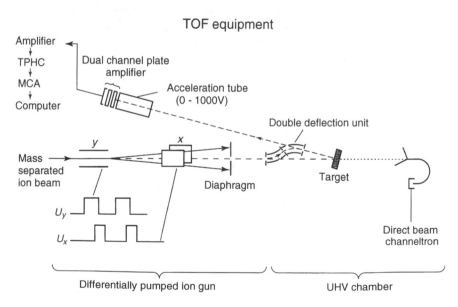

Figure 6.21. Schematic of a time-of flight (TOF) equipment for ion and neutral detection (from [66])

pairs of deflection plates. Bunches of primary particles thus impinge on the surface and, after scattering, they pass through a drift tube at ground potential until they hit the particle detector. The optional double deflection unit helps to switch electronically between two scattering angles, i.e. 165° and 180°. The time distribution of the chopped-ion beam is measured after scattering and this distribution can be transformed into an energy spectrum through the relation

$$E = \tfrac{1}{2} M_1 L^2 / t^2 \tag{39}$$

where L is the length of the flight path. The counts per constant time increment $\Delta N(t)$ are converted into constant energy increments via the expression

$$\Delta N(E) = (t^3 / M_1 L^2) \Delta N(t). \tag{40}$$

For distances of the order of 1 m, flight times of microseconds are obtained and the corresponding electronics have to chop the beam with rise times of some ten nanoseconds in order to obtain an energy resolution of about 1%.

By applying an appropriate potential to the drift tube, the scattered ions can be separated from the neutrals and charge fractions in corresponding parts of the energy spectrum can be determined. Particle detection can be achieved by channeltrons or open multipliers, just as in the case of electrostatic analysers.

If the particle energies are sufficiently high, secondary electron multipliers respond equally well to neutral particles as to ions (i.e. if the kinetic secondary electron emission is much higher than the potential emission with ions). Therefore the scattered particle energies should be above about 1 keV [68] and consequently TOF experiments are usually carried out with primary energies of 2 keV or more, up to 10 keV (thus reducing the surface specificity). Neutral particle spectra exhibit, in general, similar features to as alkali ion spectra: quantification is easier in principle but large contributions from multiple scattering can make the interpretation of spectra more complicated and computer simulations are sometimes necessary for this purpose. Since the neutral particle fraction is usually more than 90% and due to the simultaneous recording of the entire spectrum (rather than scanning with an energy window), primary ion fluences for one spectrum are two orders of magnitude or more lower than for noble gas ISS with an electrostatic analyser, i.e. 10^{11} ions/cm^2 or less. Therefore, the surface damage is correspondingly low and thus this method is virtually non-destructive.

4.3 SURFACE COMPOSITION ANALYSIS

It is pointed out in the preceding sections that ISS has the capability of analysing the elemental composition of the outermost atomic layer of a solid surface. The method therefore appears well suited for routine surface composition analysis. This, however, is true only within certain limitations which are inherent to the method:

(1) 'Technical' surfaces which are not prepared in a well defined way generally have a hydrogen-rich surface contamination layer (hydro-carbons, water) and the light H atoms do not contribute to the scattering signal under many common scattering conditions. Useful spectra are therefore only obtained after removing the contamination layer by sputtering with the analysing ion beam (see Figure 6.22).

(2) Mass resolution for heavier masses and consequently mass identification is restricted by collision kinematics (cf. Section 2.2) and therefore unambiguous determination of heavier masses without any pre-knowledge is not guaranteed.

The areas of research in which ISS has been shown to be particularly successful are surface composition analysis of adsorbates, analysis of catalyst and of metal

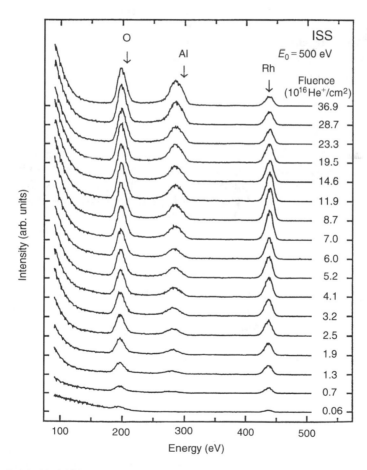

Figure 6.22. He$^+$-ISS spectra from an Al$_2$O$_3$ surface covered with about one monolayer of Rh (compare Figure 6.12 for the RBS case) (from [23])

alloy surfaces. The especially useful combination with structure determination is treated in Section 4.4.

4.3.1 Adsorbates

According to equation (37) the ion scattering signal should ideally increase linearly with the surface density N_i of the adsorbed species i. This has in fact been demonstrated for a number of systems such as S, O, CO and Pb adsorbed on Ni surfaces in which the surface coverage could be independently calibrated using other methods (neutron activation analysis, work function change, RBS, LEED) [69]. An example is given in Figure 6.23 for CO adsorption on Ni (100) [70]. The oxygen signal from CO increases linearly with coverage (and parallel to the work function change $\Delta\phi$) up to a saturation value of one monolayer (showing a $c\,(2 \times 2)$ structure). The linearity is maintained despite the large work function change of 0.9 eV which obviously does not influence the neutralization probability (P in equation (37)). This linearity is lost if the work function is drastically decreased by adsorption of Cs [46].

Adsorbed species cover the substrate atoms, whose scattering signal should then decrease accordingly. This is also demonstrated in Figure 6.23 by the Ni intensity which shows a drastic linear decrease with CO adsorption. The substrate signal I_s can therefore be expressed in analogy to equation (37) by

$$I_s = I_O{}^+ T(N_s - \alpha N_i)(d\sigma_s/d\Omega)\Delta\Omega P_s. \tag{41}$$

The shadowing factor α indicates how many substrate atoms are excluded from scattering by one adsorbate molecule; an initial value of 4 is found for the example of Figure 6.23, it decreases with increasing coverage due to overlap of the shadows.

The shadowing can also yield information on the orientation of adsorbed molecules. For the case of CO on Ni shown in Figure 6.23, the energy spectra only exhibit an oxygen peak, no scattering from C can be observed. This gives a very direct and simple indication that CO is adsorbed on Ni in a vertical orientation, the O pointing away from the surface, a result which is also deduced from data of other techniques in a more indirect manner.

Information about the geometric surface structure of adsorbates, i.e. exact adsorbate positions and bond lengths are of major importance in adsorbate studies. These have been carried out with ion scattering techniques, e.g., for H, S, and O adsorption on a number of high-symmetry metal surface (Cu, Ni, W) and are further discussed in Section 4.4. For light adsorbates (H, D) direct recoil spectroscopy is a very useful variety of ion beam techniques.

The analysing ion beam causes sputtering or ion-induced desorption of adsorbed layers. This has to be taken into account in adsorbate structure investigations, but it can also be used for determining desorption cross-sections and for obtaining near-surface concentration profiles. There is a fundamental and practical interest in desorption cross-sections for the basic understanding

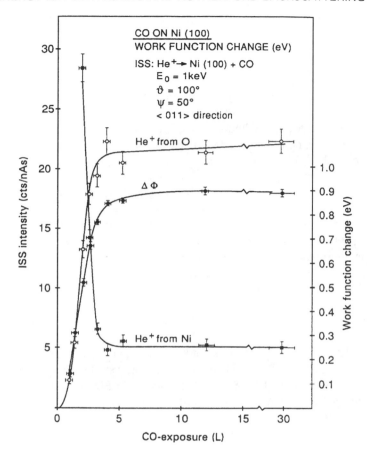

Figure 6.23. Helium ISS intensities and work function change for a Ni (100) surface as a function of CO exposure, see text (from [70])

of multicomponent material sputtering, for surface cleaning in UHV systems and in large vacuum vessels such as storage rings or fusion devices [56]. In the monolayer coverage region, the desorption cross-section σ_D can be easily determined in an ISS apparatus by simultaneously using the ion beam for monitoring the surface coverage and for the ion impact desorption. The signal from the adsorbed species I_i then decreases with time t or fluence it, where i is the primary current density:

$$I_i^+/I_0^+ = \exp(-it\sigma_D) \tag{42}$$

σ_D can be connected with the sputtering yield Y through the relation $Y = \sigma_D N_{ML}$ where N_{ML} is the monolayer areal density. Using ion scattering these cross-sections have been determined for a number of ion-adsorbate–substrate combinations [71, 72]. Their values range from 10^{-16} cm^2 (e.g. 500 eV He-Ni-O) up to such large values as 7×10^{-15} cm^2 (500 eV Ne-Ni-CO).

4.3.2 Catalysts

Supported catalysts as commonly used in heterogeneous catalysis are adsorption systems of a particular kind and represent a research field in themselves due to their enormous technical importance [73]. These catalysts generally consist of a high surface area (about 100 m^2/g) support material and one or more finely dispersed adsorbed components ('active components' and promotors) that determine the efficiency, selectivity and stability of the catalyst. The support material usually consists of highly insulating metal oxides (Al$_2$O$_3$, TiO$_2$, SiO$_2$) and the active components can be metal oxides such as MoO$_3$, WO$_3$ or V$_2$O$_5$, at least in a precursor state. These insulating materials are not accessible to the electron spectroscopies generally used in surface analysis, because of charging effects on the sample. In the case of ISS, charging can be compensated by flooding the specimen with electrons from a filament. And since ISS monitors the composition of the outermost atomic layer — which is most important for the performance of the catalyst — with this technique useful contributions for understanding the composition and structure of supported catalysts can be made [74, 75].

An example of such studies is shown in Figure 6.22 [23] which shows He$^+$-ISS spectra of Rh on Al$_2$O$_3$ as a model catalyst. These spectra are taken from the same sample as the RBS spectrum shown in Figure 6.12 and so a direct comparison of both techniques is possible on this basis. The ISS spectra show almost no scattering signal in the beginning due to surface contamination, as discussed above. With increasing He$^+$ ion fluence distinct O, Al, and Rh peaks are observed. The sequence of spectra represents a composition depth profile and the increase in Al intensity when the Rh adlayer is sputtered away can clearly be seen. By these means, the layering sequence and the spreading for a number of real and model catalysts have been studied [75]. Results from a comparative investigation of MoO$_3$ adsorption (from a solution) on Al$_2$O$_3$ are shown in Figure 6.24 [76]. The increase in the amount of adsorbed Mo species with adsorption time is plotted from RBS, AES and ISS results. ISS, being most surface-sensitive, shows the flattest slope while RBS detects nearly 80% of the final amount already after 1 minute of adsorption. AES is in between, according to its information depth of about five monolayers. This result means that, in the beginning, Mo is adsorbed in pores below the surface in less than 1 minute, afterwards additional molybdate is adsorbed on the external surface in a timescale of hours.

In these cases, model catalysts were prepared by oxidizing metal sheets and the oxide layers were thin enough to allow the use of AES. If real catalysts are studied, wavers have to be pressed from powder material and for these samples,

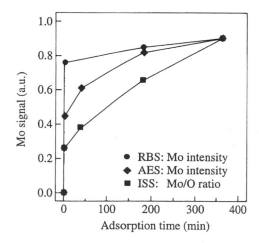

Figure 6.24. Normalized Mo intensities for RBS, AES and ISS from a Al_2O_3 surface impregnated with MoO_3 from a solution as a function of impregnation time (from [76])

charging effects exclude electron spectroscopies. The surface of these wavers is very rough (on a µm scale) and poses the question of the influence of surface roughness on ISS results. It has been found [77] that surface roughness can decrease the ion-scattering signal by more than a factor of six compared to polished samples. If, however, intensity ratios are taken as a relative measure of surface coverage, the results from smooth and rough samples are very similar.

4.3.3 Alloys

Low-energy ion scattering is also extremely valuable for the surface composition analysis of metallic alloys, again due to its 'monolayer' sensitivity. If surface segregation of one component occurs — and this is very generally the case for alloy systems — there is a discontinuity in the composition of the first and second atomic layer [78, 79]. Consequently this requires an analysis method which is able to discriminate between the first and the second layer. ISS has therefore been used for analysing a number of alloy systems and for obtaining data necessary for the fundamental understanding of surface segregation, often in combination with radiation enhanced diffusion and differential sputtering effects [80–83]. As an example Figure 6.25 shows ion scattering spectra taken during the segregation of Al on the surface of a $Fe_3Al(110)$ single crystal surface [84]. At a temperature of 700 K, already after about 10 minutes, equilibrium is reached with a coverage of 95% Al in the top layer.

It has been proven in many cases that, with these metal surfaces, quantification is possible by using elemental standards. In a binary system with components A

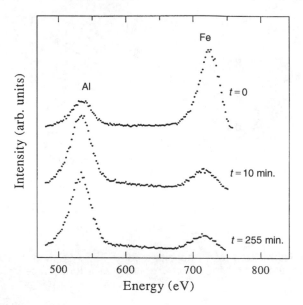

Figure 6.25. He$^+$ ISS spectra ($E_0 = 1$ keV) from a Fe$_3$Al (110) alloy surface held at 700 K, showing the surface segregation of Al with time [84]

and B the surface concentration A can then be calculated to be

$$X_A = 1/ \left(1 + \frac{I_B}{I_A} \frac{S_A}{S_B} \frac{N_B}{N_A} \right) \tag{43}$$

with $X_A + X_B = 1$. Here I refers to the scattering signal from the alloy and S to the signal from the elemental standard with surface density N, the subscripts denoting these values for the components A and B, respectively. If several data are taken from a surface with varying composition, then a calibration without standards is also possible as long as all signals vary linearly with concentration and the sum of the concentration is unity, as stated above.

An extensive study was carried out for the Cu$_3$Au(100) surface using 5 keV and 9.5 keV Ne$^+$ in a TOF technique which allowed a quantitative analysis of the top three crystal layers [80]. This alloy has an order–disorder transition (similar to the Fe$_3$Al alloy mentioned above in Figure 6.25) and upon heating above the transition temperature, the most significant composition changes were found in the second layer where the Au concentration increased from (ideally) zero to 0.25. Such layer-selective and mass-sensitive measurements are a unique feature of ISS.

Under the influence of ion bombardment, the surface composition of an alloy can be changed in addition to the thermally activated segregation by preferential sputtering and radiation enhanced diffusion. For a complete analysis then, data from the first and subsequent layers are necessary. They can either all be

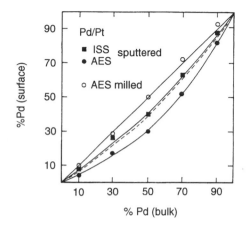

Figure 6.26. ISS and AES results for PdPt alloys with various compositions. The mechanically milled surface shows bulk composition, ISS concentrations are close to those calculated for mass conservation in steady state sputtering, AES indicates subsurface Pd depletion due to preferential sputtering [83]

based on ISS results or on a combination with a method of larger information depth, e.g. AES. An example is given in Figure 6.26 which shows the surface composition of a series of PdPt alloy samples [83]. The combination of segregation and preferential sputtering of Pd leads to a moderate depletion in the first layer (as detected by ISS), but AES averages also over the depleted sub-surface layers and therefore shows more deviation from the bulk composition. From these measurements, the segregation energies and diffusion coefficients can be determined as a function of alloy composition [85].

4.4 STRUCTURE ANALYSIS

4.4.1 Principles (ICISS)

One of the outstanding strengths of low-energy ion scattering is the possibility of obtaining mass-selective structural information in real space about the topmost surface layers of a crystal. This means that the distances between neighbouring atoms in selected crystallographic directions can be measured. In order to determine the geometry of a surface, these distances are used in connection with structure models that are generally deduced from LEED (low energy electron diffraction) results. Diffraction studies yield data in the reciprocal space from which the symmetry of the surface lattice is easily deduced but the exact position of surface atoms, particularly also for multicomponent crystals, requires substantial computational effort [86].

The specific feature of ISS used for structure determination is the shadow cone (see Section 2.4) with the peaked ion flux at the cone edge (cf. Figures 6.8

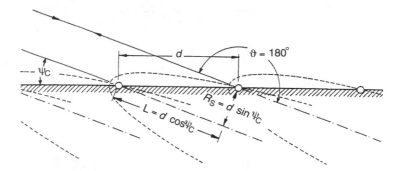

Figure 6.27. Scattering geometry illustrating the ICISS technique (after Aono *et al* [87]

and 6.10). It was applied most successfully by Aono *et al.* [87] in the so called ICISS (impact collision ion-scattering spectroscopy) technique. Its principle can be explained by Figure 6.27 [87]: A parallel flux of ions is incident along a chain of surface atoms with constant distance d. For small angles of incidence Ψ each chain atom is in the shadow of the preceding one and therefore cannot contribute to scattering. By increasing the angle of incidence, a critical angle Ψ_c is reached at which the edge of the shadow cone (with its high ion flux) exactly hits the neighbouring atom. This results in a high backscattered intensity that, upon further increasing Ψ, falls to the average value of the primary flux. Consequently a scattered ion flux distribution is expected, which represents the primary flux distribution shown in Figure 6.10. An example of an ICISS distribution is shown in Figure 6.28 [70]. The critical angle of Ψ_c is related to the shadow cone radius R_s at a distance L from the scattering centre through the atomic distance d as indicated in Figure 6.27. If the form of the shadow cone $R_s(L)$ is known (cf. equation (18)) the interatomic spacing d can be determined by measuring Ψ_c. The shadow cone radius $R_s(L)$ can be calibrated by ICISS measurements at a surface of known structure, e.g., a non-reconstructed surface which shows bulk termination (many clean low-index transition metal surfaces). This possibility of self-calibration is of course a very useful feature of ICISS and helps to avoid ambiguities arising from insufficiently well known interaction potentials. However, numerical calculations are generally useful and sometimes necessary for data interpretation. Shadow cones can be calculated using equation (18). By numerically solving this equation for Ψ_c the experimentally accessible quantity, a formula for a TFM potential was obtained [88]:

$$\ln \Psi_c = 4.6239 + \ln(d/a)(-0.0403 \ln B - 0.6730) \qquad (44)$$
$$+ \ln B \, (-0.0158 \ln B + 0.4647)$$

with $B = Z_1 Z_2 e^2 / (E_0 a 4 \pi \varepsilon_0)$. For a given projectile — target combination it reduces to $\Psi_c \approx d^{-\gamma}$ with γ between 0.7 and 0.8 [88].

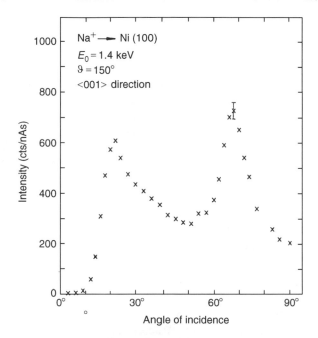

Figure 6.28. ICISS distribution: Na^+ scattering of a Ni(100) surface as a function of the angle of incidence. The first peak corresponds to neighbouring atoms in the surface [100] direction, the second peak to scattering from the second-layer neighbour atoms in [10$\bar{1}$] direction [70]

In some cases, two-dimensional numerical codes were able to reproduce experimental results quite well [90, 91, 84] and thus improved data interpretation. In detailed investigations, particularly for cases of lower symmetry, three-dimensional codes such as the MARLOWE program [92] are useful. This has been applied for determining the position of H adsorbed on Ru(001) [93] or for studies of the Au(110) reconstruction [94].

Ideally, the backscattered flux is measured at a scattering angle of $\theta = 180°$ which corresponds to an impact parameter $p = 0$ and therefore relates Ψ_c directly to the mean lattice position of the scattering atoms. Experimental designs with very nearly [89] or true 180° [66] scattering are reported in the literature, see also Figure 6.21. However, it turns out that scattering angles $\theta > 145°$ are generally large enough to allow a determination of d with an accuracy of about 0.1 Å. This limitation also arises from the thermal motion of surface atoms which results in a broadening of the ICISS intensity compared to the theoretical singularity at $\Psi = \Psi_c$ (equation (17)).

The ICISS method can, of course, not only be used with noble gas ions but also with alkali ions and with neutral particle TOF techniques. The two latter varieties have been named ALICISS and NICISS, respectively [66]. Since they are not strongly influenced by neutralization, they also provide information about atomic positions in the second or deeper layers, see Figure 6.28.

The ICISS techniques have so far been very successfully applied to structure determinations in connection with surface reconstruction (clean surfaces and adsorbate induced reconstructions), for locating the position of adsorbed species, for determining the positions of the components on ordered alloy surfaces (in connection with surface segregation) and for determining thermal vibrational amplitudes of surface atoms (for recent reviews see e.g. [2]). A closely related technique is direct recoil spectroscopy (DRS) in which directly recoiling surface atoms can be identified by their energy (equation (3)). This technique is particularly useful for light adsorbates (e.g. hydrogen isotopes) which are difficult to detect by scattering.

4.4.2 Surface Reconstruction

As an illustrative example of structure determination, we consider an ALICISS study of the reconstruction of the Ni (110) surface upon oxygen adsorption [95, 96]. The oxygen covered surface shows a (2×1) superstructure in the LEED pattern. The questions to be answered are: (i) Does the superstructure represent the oxygen overlayer or has a rearrangement of Ni surface atoms taken place? (ii) What is the reconstructed surface structure in the latter case? (iii) Where are the oxygen atoms positioned? The situation is depicted in Figure 6.29 [95].

The upper part shows the clean surface and the corresponding ICISS spectrum in [112] direction taken with 2 keV Na$^+$ ions at the energy of the binary Na — Ni collision, $E = 0.216\, E_0$. The slopes 1 and 2 represent the shadow cone enhanced scattering from first- and second-layer atoms, respectively, as indicated in the insert. The spectrum changes drastically for the reconstructed surface showing three distinct peaks. The following conclusions were drawn from this result, answering some of the questions posed above. The shift in the leading peak with a slope at $10°$ must be due to scattering from first-layer Ni atoms and thus demonstrates that the (2×1) superstructure is a result of rearrangement of Ni atoms (and not only due to the oxygen overlayer). The absence of the previous first-layer peak proves that the reconstruction is complete. Among the reconstruction models derived from the LEED pattern, the saw tooth (ST) structure and the missing row (MR) structure shown in Figure 6.29 remained as possible candidates. The ALICISS result unambiguously excludes the ST model, since the intensity increases expected from this model were not found, whereas all the expected MR features can be detected in the intensity distribution. The MR structure in which every second ⟨001⟩ row is missing was also confirmed by measurements in other azimuthal directions, and meanwhile confirmed by scanning tunnelling microscopy (with an added row growth mechanism) [97].

The ALICISS distributions can only be taken for Na$^+$–Ni scattering, no backscattering from oxygen is possible. The indicated long bridge O-position (in the [001] direction) cannot be directly demonstrated here, but it is compatible

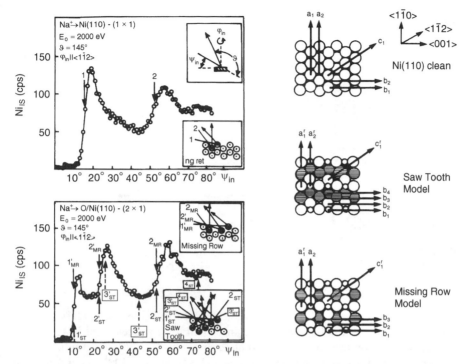

Figure 6.29. Na^+–ICISS distribution from a clean Ni(110) surface (upper left panel) and after oxygen induced (2 × 1) reconstruction (lower left panel). The surface structure models and scattering geometries are plotted on the right side (from [95]

with the ALICISS results and it is directly deduced from recoil spectroscopy, as shown in the following section.

4.4.3 Direct Recoil Spectroscopy (DRS)

DRS is a technique which is closely related to ion-scattering techniques (in the high-energy RBS regime it is usually called 'elastic recoil detection', ERD). In this method, atoms or ions are detected which are removed from the surface by one single collision with an incoming projectile. They can therefore be identified by their kinetic energy according to equation (3). An example is given in Figure 6.30 which shows a Ne^+–Ru scattering peak (at 1200 eV) and a Ne^+–H^+ recoil peak (at 68 eV) from a H covered Ru(001) surface [93]. In a TOF experiment, the corresponding flight time for a path length L is given by

$$t = L \, (M_1 + M_2) \cos \theta_2 / (8 \, M_1 E_0)^{1/2}. \tag{45}$$

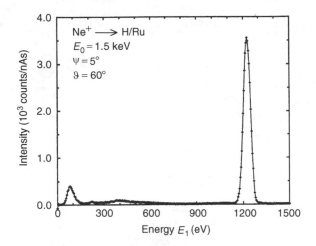

Figure 6.30. Ion energy spectrum from a hydrogen covered Ru(001) surface showing a Ne^+–H^+ recoil peak at 68 eV and a Ne^+–Ru scattering peak at 1200 eV (from [93])

Directly recoiling particles are therefore different from those secondary particles that originate from a collision cascade in a sputtering process and have a broad energy distribution around one to two electron volts. They are used in secondary ion or neutral mass spectroscopy (SIMS and SNMS, see [98] and Chapter 5). According to the collision kinematics (Section 2.2) DRS is a forward-scattering technique. It can be easily combined with ISS or ICISS studies if small and large scattering angles are available in the same apparatus (for examples see [66] and [93]). Recoil cross-sections are generally of the same order as scattering cross-sections. They are largest towards $\theta_2 = 90°$, but then the recoil energy approaches zero. A useful energy range is obtained between 30° and 60°.

The shadow-cone concept can also be applied with DRS and therefore, in principle, analogous measurements are possible as with the scattering techniques, i.e. elemental surface analysis, surface structure analysis and analysis of charge exchange processes (by comparing recoiling ions and neutrals).

As an example of structural sensitivity, we consider the oxygen recoil intensity distribution from a Ni (110)–(2 × 1) reconstructed surface, the same surface as discussed in the previous section. The azimuthal intensity distribution due to 4 keV Ar^+ bombardment is shown in Figure 6.31 together with the structural model [99]. Oxygen recoils are shadowed in certain crystallographic directions either by surface Ni atoms or by other O adsorbates, which results in a distinctly structured intensity distribution. The correlation with the structure model reveals a perfect correspondence with the missing-row reconstruction, with the oxygen in the long-bridge position along the ⟨001⟩ rows.

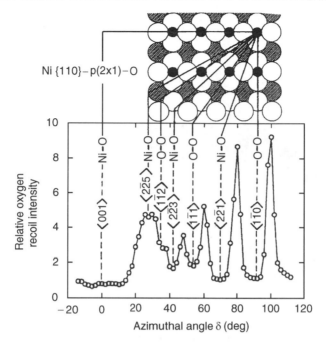

Figure 6.31. Oxygen recoil intensity distribution as a function of the azimuthal angle of the scattering plane for a Ni(110) surface with an oxygen induced (2 × 1) reconstruction (compare Figure 29). The blocking minima demonstrate the 'long bridge' position of the adsorbed oxygen (from [99])

The possibility of investigating hydrogen adsorbates by scattering is very limited, ^{4}He for example, can be scattered by H only into angles below 15°. Here DRS can be applied with advantage as demonstrated by a study determining the position of H on a Ru (001) surface [93]. A corresponding energy spectrum is shown in Figure 6.30, taken at a sample temperature of 138 K where a (1 × 1) structure is formed. Variation of Ψ yields ICISS-like distributions resulting in critical angles between 7.1° and 10.8°, depending on the azimuth. The Ne–Ru scattering critical angle is at 22°. From this a H position of 1.01 ± 0.07 Å above the top Ru layer was deduced. The lateral position of adsorbed H was determined by recording the azimuthal recoil intensity distribution which is shown in Figure 6.32 together with the Ne$^+$–Ru scattering data. H$^+$ recoils have intensity maxima at ±7.5° from the [100] direction which exhibits a broad Ne–Ru scattering minimum. Evaluation of these data and comparison with numerical MARLOWE calculations proved that H is positioned in a three-fold hollow site, 1 Å above the Ru surface. Investigations like this nicely demonstrate the usefulness of DRS for structure studies of light adsorbates.

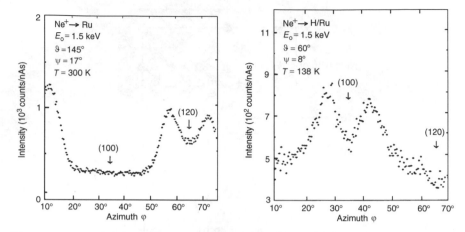

Figure 6.32. Ne$^+$–Ru scattering intensity (top) and Ne$^+$–H$^+$ recoil intensity (bottom) as a function of azimuthal angle of the scattering plane. The H$^+$ recoil intensity confirms the hydrogen position in a three-fold hollow site (from [93])

4.5   CONCLUSIONS

Low-energy ion scattering is one of the many existing surface analytical techniques, all of which have their particular advantages and limitations. The outstanding features of ISS are the possibilities of obtaining mass-selective signals exclusively from the topmost atomic surface layer and to determine interatomic distances on the surface by using straightforward simple concepts. The main limitations consist of the problems which are associated with absolute quantification (due to uncertainties in neutralization and interaction potentials) and in the limited mass identification (caused by collision kinematics and inelastic energy losses). Nevertheless, ISS has been proven to be extremely useful for surface composition analysis, e.g., of catalysts and alloy surfaces and for structure analysis of a large number of metal, semiconductor and metal oxide surfaces and also adsorbates on these surfaces. In Comparison with other common surface spectroscopies, AES is probably more generally applicable (except for insulating material) because it has an information depth of several atomic layers. Therefore, in many cases they cannot replace each other but rather should be used in a complementary way. SIMS certainly yields definite mass identification and often much higher sensitivity, but quantification problems are much larger.

Considering structural analysis, the relation of ISS techniques to LEED and scanning tunnelling microscopy (STM) is important. LEED provides the bulk of structural information from surfaces, i.e. the basic crystallographic structure. ISS can nicely complement these results due to its mass sensitivity and its capability to determine atomic positions unambiguously. STM, another real-space method [100], yields direct images of atomic arrangements on a microscopic scale (atomic resolution on an area of the order of 50 Å in linear dimension), without definite mass identification. Ion scattering averages over a comparatively large area (linear dimension of 1-2 mm). Again complementing measurements appear to be most useful for verifying structure models, positions of various atomic species and investigations of kinetic processes. Regarding this last point, sample temperature variations and time dependent measurements (with a resolution of about one second) are fairly straightforward with ion scattering, and much more limited with STM so far, but rapid progress is presently being made in this field.

REFERENCES

[1]   D.P. Smith, *J. Appl. Phys.* **18**, 340 (1967).
[2]   For reviews see e.g.:
      E. Taglauer in *Methods of Surface Characterization*, Vol. 2 A.W. Czanderna and D.M. Hercules, (eds.) Plenum Press, New York p. 363 (1991), H. Niehus, W. Heiland and E. Taglauer, *Surf. Science Report* **17**, 213 (1993), H. Niehus, in *Practical Surface Analysis by Ion and Neutral Spectroscopy*, Vol. 2, D. Briggs and M.P. Seah (eds.) John Wiley, Chichester (1991), J.A. van den Berg and D.G. Armour, *Vacuum* **31**, 259 (1981).
[3]   For a comprehensive description see e.g.:
      W.-K. Chu, J.W. Mayer and M.-A. Nicolet, *Backscattering Spectrometry*, Academic Press, New York (1978).
[4]   N. Bohr, *Mat.-Fys. Medd. Kgl. Dan. Vid. Selsk.* **18**, 8 (1948).
[5]   See e.g. H. Goldstein, *Classical Mechanics*, Addison-Wesley, Reading, Massachusetts (1965).
[6]   E. Rutherford, *Phil. Mag.* **21**, 669 (1911).
[7]   C.G. Darwin, *Phil. Mag.* **28**, 499 (1914).
[8]   O.B. Firsov, *JETP* **6**, 534 (1958).
[9]   J. Lindhard, V. Nielsen and M. Scharff, *Mat.-Fys. Medd. Dan. Vid. Selsk.* **36**, 10 (1968).
[10] J. Lindhard, *Mat. Fys. Medd. Dan. Vid. Selsk.* **34**, 14 (1965).
[11] A.G.J. de Wit, R.P.N. Bronkers and J.M. Fluit, *Surf. Sci.* **82**, 177 (1979).
[12] O.S. Oen, *Surf Sci.* **131**, L407 (1983).
[13] H.A. Bethe, *Ann. Phys.* **5**, 325 (1930).
[14] J.F. Ziegler, *Helium Stopping Powers and Ranges in All Elemental Matter*, Pergamon Press, New York (1977).
[15] W.H. Bragg and R. Kleemann, *Phil. Mag.* **10**, 318 (1905).
[16] J.S.-Y. Feng, W.-K. Chu and M.-A. Nicolet, *Phys. Rev.* **B10**, 3781 (1974).
[17] N. Bohr, *Phil. Mag.* **30**, 581 (1915).
[18] W.-K. Chu in *Ion Beam Handbook for Materials Analysis* J.W. Mayer and E. Rimini (eds.) Academic Press, New York (1977).
[19] J.F. van der Veen, R.G. Smeenk, R.M. Tromp and F.W. Saris, *Surf. Sci.* **79**, 212 (1986).
[20] R. Behrisch (ed.) *Sputtering by Particle Bombardment I* Springer, Berlin (1982).
[21] E. Taglauer, *Nucl. Fusion Special Issue* p. 43 (1984).
[22] L.C. Feldman in *Methods of Surface Characterization*, Vol. 2, A.W. Czanderna and D.M. Hercules (eds.) Plenum Press p. 311 (1991).
[23] Ch. Linsmeier, H. Knözinger and E. Taglauer, *Surf. Sci.* **275**, 101 (1992).
[24] P. Berberich, W. Dietsche, H. Kinder, J. Tate, Ch. Thomsen, and B. Scherzer, *Proc. Int. Conf. High T Supercond. Mater.* Interlaken, CH, (1988).

[25] L.C. Feldman, J.W. Mayer, S.T. Picraux, *Materials Analysis by Ion Channeling* Academic Press, New York (1982).

[26] D.S. Gemmel, *Rev. Mod. Phys.* **46**, 129 (1974).

[27] J.F. van der Veen, *Surf. Sci. Rep.* 5, **199** (1985).

[28] W.C. Turkenburg, W. Soszka, F.W. Saris, H.H. Kersten and B.G. Colenbrander, *Nucl. Instr. Meth.* **132**, 587 (1976).

[29] L.C. Feldman, *Nucl. Instr. Meth.* **191**, 211 (1981).

[30] J.A. Davies, D.P. Jackson, J.B. Mitchell, P.R. Norton and R.L. Tapping, *Phys. Lett.* **54A**, 239 (1975).

[31] T.E. Jackman, K. Griffiths, J.A. Davies and P.R. Norton, *J. Chem. Phys.* **79**, 3529 (1983).

[32] L.M. Howe, M.L. Swanson and J.A. Davies, *Methods of Experimental Physics* **21**, 275 (1983).

[33] F. Besenbacher, I. Stensgard and K. Mortensen, *Surf. Sci.* **191**, 288 (1987).

[34] J.W.M. Frenken, P.M.J. Marée and J.F. van der Veen, *Phys. Rev.* **B34**, 7506 (1986).

[35] B.L. Doyle, D.S. Walsh and S.R. Lee, *Nucl. Instr. Meth. Phys. Res.* **B54**, 244 (1991).

[36] S. Raman in *Applied Atomic Collision Physics*, Vol. 4, S. Datz, (ed.) Academic Press, Orlando p. 407 (1983).
S.A.E. Johansson and J.L. Campbell, *PIXE-A Novel Technique For Elemental Analysis*, John Wiley, New York, (1988).

[37] E. Taglauer and G. Staudenmaier, *J. Vac. Sci. Technol.* **A5**, 1352 (1987).

[38] R. West, this Volume.

[39] M.J. Kelley, D.G. Swartzfager and V.S. Sundaram, *J. Vac. Sci. Technol.* **16**, 664 (1979).

[40] P. Novacek, E. Taglauer and P. Varga and *Fresenius J. Anal. Chem.* **341**, 136 (1991).

[41] D.J. Godfrey and D.P. Woodruff, *Surf. Sci.*, **105**, 438 (1981).

[42] G. Engelmann, E. Taglauer and D.P. Jackson, *Nucl. Instr. Meth. Phys. Res.* **B13**, 240 (1986).

[43] W. Englert, E. Taglauer, W. Heiland and D.P. Jackson, *Physica Scripta* **T6**, 38 (1983).

[44] H.D. Hagstrum, in *Inelastic Ion-Surface Collisions*, N.H. Tolk, J.C. Tully, W. Heiland and C.W. White (eds.) Academic Press, New York p. 1 (1976).

[45] D.J. O'Connor, Y.G. Shen, J.M. Wilson and R.J. MacDonald, *Surf. Sci.* **197**, 277 (1988).

[46] M. Beckschulte and E. Taglauer, *Nucl. Instr. Meth. Phys. Res.* (1993).

[47] M. Aono and R. Souda, *Nucl. Instr. Meth. Phys. Res.* **B27**, 55 (1987).

[48] A.W. Czanderna and J.R. Pitts, *Surf Sci.* **175**, L737 (1986).

[49] W. Heiland and E. Taglauer, *Nucl. Instr. Meth.* **132**, 535 (1976).

[50] J. Los and J.J.C. Geerlings, *Phys. Rep.* **190**, 133 (1990).

[51] R. Brako and D.M. Newns, *Rep. Prog. Phys.* **52**, 655 (1989).

[52] R.L. Ericksen and D.P. Smith, *Phys. Rev. Lett.* **34**, 297 (1975).

[53] See e.g., A. Zangwill, *Physics at Surfaces*, Cambridge University Press, New York (1988).

[54] H.J. Kreuzer and Z.W. Gortel, *Physisorption Kinetics*, Springer, Berlin (1986).

[55] H.H. Andersen and H.L. Bay in *ref 20*, p. 145.

[56] E. Taglauer, *Appl. Phys.* **A51**, 238 (1990).

[57] E. Taglauer, W. Melchior, F. Schuster and W. Heiland, *J Phys* **E8**, 768 (1975).
F. Huussen, J.W.M. Frenken and J.F. van der Veen, *Vacuum* **36**, 259 (1986).
H. Dürr, Th. Fauster and R. Schneider, *J. Vac. Sci. Technol.* **A8**, 145 (1990).

[58] E. Taglauer, *Appl. Phys.* **A38**, 161 (1985).
[59] P.A.J. Ackermans, P.F.H.M. van der Meulen, H. Ottevanger, F.E. van Straten and H.H. Brongersma, *Nucl. Intr. Meth. Phys. Res.* **B35**, 541 (1988).
[60] H.A. Engelhardt, W. Bäck, D. Menzel and H. Liebl, *Rev. Sci. Instr.* **52**, 835 (1981).
[61] Y.-S. Chen, G.L. Miller, D.A.H. Robinson, G.H. Wheatley and T.M. Buck, *Surf. Sci.* **62**, 133 (1967).
[62] T.M. Buck, G.H. Wheatley, G.L. Miller, D.A.H. Robinson and Y.-S. Chen, *Nucl. Instr. Meth.* **149**, 591 (1978).
[63] S.B. Luitjens, A.J. Algra, E.P.Th. Suurmeijer and A.L. Boers, *Appl. Phys.* **A21**, 205 (1980).
[64] J.W. Rabalais, J.A. Schultz and R. Kumar, *Nucl. Instr. Meth. Phys. Res.* **218**, 719 (1983).
[65] D. Rathmann, N. Exeler and B. Willerding, *J. Phys.* **E18**, 17 (1985).
[66] H. Niehus and G. Comsa, *Nucl. Instr. Meth. Phys. Res.* **B15**, 122 (1986).
[67] R. Aratari, *Nucl. Instr. Meth. Phys. Res.*, **B34**, 493 (1988).
[68] H. Verbeek, W. Eckstein and F.E.P. Matschke, *J. Phys.* **E10**, 944 (1977).
[69] E. Taglauer and W. Heiland, *Appl. Phys. Lett.* **24**, 437 (1974).
[70] M. Beckschulte, D. Mehl, and E. Taglauer, *Vacuum* **41**, 67 (1990)
[71] E. Taglauer, W. Heiland, and J. Onsgaard, *Nucl. Instr. Meth.* **168**, 571 (1980)
[72] A. Koma, IPPJ-AM, 22 (1982)
[73] J. Knözinger, in: Fundamental Aspects of Heterogeneous Catalysis Studied by Particle Beams (H.H. Brongersma and R.A. van Santen, eds.) Plenum Press, New York (1991) p. 7
[74] H.H. Brongersma and G.C. van Leerdam, ibid. p. 283
[75] E. Taglauer, ibid. p. 301
[76] K. Josek, Ch. Linsmeier, H. Knözinger, and E. Tanglauer, *Nucl. Instr. Meth. Phys. Res.* **B64**, 596 (1992)
[77] R. Margraf, H. Knözinger, and E. Tanglauer, *Surf. Sci.* 211/212, 1083 (1989)
[78] R. Kelly. *Surf. Interf. Anal.* **7**, 1 (1985)
[79] J. du Plessis, Surface Segregation, in: Solid State Phenomena, Vol. 11, *Sci.-Tech. Pub.*, Vaduz (1990)
[80] T.M. Buck, in: Chemistry and Physics of Solid Surfaces IV (R. Vanselow and R. Howe, eds.) Springer, Berlin (1982) p. 435
[81] N.Q. Lam and H. Wiedersich, *Nucl. Instr Meth. Phys. Res.* **B18**, 471 (1987)
[82] R. Shimizu, *Nucl. Instr Meth. Phys. Res.* **B18**, 486 (1987)
[83] J. du Plessis, G.N. van Wyk, and E. Taglauer, *Surf. Sci.* **220**, 381 (1989)
[84] D. Voges, E. Taglauer, H. Dosch, and J. Peisl, *Surf. Sci.* 269/270, 1142 (1992)
[85] J. du Plessis and E. Taglauer, *Nucl. Instr. Meth. Phys Res.* (1993)
[86] M.A. van Hove, W.H. Weinberg, and C.-M. Chen, Low-Energy Electron Diffraction, Springer Verlag, Berlin (1986)
       W. Flavel, this Volume
[87] M. Aono, C. Oshima, S. Zaima, S. Otani, and Y. Ishizawa, *Japn. J. Appl. Phys.*, **20**, L829 (1981)
[88] Th. Fauster, *Vacuum* **38**, 129 (1988)
[89] I. Kamiya, M. Katayama, E. Nomura, and M. Aono, *Surf. Sci.* **242**, 404 (1991)
[90] R.S. Daley, J.H. Huang, and R.S. Williams, *Surf. Sci.* **215**, 281 (1989)
[91] R. Spitzl, H. Niehus, and G. Comsa, *Rev. Sci. Instr.*, **61**, 1275 (1990)
[92] M.T. Robinson and I.M. Torrens, *Phys. Rev.* **B9**, 5008 (1974)
[93] J. Schulz, E. Taglauer, P. Feulner, and D. Menzel, *Nucl. Instr. Meth. Phys. Res.* **B64**, 588 (1992)
[94] H. Hemme and W. Heiland, *Nucl. Instr. Meth. Phys. Res.* **B9**, 41 (1985)

[95]   H. Neihus and G. Comsa, *Surf. Sci.* **151**, 1171 (1985)
[96]   J.A. van den Berg, L.K. Verheij, and D.G. Armour, *Surf. Sci.* **91**, 218 (1980)
[97]   L. Eierdal, F. Besenbacher, E. Laegsgaard and I. Stensgard, *Ultramicroscopy*, **42**-**44** (1992).
[98]   D.J. Surman, J.A. van den Berg and J.C. Vickerman, *Surf. Interf. Anal.* **4**, 160 (1982).
[99]   H. Bu, M. Shi, K. Boyd and J.W. Rabalais, *J. Chem. Phys.* **95**, 2882 (1991).
[100] G. Binnig, H. Rohrer, Ch. Gerber and E. Weibel, *Phys. Rev. Lett.* **49**, 57 (1982).

QUESTIONS

1. Why can thermal vibrations be neglected in the treatment of collision kinematics (Section 2.2)? Compare relevant timescales. Estimate the energy broadening resulting from momentum transfer due to the thermal motion of target atoms.

2. Which energy and angular resolutions are required to separate the stainless steel components (Fe, Cr, Ni)? At which experimental parameters (projectile mass, scattering angle) can these requirements be met in ISS or RBS?

3. Estimate the Y:Ba:Cu concentration ratios from the spectrum of the high T_c film displayed in Figure 6.13. How close is it to the nominal 1:2:3 ratios?

4. What are the accessible angular and energy regions for the detection of hydrogen isotopes by backscattering or by direct recoil detection? Which projectile masses and energies should be favoured (assuming Rutherford cross-sections)?

5. Estimate the interatomic distances in a Ni crystal from the ICISS distribution shown in Figure 6.28, using the appropriate Na^+ shadow cone radius.

6. Estimate the amount of damage to a Ni target surface which will occur when taking spectra with 1 keV He^+ ISS and 1 MeV He^+ RBS where the sputtering yields are 0.15 and 0.01 atoms/ion, respectively. Compare these with the damage resulting from the ISS analysis of a CO adlayer with a desorption cross-section of 10^{-15} cm^{-2}.

CHAPTER 7

Vibrational Spectroscopy from Surfaces

MARTYN E. PEMBLE

University of Salford, UK

1 INTRODUCTION

There can be few techniques as versatile to the chemist as vibrational spectroscopy. This also holds true for the surface chemist. This chapter reviews some of the techniques employed by surface chemists to measure vibrational spectra from molecules adsorbed at a variety of surfaces and gives specific examples of their use. The techniques may be summarised as involving the interactions of photons or particles with surfaces which result in energy transfer to or from the surface adsorbed species via vibrational excitation or de-excitation. The discrete energies transferred of course correspond to vibrational quanta, and analysis of these energies provides the means of determining the structure of the surface species. The basic theory of vibrational spectroscopy is not covered here, but rather the reader is referred to the numerous undergraduate texts that deal with this material.

Let us begin by asking the question, 'Why use vibrational spectroscopy as a surface probe? To answer this we must consider the usual methodology adopted by the surface chemist, who often notes that there is no one technique that provides all of the information required in order to solve a particular problem in surface chemistry. Many of the techniques available to the surface chemist involve the use of particle beams and hence high-vacuum systems, such that it becomes difficult to examine real surfaces under operating conditions, e.g., pertaining to corrosion, catalysis etc. Yet as will be seen, there is a form of surface vibrational spectroscopy which is capable of dealing with almost any surface one cares to imagine, under almost any conditions!

For the research and development chemist interested in the fundamental chemical mechanisms occurring on surfaces, vibrational spectroscopy is one of the most important methods of studying surface chemical reactions. However, we must introduce a note of caution at this point in that, although vibrational spectroscopy may allow us to identify intermediate species forming on surfaces, they

Surface Analysis — The Principal Techniques. Edited by John C. Vickerman
© 1997 John Wiley & Sons Ltd

may not be involved in the rate determining step of the reaction. Such species, although of interest, may thus provide little insight into the nature of true surface processes — they are so called 'spectator species'. 'Despite this' since any mechanistic study requires not only the measurement of reaction rates but also the identification of possible intermediates, any technique which reveals intermediates has a part to play.

This chapter begins with a description of the most widely-used form of surface vibrational spectroscopy, infrared spectroscopy.

2 INFRARED SPECTROSCOPY FROM SURFACES

IR spectroscopy is attractive as a means of studying surface species because of its enormous versatility, being applicable to almost any surface, capable of operating under both high and low pressure conditions and having a relatively low cost compared to, say, a technique which requires high vacuum for operation. Most modern surface IR facilities, whichever sampling mode is used, will utilise a Fourier transform IR spectrometer (FTIR) The essential feature of an FTIR spectrometer as compared to a dispersive instrument is that all of the light from the source falls onto the detector at any instant. Since this inherently leads to increased signal levels this automatically improves the signal-to-noise ratio at any point in the spectrum. Wavelength or frequency identification is not achieved using monchromators but rather through a careful frequency analysis (Fourier analysis) of the periodic signal at the detector produced by the Michelson interferometer or similar device. This device induces a periodically varying path length difference between two, usually equally intense, beams from the source produced by a simple optical device designed to split the beam intensity — a so called beam splitter. The two beams, which contain all wavelengths emanating from the source, are recombined and detected. The intensity measured depends on the overall effects of the phase difference for each component wavelength. The phase difference of course varies with each component wavelength. The mathematical operation of converting or transforming a signal which varies with path length to a spectrum in which intensity varies with wavelength is known as Fourier transformation.

The advantages that FTIR methods brings are usually described in terms of the Jaquinot or multiplex advantage which arises since all channels are samples at once, the Fellgett or throughput advantage, which arises because — as compared to a dispersive instrument — the signal level at the detector is always higher, and the Connes advantage, which represents enhanced photometric accuracy arising from the in-built electronic calibration resulting from a single wavelength interferogram produced by the interaction of an alignment laser, usually a HeNe, with the beam splitter. For further information on the operation of an FTIR instrument the reader is directed to references [1– 3].

There are both routine and complex research applications of surface IR spectroscopy and this section aims to cover briefly, most of these methods.

2.1 TRANSMISSION IR SPECTROSCOPY

As the name suggests, this mode of sampling involves the passage of the IR beam through the sample, which must therefore be at least partially transmissive in the IR region of the spectrum, Figure 7.1.

Figure 7.1(a) depicts the case of an IR transmissive substrate with an adsorbate layer on both its exposed surfaces while Figure 7.1(b) depicts the case for a species adsorbed at the surface of a highly particulate medium. In the case of sampling mode (b), samples are commonly prepared by pressing the particulate sample under high pressure, often with some additional support material such as KBr, such that it forms a self-supporting disc. This basic arrangement employing transmission techniques was used in the early pioneering experiments of Eischens *et al.*, who demonstrated for the first time, the utility of IR methods for the study of adsorbed species [4, 5].

Using transmission IR spectroscopy as an example, it is possible to see how IR measurements may be related to the amount of material present and its refractive index: If the incident IR beam has intensity I_0 and the transmitted beam has intensity I_t, then the relationship between these two quantities is,

$$I_0/I_t = \exp[-kcl] \tag{1}$$

where the dimensionless quotient I_0/I_t is known as the *transmittance*, or, when multiplied by 100, the *percentage transmittance* and k is the absorption coefficient, which corresponds to the imaginary part of the refractive index of the

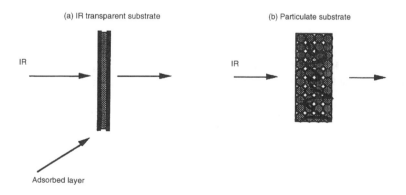

Figure 7.1. (a) An IR transmissive substrate with an adsorbate layer on both its exposed surfaces. (b) depicts the case for a species adsorbed at the surface of a highly particulate medium

medium n, such that

$$n = n + \mathrm{i}k \tag{2}$$

The use of an imaginary function is a convenient way of representing an absorption process since the intensity of the transmitted radiation is proportional to the square of the refractive index, which therefore results in a 'negative reflectance' (absorption) for the imaginary term. Although equation (1) is a common way of depicting the 'strength' of an IR absorption, it also shows that transmittance is an exponential function of concentration. A linear form of this expression may be obtained by taking logarithms to base 10 and inverting such that we obtain,

$$\log_{10}(I_\mathrm{t}/I_0) = \varepsilon cl \tag{3}$$

where $\varepsilon = k/\ln 10$ and is a constant known as the absorption cross-section. The term $\log_{10}(I_\mathrm{t}/I_0)$ is known as the *absorbance* and is a linear function of concentration. Two things emerge from this simple treatment which are worthy of note. Firstly, the relationship between the IR absorption cross-section and the refractive index of the material has been established. Secondly, in order to have a meaningful measurement, at least two intensity measurements are required which in practice means that one must record both a sample and reference spectrum.

Let us now examine a typical transmission IR spectrum of interest to the surface chemist: In order to record such a spectrum, the substrate may be any material that allows at least partial transmission of the IR beam. For example, this mode of IR spectroscopy is often used to study reactions that occur upon so-called 'supported' metal catalysts. The term supported, refers to the fact that the metal particles are chemically impregnated onto a support material, usually a high specific surface area oxide, which serves to increase the active working area of the catalysts and also prevents high temperature agglomeration of metal particles or 'sintering'. Such materials are usually black in colour and not at all the obvious choice of material for a transmission IR experiment. However, when pressed into a thin flat disc, it is found that up to ten percent of the IR radiation falling onto the disc may be transmitted and this is more than sufficient to obtain an IR spectrum. The transmission process for these materials involves partial absorption by the supporting oxide but has also been shown by the group of Sheppard [6, 7] to involve a series of complex reflections from the metallic surfaces and for this reason the spectra obtained are often subject to the so called 'surface selection rule' which applies for species adsorbed on flat substrates where the spectrum is measured in reflection. This latter method, known as reflection–absorption IR spectroscopy or RAIRS, is described in detail in a later section. An example of a comparison between the application of RAIRS and transmission IR methods is shown in Figure 7.2.

This figure demonstrates the similarity of the data obtained by the two methods of transmission and reflection. Note that the transmission spectra do not reveal any information in the spectral region below ca. 1300 cm^{-1}. This is due to the

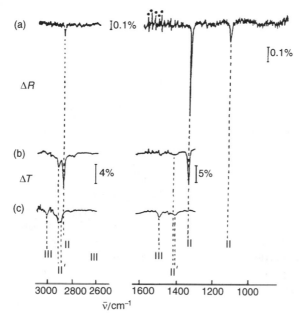

Figure 7.2. Comparison of RAIR data recorded from a flat Pt (111) substrate and data recorded in transmission from a Pt/SiO$_2$ catalyst sample, for the adsorption of ethene [7]

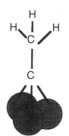

Figure 7.3. Schematic representation of the ethylidyne species formed via reaction of ethene over Pt surfaces

so called oxide 'blackout' below 1300 cm^{-1} which obscures this region of the spectrum. These data also reveal the wealth of information that may be obtained even from a 'simple' system such as this. Comparison of these data with data from other techniques such as electron energy loss spectroscopy and from IR spectra of inorganic cluster compounds enabled Sheppard and co-workers to ascertain that ethene undergoes a rearrangement on Pt surfaces near 300 K to produce the ethylidyne species, Figure 7.3.

Not too surprisingly from purely geometric considerations, this species is formed most readily on the (111) surfaces of face-centred cubic metals since these surfaces necessarily have the three-fold sites depicted above.

2.2 PHOTOACOUSTIC SPECTROSCOPY

Some particulate materials are unsuitable for transmission IR studies because they either absorb too much radiation, or they divert the path of the radiation through scattering. It is still possible to record IR spectra from such materials using alternative sampling modes. For highly absorbing materials, the technique *photoacoustic IR* spectroscopy has been developed. This method relies upon the fact that when IR radiation is incident upon a highly absorbing material, the material effectively heats up. If the material is in contact with some inert gas, then the local heating caused by absorption of a specific wavelength of IR radiation gives rise to a minute thermal shock wave which propagates out into the gas. This shock wave is directly analogous to a sound wave and, as such, may be detected using a sensitive diaphragm, i.e. the sampling cell forms a microphone unit. This description makes the photoacoustic detector sound somewhat esoteric but this is not the case. Photoacoustic units are commonly available as off-the-shelf items. A schematic diagram of a photoacoustic detector is shown in Figure 7.4.

The particular arrangement shown in Figure 7.4 corresponds to absorbance sampling, although it is possible to obtain photoacoustic sampling accessories that will also work in diffuse reflectance (see Section 2.3.2) or transmittance modes. In absorbance sampling, the photoacoustic signal arises from the sample heating produced by absorption of the IR radiation. Selective absorption such as this is obscured when using black highly absorbing samples and so transmittance sampling is employed, in which the radiation passing through the sample is absorbed into a gas which then produces the photoacoustic signal at the microphone. It is thus possible to obtain both absorbance and transmittance spectra using this sampling mode. In absorbance mode, the photoacoustic signal is, as expected, directly proportional to the amount of substance present. It is widely believed that photoacoustic detection only provides an advantage over conventional methods when dealing with black samples. This is not the case,

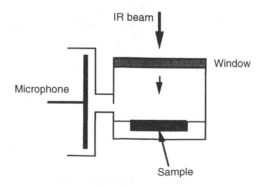

Figure 7.4. A schematic representation of an absorbance photoacoustic sampling system

although it is true that for such materials it is hard to match the quality of a photoacoustic spectrum using another sampling technique.

2.3 REFLECTANCE METHODS

Many samples are effectively opaque to the IR radiation and thus cannot be studied in transmission mode. Reflectance methods are particularly useful here and have found wide application in both routine and research oriented surface analysis. In order to be able to optimise the efficiency of a reflectance experiment it is necessary to understand the equations governing the reflectance process.

The processes of reflection, refraction and absorption are all related. The purely directional changes that occur when electromagnetic radiation encounters an interface may be explained by considering the refractive indices of the interfacial system. Figure 7.5 depicts the processes of reflection and refraction at an interface.

From this figure we can establish a relationship between the angles of reflection and refraction which was first established by Snell and then later developed mathematically by Descart,

$$n_1/n_2 = \frac{\sin \theta_1}{\sin \theta_2} \tag{4}$$

where n_1 and n_2 are the refractive indices of the two media forming the interface. This simple ray diagram also defines the so called plane of incidence, which is the plane containing the incident and reflected rays and the surface normal. To determine the intensity of the reflected and refracted rays, the treatment of Fresnel is employed. Fresnel, in the late eighteenth century, considered the interaction of the electromagnetic wave with the surface between the two media as involving two extreme orientations or *polarisations* of the electric vector — light polarised in the plane of incidence, p-polarisation and light polarised perpendicular to the

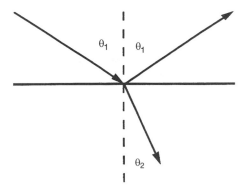

Figure 7.5. The relationship between reflection and refraction

plane of incidence, s-polarisation. For a more detailed description of these polarisation states see Figure 7.9. Fresnel determined that the fraction of p-polarised
monochromatic light reflected from the interface, R_p, is given by,

$$R_p = r_p r_p{}^*$$ (5)

$$\text{where } r_p = \frac{n_2 \cos \theta_1 - n_1 \cos \theta_2}{n_2 \cos \theta_1 + n_1 \cos \theta_2}$$ (6)

and $r_p{}^*$ is the complex conjugate of r_p. The use of the complex conjugate may
be attributed to the nature of the refractive index of the media, which as was
shown in section consists of a real and imaginary part, the imaginary part being
associated with absorption in the medium. Similarly for s-polarised light,

$$R_s = r_s r_s{}^*$$ (7)

$$\text{where } r_s = \frac{n_1 \cos \theta_1 - n_2 \cos \theta_2}{n_1 \cos \theta_1 + n_2 \cos \theta_2}$$ (8)

and again we note that the true form of the refractive index contains a complex
part, such that for a vacuum solid interface, where $n_1 = 1.0$ and $n_2 = n + ik$,
the reflectivity equations reduce to

$$R_p = \frac{\cos^2 \theta_2 - 2n \cos \theta_1 \cos \theta_2 + (n^2 + k^2) \cos^2 \theta_1}{\cos^2 \theta_2 + 2n \cos \theta_1 \cos \theta_2 + (n^2 + k^2) \cos^2 \theta_1}$$ (9)

and

$$R_s = \frac{\cos^2 \theta_1 - 2n \cos \theta_1 \cos \theta_2 + (n^2 + k^2) \cos^2 \theta_2}{\cos^2 \theta_1 + 2n \cos \theta_1 \cos \theta_2 + (n^2 + k^2) \cos^2 \theta_2}$$ (10)

Equations such as (9) and (10) allow us to predict the variation in reflectivity as
a function of angle of incidence for a particular system and thus determine the
optimum angle of incidence to use in obtaining a specular reflection spectrum.
Two extreme cases can be visualised:

(i) **Near normal incidence**
    This technique is commonly employed for the study of relatively thick
    films of material adsorbed onto highly reflecting substrates such as metallic
    surfaces. For films of sufficient thickness, this method is effectively a
    double-pass transmission experiment.
(ii) **Grazing incidence**
    This technique is of most use when studying very thin films adsorbed at
    conducting surfaces. Known as reflection-absorption infrared spectroscopy
    (RAIRS), this technique is discussed in more detail in Section 2.4.

2.3.1 Attenuated Total (Internal) Reflection (ATR)

Here the substrate to be analysed in pressed into intimate optical contact with
a prism which is transparent over the range of IR wavelengths to be studied.

Incident IR beam

Substrate material

IR transparent prism

Figure 7.6. The path of the sampling IR beam as it passes through the waveguide prism in contact with an opaque sample in the ATR experiment

The IR radiation enters the prism and is incident on the surfaces of the prism at angles greater than the critical angle. If the geometry of the experiment is arranged correctly, then multiple internal reflection occurs, Figure 7.6.

From Figure 7.6 it may be seen that, by varying the angle of incidence, it is possible to vary the number of internal reflections within the ATR element. In practice, up to 100 internal reflections may be employed. The substrate surface is pressed against the ATR prism and at each reflection, the electric vector of the IR radiation samples the surface in contact with the prism via the so called 'evanescant' wave, which extends beyond the boundary of the prism. To obtain internal reflectance, the angle of incidence must exceed the so-called 'critical' angle. This angle is a function of the real parts of the refractive indices of both the sample and the ATR prism:

$$\theta_c = \sin^{-1}(n_2/n_1) \tag{11}$$

where n_2 is the refractive index of the sample and n_1 is the refractive index of the prism. The evanescent wave decays into the sample exponentially with distance from the surface of the prism over a distance on the order of microns. The depth of penetration of the evanescent wave d_{ev} is determined by

$$d_{ev} = \lambda/\{2\pi n_1[\sin^2\theta - (n_2/n_1)^2]^{0.5}\} \tag{12}$$

where λ is the wavelength of the IR radiation. Comparisons can be drawn between ATR spectra and transmission spectra by estimating the path length through the sample in the ATR experiment. This 'effective pathlength' may be taken as simply the penetration depth $\times$ the number of reflections.

2.3.2 Diffuse Reflectance

For highly scattering particulate samples such as white powders, photoacoustic detection is not efficient. An alternative method is available which collects the scattered light from the substrate surface and directs it onto the IR detector. This method of sampling is known as diffuse reflectance, Figure 7.7.

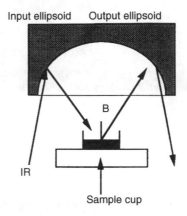

Input ellipsoid   Output ellipsoid

B

IR

Sample cup

Figure 7.7. A schematic representation of the arrangement for diffuse reflectance sampling, B = beam blocker

In this mode of analysis, only that radiation which undergoes diffuse scattering is considered to have penetrated into the surface of the particulate material, i.e., true specular scattering does not interact with surface species and is not absorbed. In order to maximise signal-to-noise performance it is necessary to prevent unwanted light from reaching the detector, this is partially achieved using a specular beam blocker, labelled B in the diagram. The form of the spectrum obtained in this manner does not directly coincide with that of a spectrum obtained in the conventional manner, since the spectral profile is dominated by the relationship between wavelength and scattering efficiency. This feature is eliminated by transforming the spectrum using a mathematical process known as a Kubelka–Munk transformation which compensates for the wavelength: scattering power relationship. Specific instruments have this routine built in to the permanent memory of the computational facility in addition to the normal Fourier routines. This is obviously a highly convenient method of obtaining a surface spectrum from a particulate material that involves almost no sample preparation! It is particularly useful for the study of solid catalysts, where almost any material may be studied and via the use of environmental chambers, gas pressures of up to 100 atmospheres may be introduced over the catalyst surface. For such purposes it is possible to obtain high pressure and temperature environmental chambers designed specifically for diffuse reflectance or transmission work. The resulting spectra, recorded under 'real' conditions, may reveal the presence of species not previously postulated to be intermediates in the catalytic process. An example of this is shown in Figure 7.8 in which Holmes [8] and co-workers in Edinburgh have obtained spectra from a supported Ni catalyst under dynamic conditions.

This figure depicts the time-dependent evolution of an ethylidyne species similar to that detected on supported Pt catalysts (see Figure 7.2) which was previously not believed to be formed on Ni surfaces. It can be seen that this

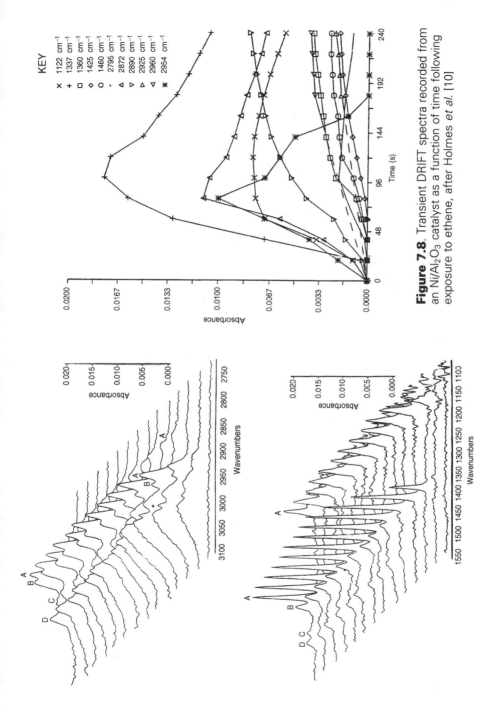

Figure 7.8. Transient DRIFT spectra recorded from an Ni/Al$_2$O$_3$ catalyst as a function of time following exposure to ethene, after Holmes et al. [10]

species is a true transient species, present for a limited time only during the reaction. The ability to record batches of spectra and to display them in this manner is also a good indication of the usefulness of the computer associated with a modern FTIR instrument.

2.4 REFLECTION–ABSORPTION IR SPECTROSCOPY (RAIRS)

This technique has proved to be a particularly powerful research tool for the study of adsorbed layers on metal surfaces. For adsorbates on metallic or any conducting film it was Greenler [9] who first demonstrated that the absorption of IR radiation by the adsorbate overlayer is enhanced at high angles of incidence (near grazing) and involves only one polarisation of the incident IR beam, Figure 7.9.

Figure 7.9 illustrates the incident and reflected electric vectors of the so called s and p components of the radiation where p refers to *parallel* polarised radiation and s to *perpendicular* polarised radiation with respect to the plane of incidence. Greenler [9] highlighted the fact that at the point of contact with the surface, the p-polarised radiation has a net combined amplitude that is almost twice that of the incident radiation via the vector summation of E_p and $E_{p'}$. However, for the s-polarised radiation, the incident and emitted electric vectors E_s and $E_{s'}$ undergo a 180 degree phase-shift with respect to each other and so the net amplitude of the IR radiation parallel to the surface plane is zero, Figure 7.10.

Thus only radiation having a p-component may possess a finite interaction with the surface and hence the only active vibrations that may be observed in RAIRS must have a component of the dynamic dipole polarised in the direction normal to the surface plane. This is a statement of the so called 'surface selection rule' for reflection IR spectroscopy. The surface selection rule is discussed in greater detail in Section 4 where the group theory of analysis of surface vibrations is presented.

Obviously the same consideration would apply for all angles of incidence. However, it is possible to show via the use of Maxwell's equations, that the resultant amplitude of the p-polarised component of electromagnetic radiation that

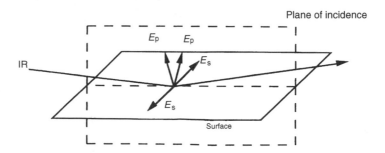

Figure 7.9. Schematic representation of the plane of incidence and the definition of s and p polarised radiation in the RAIRS experiment

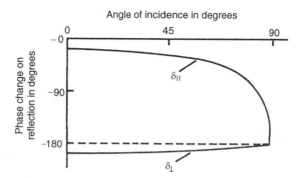

Figure 7.10. The phase shift for light reflected from a metal surface as calculated for light polarised both parallel to (p) and perpendicular to (s) the plane of incidence, after Greenler [9]

is oriented perpendicular to the surface plane reaches a maximum near grazing incidence, while the amplitude of the p-component oriented parallel to the surface plane is low and relatively structureless as a function of angle of incidence. Figure 7.10 shows that, as expected, the net amplitude of the s component upon reflection is zero. In order to determine the variation in *band intensity* expected over a range of angles of incidence, it is necessary to note that the spectral intensity will vary as a function of the *square* of the amplitude of the electric vector and also the variation in sampled surface area. It may be shown by trigonometry that the surface area sampled for a given incident beam diameter varies as a function of $1/\cos\theta$, i.e. $\sec\theta$. Thus the band intensity will vary as a function of $E^2 \sec\theta$. This function was first calculated for a series of model overlayers with varying refractive indices by Greenler [9]. A more recent summary of the behaviour expected has been produced by Chesters [10], Figure 7.11.

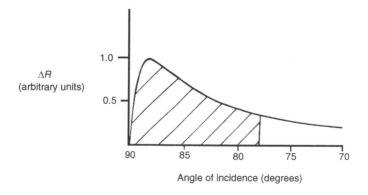

Figure 7.11. Schematic representation of the variation in band intensity with angle of incidence, after Chesters [10]

From this figure it may be seen that the anticipated variation in band intensity reaches a maximum near grazing incidence. We are now in a position to state the optimum conditions for the measurement of RAIR spectra from species adsorbed at metallic surfaces:

'Maximum spectral intensity will be observed when using p-polarised radiation oriented perpendicularly to the surface plane at grazing incidence'

2.4.1 The RAIRS Experiment

At the beginning of the section on IR methods, it was stated that most modern instrumentation utilises FTIR techniques. This is also true for the RAIRS experiment, which is therefore termed FT-RAIRS by some, although some of the highest quality RAIRS data have been obtained using purpose-built dispersive instruments designed to maximise sensitivity in one small spectral region such as that containing the C−O stretching vibration for molecularly adsorbed CO on certain metal surfaces. The particular methods employed to enhance signal-to-noise levels here are discussed in Section 2.4.2. However, a typical layout for the FT-RAIRS experiment is shown in Figure 7.12.

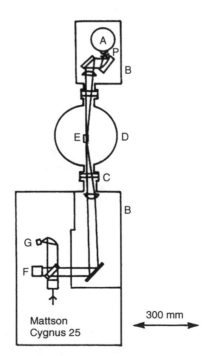

Figure 7.12. A schematic representation of the FT-RAIRS experiment as described by Chesters [10]

The resolution of IR spectroscopy is such that, for the analysis of surface species, the bandwidths are determined entirely by the heterogeneity of the surface and the nature of the surface-molecule interactions rather than any experimental artifacts (unlike electron energy loss spectroscopy, see Section 3). A typical RAIRS spectrum from a strong IR absorber, CO, adsorbed at a copper single crystal surface is shown in Figure 7.13 along with analogous data obtained using the technique of electron energy loss spectroscopy (see Section 3).

The RAIR data presented in Figure 7.13 reveal that at least two types of adsorbed CO are present on the saturated Cu (111) surface as indicated by two discrete bands in the C−O stretching region. It would be tempting to apply a semi-quantitative analysis to the band intensities depicted in Figure 7.13 and suggest that the more intense band corresponds to the majority species. However, it may be shown that for ordered overlayers such as this, where the dipole moments of the adsorbed species are also arranged in an ordered array, that the phenomenon known as dipole coupling may give rise to variations in band intensity such that intensities alone are not a reliable guide to quantity [11, 12]. Although the theory of dipole coupling is beyond the scope of this text, the effects of dipole coupling may be isolated via the use of dilute mixtures of isotopically substituted molecules which, as a result of the frequency shift arising from the variation of masses, do not take part in the collective dipole coupling process. Figure 7.13 also reveals the high signal-to-noise levels achievable using FT-RAIRS. The advent of FTIR techniques has opened up the use of RAIRS beyond the study of molecules which have high IR absorption coefficients, to species such

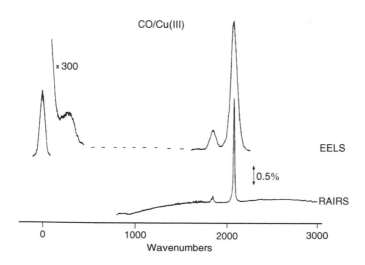

Figure 7.13. RAIRS spectrum of a monlayer of CO adsorbed on a Cu (111) surface at 95 K in comparison with analogous data obtained using electron energy loss spectroscopy, after Chesters *et al*. [10]

Cyclohexane/Cu(III)

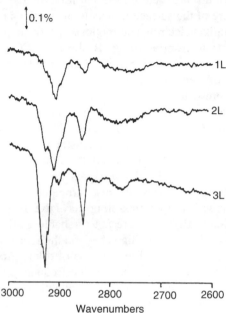

Figure 7.14. RAIRS spectrum of cyclohexane adsorbed on a Cu (111) surface at 95 K as a function of exposure in Langmuirs, after Chesters, Parker and Raval [13]

as simple organics. Although not a routine method by any means, it is now relatively easy to record RAIR spectra from sub-monolayer amounts of such weak absorbers on metallic surfaces. Figure 7.14 illustrates this by presenting data for cyclohexane adsorption at a Cu (111) single crystal surface.

Figure 7.14 demonstrates well that hydrocarbon species may be detected at the submonolayer level when adsorbed on metal surfaces using RAIRS. The particular spectral region described by Figure 7.14 depicts the C−H stretching region. The weak broad features near 2600 cm^{-1} are particularly worthy of note. These features arise due to the close proximity of C−H bonds to surface metal atoms, resulting in the formation of partial hydrogen bonds to the surface and generating so called 'softened' C−H modes. Obviously such states may well play a part in the mechanism of dehydrogenation of alkanes which occur over such surfaces.

2.4.2 Signal Enhancement Techniques in RAIRS

Although RAIRS has become associated with the use of high-vacuum surface science techniques, several studies have shown that it is also possible to obtain

RAIRS data under high overpressures of reactants. This is of course a tremendous advantage over many high-vacuum surface analysis methods. To understand how the gas-phase and surface-phase spectra may be decoupled we must consider the result shown in Figure 7.9. This figure demonstrates that radiation polarised parallel to the plane of incidence will not interact with surface species. This radiation will, however, interact with gas phase species. Thus if the s-polarised light is sensitive to gas phase species and the p-polarised light is sensitive to gas-phase and surface-phase species, then by utilising a polariser and gating the response of the detector accordingly, it is possible to extract a surface spectrum. The use of such 'polarisation modulation' methods has met with some success, particularly if further signal-to-noise improvements have been made by concentrating on a small spectral region and employing some form of wavelength modulation as well.

The use of such methods has enabled not only traditional surface chemists to use RAIRS as a surface probe, but also electrochemists. Bewick and co-workers at Southampton were able to obtain RAIR spectra from electrode surfaces immersed in aqueous solution by using thin layer cells coupled with electrochemical modulation methods [14]. In these experiments, two electrode potentials are selected — one at which the adsorbate is adsorbed at the electrode surface and the other where there is no adsorption. By recording spectra and rapidly switching the electrode potential, the difference spectrum may be obtained. This method is known either as EMIRS (electrochemically modulated infrared spectroscopy)

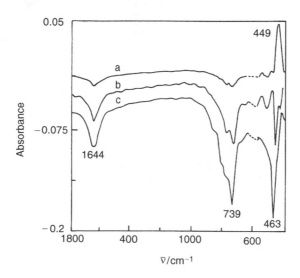

Figure 7.15. Modified SNIFTIRS spectra recorded from the surface of a Ni electrode in aqueous hydroxide media showing via the use of the time-dependent difference spectra the long-term evolution of a layer of α-Ni(OH)$_2$, after Beden and Bewick [15]

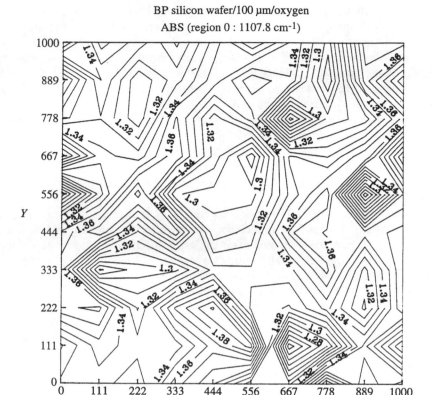

BP silicon wafer/100 μm/oxygen
ABS (region 0 : 1107.8 cm⁻¹)

Figure 7.16. Absorbance map recorded from a Si wafer covered with oxide using a novel microscope attachment, courtesy of Spectra Tech Europe Ltd and BP Research Sunbury. The data refer to the detection of the 1108 cm⁻¹ band of the SiO_2 overlayer present on the wafer

or the more flexible SNIFTIRS (subtraction and normalisation of interferrograms Fourier transform infrared spectroscopy) which is capable of compensating for high background absorptions such as those due to water. These techniques work well but do require that at the 'potential of no adsorption' there are no other processes occurring which may alter the state of the surface. For many electrochemical systems of interest this is not the case. An example of this concerns the formation of a corrosion layer at a nickel electrode surface in aqueous hydroxide media. This system is believed to involve the formation of a variety of hydroxides and oxy-hydroxides over a wide potential range and as such it is not possible to define a 'potential of no adsorption'. In order to be able to investigate this system the group at Southampton developed a method by which sets of interferrograms could be recorded at short time interval separations, while the nickel electrode was also subjected to continuous potential cycling in order to stabilise the surface towards large changes over the timescale of the experiment. An example of the data arising from this 'modified SNIFTIRS' approach is presented in Figure 7.15.

Figure 7.15 shows that, despite the fact that the sample was immersed in aqueous solution, the modified SNIFTIRS approach allows the detection of the a-Ni species via the O—H stretching mode near 1644 cm^{-1}, amongst other bands. Thus there can be no doubt that these are very powerful methods for studying electrode surface processes *in situ*. A later section describes the application of the complementary technique of Raman scattering to the *in-situ* study of electrode surfaces.

2.5 SPATIAL RESOLUTION IN SURFACE IR SPECTROSCOPY

FTIR microscope attachments are commonly available accessories for most instruments. Using these, which operate as beam condensing devices, it is possible to analyse by transmission methods, samples as small as 10–20 microns. While suitable for certain surface chemical applications, this method is better thought of as a bulk analysis method. For specific surface chemical applications, small area reflection accessories may be obtained, which permit the IR 'mapping' of a surface. Figure 7.16 shows a surface absorbance map obtained from an area of 100 μ × 100 μ on a silicon wafer covered with a layer of oxide. The absorbance data obtained corresponds to the 1108 cm^{-1} Si—O mode of the oxide film.

Such sampling methods have obvious application in, say, the semiconductor industry as part of the quality control of large area wafers.

3 ELECTRON ENERGY LOSS SPECTROSCOPY (EELS)

The development of this method as a surface probe by Ibach and coworkers [16] in the early 1970's effectively revolutionised surface science. Up to this time, IR methods such as RAIRS were the only viable means of recording surface IR spectra and due to limitations with equipment and detectors, these early IR experiments were limited to the study of molecules with large dynamic dipole moments such as CO, NO etc. In contrast, it was soon demonstrated that EELS was sensitive to submonolayer amounts of adsorbates possessing relatively weak dynamic dipoles which could not be studied using IR methods at the time [17–19].

The EELS experiment was developed from gas-phase electron scattering experiments which probed electronic states within molecules. For the analysis of surface vibrations, the technique utilises the interaction of very low energy electrons (1–10 eV) with the surface electric fields produced by adsorbate molecules and substrate atoms, Figure 7.17.

Two types of scattering of electrons may be considered, 'elastic' and inelastic (a third mechanism in which the incoming electron is trapped for a finite time within

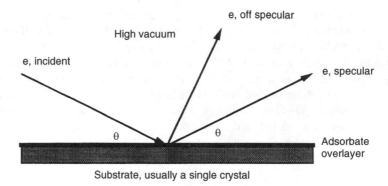

Figure 7.17. A schematic representation of the electron energy loss spectroscopy experiment

the surface forming a so called 'negative ion resonance' is not treated here). Electron scattering is directly analogous to neutron diffraction. This analogy with a diffraction experiment is a useful one, since it illustrates why the best energy resolution in EELS, as measured by the full width at half maximum height (FWHM) of the elastically scattered electron peak, is obtained when using a single crystal substrate. Both elastic and inelastic scattering are broadened in energy terms when the scattering potential of the surface is poorly defined, e.g. a polycrystalline sample. Since the resolution of the experiment is determined largely by the efficiency of the energy selection system and the sample surface, typical values obtained are of the order of 30–40 cm^{-1}, which is very poor in comparison to IR or laser Raman methods but more than adequate to record good quality vibrational spectra from single crystal surfaces. Some modern EEL spectrometers are able to obtain resolution of <8 cm^{-1} from well prepared surfaces.

3.1 INELASTIC OR IMPACT SCATTERING

Inelastic or 'impact' electron scattering can be described in the following way: impact scattered electrons have effectively 'forgotten' their initial angle and plane of incidence and have been scattered by short-range interaction with surface atomic potentials, which are, of course, modulated at vibrational frequencies. If a net surface dipole exists, these electrons may lose energy into the corresponding vibrational mode of the surface-adsorbate complex. The efficiency of the scattering process, which is known as the scattering cross-section, depends upon the net momentum of the electrons *at the point of interaction to*gether with the 'magnitude'(amplitude and direction) of the dipole moment accompanying the surface vibration. Rather than deal with electric field strength as in the case of photons, for electrons we consider electron momentum sometimes known as wave vector for which we use the symbol k. Using these terms, the impact scattering experiment may be described as shown in Figure 7.18.

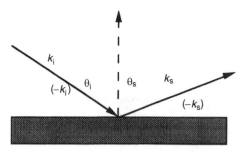

Figure 7.18. A schematic representation of the impact scattering process indicating the change in electron wave-vector which is observed

The wave vector in brackets represents the electron wave under conditions of *time-reversal*, illustrating that both energy and momentum of the electron must be conserved in the complete scattering event, the negative signs arising from the imposition of a fixed coordinate frame. At the point of interaction, the net wave-vector is given by the difference between the incident and scattered wave vectors, $k_s - k_i$. The scalar product of this function with a function describing the time-dependent electric field of the adsorbate vibration is the inelastic scattering cross-section. It may be seen that for the situation where $\theta_i = \theta_s$, i.e. specular scattering, the net wave-vector will be oriented perpendicular to the surface. Since under specular scattering conditions the quantity $k_s - k_i$ has no component parallel to the surface plane, the cross-section for scattering from those vibrational modes oriented parallel to the surface must necessarily be zero. This then represents a statement of the surface selection rule that applies for impact scattering in the specular position. This may also be seen from a more general treatment of vibrational excitation:

Quantum mechanically the probability (cross-section) for excitation of a particular vibration in state Ψ_0 to state Ψ_1 is given by the overlap integral denoted as $< \Psi_0 |V| \Psi_1 >$, where V is the electron–vibration interaction potential. The symmetry of this function may be understood by considering a plan-view projection of Figure 7.19.

This figure reveals that, providing there is little change in the trajectory for the electron upon interaction with the surface, i.e. $k_i \approx k_s$ the shape of the electron potential as seen by the adsorbate is independent of the direction of propagation, i.e it does not matter whether one considers the scattering event as occurring in the forward or reverse (time reversal) directions. The electron-vibration interaction potential V is said to be an even function. Since the ground vibrational state Ψ_0 must by definition be even with respect to the symmetry operations for the point group of the adsorbed complex, the only allowed excited states must be those which are also even with respect to this point group. For a vibration which has its dynamic dipole oriented parallel to the surface as shown in Figure 7.19 it is clear that such a dipole transforms as odd with respect to the symmetry plane that

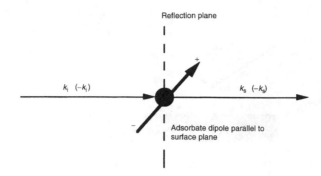

Figure 7.19. Schematic representation of the symmetry factors involved in the consideration of whether a vibrational mode will be observed via the impact scattering mechanism

represents time-reversal inversion and therefore will give rise to a zero scattering cross-section. The only orientation of the adsorbate dynamic dipole that does not violate this rule is where the dipole is parallel to the plane of reflection, but since this direction is orthogonal to both k_s and k_i, there can be no interaction anyway.

Thus it may be stated that, for impact scattering on specular, where the condition $k_i \approx k_s$ holds, a vibration which transforms as a vector parallel to the surface plane will not be observed in the energy loss spectrum. Obviously this rule may only be strictly observed where $k_i = k_s$ but in practice it is found that the cross-section for excitation of parallel modes fall rapidly as one approaches the true specular position.

To determine whether a vibrational mode, having a component of its dynamic dipole parallel to the surface plane will give rise to an impact scattered loss feature, the orientation of the dynamic dipole with respect to the plane of incidence of the electron beam must be considered, Figure 7.20.

Figure 7.20 depicts a particular orientation of an adsorbate dipole towards symmetry elements determined by the plane of incidence of the electron beam, which is usually chosen, in the case of single crystal substrates, to be coincident with a direction of symmetry of the surface. The symmetry elements in question are a two-fold axis of rotation and planes of reflection parallel to and perpendicular to the plane of incidence. The following rules may now be stated:

1. If a vibration transforms as antisymmetric with respect to the two-fold axis oriented perpendicularly to the surface plane, then impact scattering is only observed in the *off-specular* directions.

2. If a vibration transforms as antisymmetric with respect to a plane of reflection oriented perpendicular to both the surface plane and the plane of incidence then impact scattering is only observed in the *off-specular directions confined to the plane of incidence*.

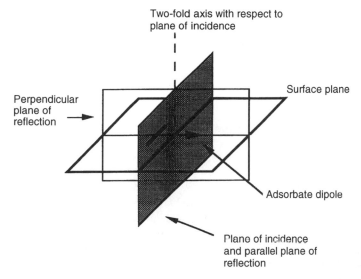

Two-fold axis with respect to
plane of incidence

Surface plane

Perpendicular
plane of
reflection

Adsorbate dipole

Plane of incidence
and parallel plane of
reflection

Figure 7.20. Construction of the symmetry elements used in the determination of whether a particular experimental geometry will permit vibrational modes oriented parallel to the surface plane to be observed via the impact scattering mechanism

3. If a vibration transforms as antisymmetric with respect to a plane of reflection oriented perpendicular to the surface plane but parallel to the plane of incidence then impact scattering is *disallowed* both on and off-specular.

Thus unlike in RAIRS, it is possible to detect vibrational modes that produce dynamic dipoles even for the case when those dipoles are oriented parallel to the surface.

3.2 ELASTIC OR DIPOLE SCATTERING

Elastic or 'dipole' scattering occurs when the change in momentum vector k of the electron upon reflection is minimal, which in practice means that small energy losses may occur such that $k_S \approx k_i$ (equal amplitudes) but the net *direction* of the momentum vector remains unchanged, i.e. $k_S \approx k_i$. The scattered electrons are therefore grouped in a small angular lobe of perhaps 2–3 degrees about the true specular position. Most of these electrons interact with surface dynamic dipoles via a long-range electrostatic interaction such that coulombic energy transfer occurs while the electron is still some 100–200 Angstroms above the surface. This 'dipole' scattering may be thought of as an entirely different scattering mechanism to that which occurs under the influence of short range interactions. The electron in vacuo sees not only the surface dipole but also the response of the metallic conduction electrons to this dipole. This response is known as the *image dipole* and may be depicted for the two extreme orientations as shown in Figure 7.21.

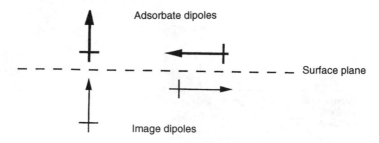

Figure 7.21. Schematic representation of the two extremes of dipole orientation with respect to the surface plane and the corresponding orientation of the image dipole formed via the response of the conduction electrons in the substrate

From figure 7.21 it may be seen that for an adsorbate dipole of magnitude p oriented parallel to the *surface normal*, the presence of the image dipole created by a redistribution of surface conduction electrons in response to the adsorbate dipole results in a net surface dipole of magnitude $\approx 2p$, i.e. approximately twice that of the adsorbate dipole alone. However, for adsorbate dipoles oriented parallel to the *surface plane*, the image dipole acts to negate the surface dipole such that the net dipole is effectively zero. Thus a surface selection rule exists that is effectively identical to that described for the RAIRS experiment in that for long-range, dipolar scattering, energy is only lost to those surface vibrations which have a component of their dynamic dipole oriented parallel to the surface normal. For this reason, a RAIRS spectrum and a dipolar EELS spectrum are entirely analogous in terms of band positions, although relative intensities may vary due to differences in scattering and absorption factors. The key features of dipole scattering, namely the narrow angular spread about the specular position and the long-range nature of the interaction are a direct consequence of an electrostatic model of the interaction of the surface dipole with the incident and outgoing electron. This treatment is beyond the scope of this text and the reader is referred to the work of Ibach and Mills [20] for a full description.

Considering both dipole and impact scattering, it may be seen that for surfaces having well defined symmetry elements such as single crystals, where the plane of incidence of the electrons may be well-defined, a comparison of dipolar and impact EELS data and application of the impact scattering selection rules often allows for the complete structural and orientational characterisation of the adsorbed species. Examples of such analyses are given in a later section.

However, it is possible to take such analyses one stage further and arrive at a typical spectroscopic intensity *pattern* for an adsorbed species which is attributable to a particular mode of bonding. For ethene adsorbed on a variety of single crystal metal surfaces this exercise has been carried out by Sheppard and co-workers [21] who have classified spectra as type I, with some di-σ character and type II, mainly π-bonding, Figure 7.22.

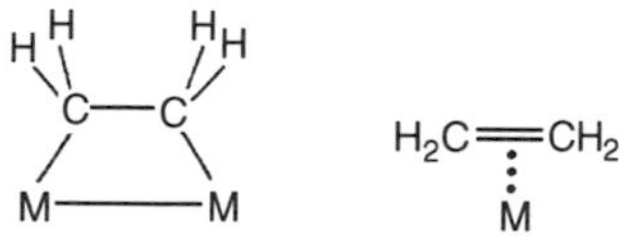

Figure 7.22. The classification of adsorbed molecular ethene species according to Sheppard and coworkers [21]

The assignment of the type I spectrum to that of a true di-σ species was made with the aid of data for the cluster compound $(C_2H_4)Os_2(CO)_8$ which is known to have this structure [22]. Similarly, the assignment of the type II spectrum to that of a π-bonded species was made with the aid of IR data for Zeise's salt [23], $K^+[(C_2H_4)PtCl_3)]^-H_2O$, which includes data for the C=C stretching mode, which does not possess a dipole moment. This apparent anomaly also occurs in the spectra, both IR and EELS of π-bonded ethene and may be explained by considering the bonding of the molecule/ligand to the surface/metal atom as via donation from the π-electron cloud, which periodically distorts at the frequency of the C=C vibration. Thus as the C=C bond vibrates, a similar periodic movement of electrons occurs in and out of the surface. This oscillation is of course entirely equivalent to a dipole oriented perpendicular to the surface plane, and thus is observed as either an absorption band (IR) or an energy loss feature (EELS).

3.3 THE EELS EXPERIMENT

If we consider the experimental requirements for EELS, low energy electrons, single crystal substrates (usually), then it is clear that this is a high vacuum technique. In practice the working pressure limit for most EELS instruments is 10^{-6} mbar due to a gas-phase scattering processes. Figure 7.23 is a schematic of the essential features of an EEL spectrometer:

Usually at least one unit, the monochromator or analyser has some ability to rotate about an axis perpendicular to the plane of the experiment while the sample may also rotate. In this way it is possible to vary the angle of incidence over a wide angular range and still maintain the ability to detect at angles away from the specular position. The monochromators and analysers are electrostatic selection units based upon either 127 degree cylindrical capacitors or hemispherical capacitors. The potentials applied to these capacitors are such that only electrons having specific pass energies may reach first the sample surface and second the detector. In practice, the 127 degree systems produce the best experimental resolution although the hemispherical systems produce a higher beam current. To avoid the influence of stray magnetic fields, the unit is normally encased in magnetic shielding. The working surfaces of the spectrometer are coated with an inert material such as graphite in order to minimise changes in spectrometer work function that may occur upon exposure to various gases.

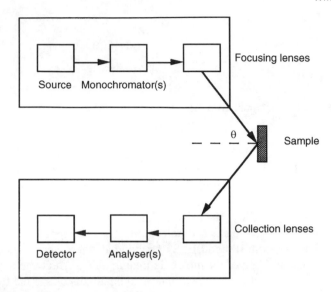

Figure 7.23. A schematic representation of the EELS experiment

3.4 SPATIAL RESOLUTION IN EELS

The design of an EEL spectrometer is such that the energy spread of the incident low energy electron beam is minimised while maintaining a reasonable electron flux. This primary consideration means that the area of surface sampled by the beam is of secondary importance. In some instruments which utilise 127 degree sector capacitors the resulting beam is ribbon shaped, having dimensions on the order of 5 mm × 0.5 mm Other instruments which utilise hemispherical sector capacitors produce a spot focus beam of diameter of the order of 1 mm. Thus the spacial resolution is poor in comparison with other methods, although the applications of EELS are such that it should not be thought of along with other more 'routine' techniques.

4 THE GROUP THEORY OF SURFACE VIBRATIONS

4.1 THE GENERAL APPROACH

In earlier sections it was noted that for particular kinds of experiment only certain modes would be expected to be observed. This point was also made during discussion of the surface selection rule. In this section we will look at the way in which vibrations are classified using molecular symmetry in an attempt to be able to predict the form of various surface spectra.

It is well known that the link between spectroscopy and molecular structure is symmetry. Molecules may be classified into symmetry types or 'point groups' by invoking rules based upon the relationship between various types of symmetry operations. The mathematical process of allowing each symmetry element to operate on a system in such a way as to produce a related system is known as Group Theory, a detailed discussion of which is beyond the scope of this text. The interested reader is referred to the text by Bishop [24]. However, even the non-mathematician may easily see that it should be possible to write down the process of symmetry operation in the form of a series of equations. Accepting this, it is then possible to note that for each point group there will be a series of such equations which will apply. Sets of equations created in this manner may be expressed in matrix notation such that the process of symmetry operation may be expressed in terms of the product of two matrices. Fortunately, the experimentalist need not be too concerned with the detailed mathematics of this process since the coefficients of the matrices for each point group are readily expressed in terms of 'Character Tables' which not only list all of the operations, but also all of the unique combinations of operation which may be created. In the language of group theory these are the possible 'permutations' which may be derived, with each possible set of permutations describing a particular mode of motion. These modes include the so called normal modes that we are used to studying in conventional vibrational spectroscopy, but also include translational and rotational motion. This factor must be accounted for when approaching the problem of predicting the form of a surface vibrational spectrum.

All of the above discussion is somewhat academic without illustration by suitable examples and so at this point we will turn to a particular example to provide some clarity.

4.2 GROUP THEORY ANALYSIS OF ETHYNE ADSORBED ON TO A FLAT SURFACE

For the purpose of this exercise let us view the surface in question as a flat, featureless plane. We realise at the outset that this is a gross assumption but it serves as a useful starting point for this analysis. We will also assume that the ethyne molecule adsorbs in a manner in which the molecular plane is parallel to the surface plane. The first step in the analysis is to determine the point group of the 'system', molecule + surface. At its simplest, this may be achieved by simply listing the symmetry elements that the system possesses and then comparing the list with those tabulated for each point group. When the list matches that for a particular point group, this is the point group of the system. For gas-phase ethyne, there is an infinite axis of rotation which corresponds to the molecular axis. This is labelled C_∞. The molecule also possesses mirror planes of reflection in both the horizontal and vertical directions. These are labelled σ_h and σ_v respectively, while the presence of orthogonal planes of this nature tell us that the point group is likely to be of symmetry higher than C, in this case a D group. Overall, the symmetry of the ethyne molecule corresponds to the point group $D_{\infty h}$. However, this is not the point group of the adsorbed ethyne molecule. We may

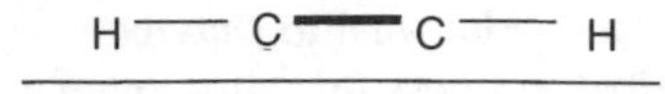

$$H \text{——} C \text{===} C \text{——} H$$

Figure 7.24. Ethyne adsorbed at a flat, featureless surface

note here a generalisation which applies to all adsorbed molecules, namely that 'the symmetry of the adsorbed species is either equal to or lower than that of the free molecule'. The inclusion of the surface can never increase the symmetry of the system. Let us now return to the adsorbed molecule, Figure 7.24.

If we list the symmetry elements of this system we may immediately note that inclusion of the surface plane removes the C_∞ element. In fact the highest axis of rotation is now the one which is normal to the surface and which passes thorough the C—C bond, a C_2 axis. The inclusion of the surface also removes the horizontal reflection plane since this would flip the surface above the molecule! Thus the only elements of symmetry are the C_2 axis and reflection planes parallel to this axis and in the plane of the paper. We add to these three elements the so-called identity element, usually given the symbol I or E, which is the allowed operation of 'leaving the system alone'. Thus we have four symmetry elements E, C_2 and 2σ. At this stage we need to be able to distinguish the two mirror planes and also to be able to interpret the meaning of each operation on the motion of the molecule. For this reason we introduce a coordinate system in which the z-direction is always normal to the surface. The molecular plane thus becomes the xy plane and we may distinguish the two mirror planes as σ_{xz} and σ_{yz}. An examination of the character tables then attributes these four elements to the point group C_{2v}. This character table appears as follows, Table 7.1:

Looking at this table it is apparent that some further explanation is required. The labels forming the first column represent the four possible permutations of each symmetry element, whereas the 1 or -1 in the rows represent the character of each operation and indicate either that operation of the element produces an indistinguishable system to that with which we started (1) or that the system changes under the operation (-1). With this in mind it may be easily seen that the highest symmetry permutation is A_1, since here operation of all elements results in an unchanged system. As we proceed down the first column we move to lower symmetry. Note that because of the special relationship between symmetry

Table 7.1. The C_{2v} point group character table

	E	C_2	σ_{xz}	σ_{yz}
A_1	1	1	1	1
A_2	1	1	-1	-1
B_1	1	-1	1	-1
B_2	1	-1	-1	1

elements that must exist, it is not possible to have a permutation in which the character of every element is -1. In other character tables the permutations may be listed in a similar fashion, although the nomenclature is readily extended to represent more subtle variations in character which may arise.

At this stage it would be possible, for simple molecules, to determine the symmetry of the vibrations which may be identified by inspection but this is not generally the case. For this reason we will proceed to determine the full nature of the motions of this system from first principles. Since each atom in the molecule possesses three degrees of freedom, it is convenient to place each atom at the centre of a coordinate system identical to that used for the whole molecule. In this way we create x, y and z axes centered on each atom. We then allow each element to operate on the molecule, giving a nominal value of 1 for an axis that remains unchanged and -1 if the atom is not moved but the axis in inverted. The easiest example is clearly the E operation, since this leaves all axes unchanged by definition. We then write onto the table, the net character for all atoms. Thus in the column under the E element, for ethyne we would write 12, being 4×3 axes unchanged, Table 7.2.

For the C_2 operation, the situation is also quite clear. The axis of rotation lies through the centre of the C–C bond and thus the rotation moves all axes because all atoms move. Thus in the column under C_2 we place zeros. For the σ_{xz} operation, since the xz plane is the plane of the paper, operation of the element leaves two axes unchanged for each atom (x and z) while the y axis is inverted. The total character per atom is therefore $+2 - 1 = 1$. Since there are four atoms, we place the number 4 in the column under the σ_{xz} operation. For the σ_{yz} operation the situation is the same as for the axis of rotation in that all atoms move. Thus we place zeros in the column under σ_{yz}. The total value accumulated across the rows, remembering that each row represents an allowed symmetry permutation, will yield the number of each possible permutation when divided by the character of the table, which is simply the number of elements, in this case, 4. Thus for the A_1 permutation, the sum across the row is $12+0+4+0 = 16$. In this case this means that there are 4 possible A_1 permutations (being 16/4). For the A_2 permutation, summing across the row yields $12 - 4 = 8$, giving in turn 2 A_2 permutations. Similarly for the B_1 and B_2 permutations, we find 4 and

Table 7.2. Inclusion of the character arising from the operation of the C_{2v} elements in the character table for adsorbed ethyne as depicted in Figure 7.25

	E	C_2	σ_{xz}	σ_{yz}
A_1	1(12)	1(0)	1(4)	1(0)
A_2	1(12)	1(0)	-1(4)	-1(0)
B_1	1(12)	-1(0)	1(4)	-1(0)
B_2	1(12)	-1(0)	-1(4)	1(0)

2 permutations respectively. Thus the total 'representation', sometimes referred to as the reducible representation, is $4A_1 + 2A_2 + 2B_2 = 12$. The total is in accord with that expected for four atoms each with three degrees of freedom. We now return to the point made earlier when we noted that this representation also includes translations and rotations of the system. To obtain just the vibrational part we must subtract those permutations which correspond to translation and rotation about x, y and z. In order to see which permutations to subtract from the reducible representation we return to the character table. In most published tables translations are marked in an additional column on the right-hand side as T_x, T_y or T_z while rotations are labelled R_x, R_y or R_z. A more complete version of the C_{2v} character table would therefore appear as in Table 7.3.

Note that in addition character tables usually Raman activity expressed in terms of the components of molecular polarisability, (see Section 5), usually written as ∞_{ij} terms, where i and j represent combinations of the cartesian coordinates x, y and z. Returning now to our example, we note that for the translational motion we must subtract $A_1 + B_1 + B_2$, while for the rotations we must subtract $A_2 + B_1 + B_2$. Removing these six permutations leaves us with the irreducible representation of the vibrational motion, $3A_1 + A_2 + 2B_1$. At this point we note an interesting observation — namely that, even if the surface is considered flat and featureless, then the number of vibrations expected is not that predicted by considering the system as a linear molecule, i.e. using the $3N - 5$ rule where N is the number of atoms. This arises because the presence of even a model surface reduces the symmetry of the system. Thus we have $3N - 6 = 6$ vibrations to consider.

Now that we have performed this exercise we may increase the complexity of the system by invoking the presence of a real surface, and necessarily introduce the symmetry of the surface.

4.3 GROUP THEORY ANALYSIS OF ETHYNE ADSORBED ON A (100) SURFACE OF AN FCC METAL

In this example we place the ethyne molecule in a two-fold bridge site on an unreconstructed (100) surface of an fcc metal such as Pt or Cu, which before adsorption is said to have 4 mm symmetry. The '4' refers to a four-fold axis of rotation while the mm refers to the presence of two orthogonal mirror planes in the surface, Figure 7.25.

Table 7.3. A general version of the C_{2v} character table

	E	C_2	σ_{xz}	σ_{yz}	
A_1	1	1	1	1	T_z
A_2	1	1	-1	-1	R_z
B_1	1	-1	1	-1	T_x, R_y
B_2	1	-1	-1	1	T_y, R_x

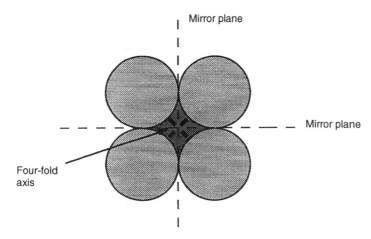

Figure 7.25. Schematic representation of an unreconstructed (100) surface for an fcc single crystal metal

Adsorption not only reduces the symmetry attributed to the ethyne molecule, but also the symmetry of the surface. Placing the ethyne molecule in the two-fold site over two Pt atoms creates a species with overall symmetry C_{2v}, i.e. the same as before. In this case we would obtain more vibrations in our analysis simply because there are more atoms involved. However, the method would be almost identical to that employed in the previous section.

Let us now make the system more complex by choosing to adsorb at the three-fold site involving two atoms from the top layer and one atom in the second layer, such that the plane formed by these three atoms is inclined with respect to the surface plane. Now since the plane of adsorption is inclined with respect to our surface plane, there can be no C_2 axis of rotation. There will still be one mirror plane in the system, but the other mirror plane found for C_{2v} disappears along with the axis of rotation. From character tables we would attribute the symmetry of this species as C_S, again noting the reduction in overall symmetry. The table for this point group is simpler than that above simply because of the reduced number of elements, Table 7.4:

In our analysis we must now also include the three surface atoms, since these contribute to the symmetry of the species. Note that all of the permutations

Table 7.4. The C_s point group character table

	E	σ
A′	1	1
A″	1	−1

have the 'A' classification. This is to be expected since reducing the symmetry increases the number of types of motion which effectively preserve this reduced symmetry. It then follows that a system with the lowest possible symmetry classification also has the highest number of A-type modes.

4.4 THE FORM OF THE RAIRS AND DIPOLAR EELS SPECTRUM

Here we are invoking the surface selection rule described earlier which states that a vibration must produce a component of its dynamic dipole perpendicular to the surface plane in order to be detectable. In our analysis so far we have not indicated any means by which we may select vibrations from our derived irreducible representation. To do this we return to the character table and study the form of our vibrations. Using the ethyne adsorbed at a flat featureless surface system as an example, we may examine the form of the vibrations and determine whether they satisfy the surface selection rule. Intuitively we may guess the form of the vibrations up to a point. There will of course be symmetric and asymmetric $C-H$ stretching modes, both in the plane of the molecule. The symmetric stretch will not alter the C_{2v} symmetry and thus has a character of 1 under each operation. It is thus assigned to an A_1 mode. The asymmetric stretch must obviously have a character of 1 under E, but will have a character of -1 under C_2 and also -1 under σ_{yz}, because of the distortion to the overall symmetry of the system. It also follows that for the σ_{xz} the character remains 1. Thus from the table we assign this to a B_1 mode. The $C-C$ stretching mode maintains the overall C_{2v} symmetry for all operations and hence must be A_1. The presence of the surface cannot be ignored and hence the vibration of the whole molecule against the surface must be considered. This is clearly an A_1 mode since the C_{2v} symmetry does not change simply if the distance between the molecule and the surface changes. Thus far we have found our three A_1 modes plus one B_1 mode. There are thus $1B_1$ and $1A_2$ modes to be found. Looking first at the B_1 mode we note that it must effectively remove the C_2 axis and the corresponding mirror plane. This is therefore an out-of-phase, out-of-plane wagging of the H atoms. The A_2 mode preserves the C_2 axis but removes both mirror planes. This is an in-plane twist of the molecule which may be thought of in part as an out-of-phase in-plane wag of the H atoms. These then are the six modes that we wished to identify. In terms of our surface selection rule we must now examine which of these modes is likely to produce a changing dipole in the z direction. Clearly the A_1 mode in which the whole molecule moves against the surface, meets this criterion. The $C-C$ stretch, which is also A_1, does not at first sight satisfy this rule but we have implicitly placed a chemical bond between the molecule and the surface (implicitly because we have assigned this bond a vibration). As the $C-C$ bond expands and contracts so the bond between the molecule and the surface must fluctuate in 'strength', at the same frequency. Thus this A_1 mode would also appear in our spectrum. No other modes have components of their dynamic dipoles perpendicular to the surface and thus of the six possible vibrations, only two appear in the RAIRS or dipolar EELS spectrum.

An interesting general conclusion emerges from this analysis:

> 'Only modes with "A" character (of the highest symmetry) may yield features in the RAIRS or dipolar EELS spectrum'.

This combined with our prediction as to the influence of symmetry upon the number of A-type modes indicates that the RAIRS and dipolar EELS spectra associated with molecules adsorbed in such a manner as to produce low-symmetry systems will contain the largest number of features. Conversely, high symmetry systems will yield the simpler spectra containing fewer features. Application of this type of analysis is clearly crucial in determining the nature of the adsorbate–substrate system.

5 LASER RAMAN SPECTROSCOPY FROM SURFACES

Raman spectroscopy is often said to be complementary to infrared spectroscopy in that it is sensitive to those vibrational modes which are either not observed via IR or give rise to only weak IR absorption bands. Predicted by Smekal in 1921 [25] and first observed by Sir C V Raman in 1928 [26] the Raman effect relies upon the polarisation of the electron cloud describing a chemical bond by the electric field of incident electromagnetic radiation, which induces a dipole moment, which in turn is time dependent due to the vibration of the atoms forming the bond. Thus it is the *polarisability* rather than the dipole moment which is the important molecular parameter in determining Raman intensities. The Raman effect is a scattering phenomenon, and thus may be thought of as directly analogous to EELS in that one analyses the energy lost by, in this case photons rather than electrons, in order to detect molecular vibrations. It is a weak effect, with perhaps only 1 in 10^{11} photons being inelastically scattered in a typical process and for this reason intense light sources are normally required. Early work utilised mercury arc lamps but modern instruments employ lasers of some description. A schematic diagram of a typical Raman experiment is given in Figure 7.26.

From this figure it is clear that the Raman technique may be applied to the study of a wide range of materials, including most solid surfaces. The use of a laser as the excitation source also means that this technique is applicable to small areas of a sample surface. This ability to map a surface or to focus onto a small area of a surface is most effectively exploited in the Raman microprobe instrument. Modern instruments may employ holographic notch filters in preference to complex expensive monochromator systems in order to achieve the required elastically-scattered light rejection capability.

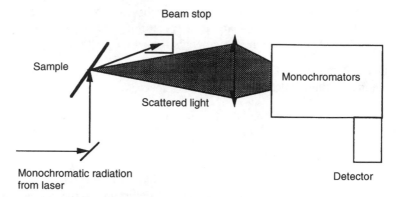

Figure 7.26. Schematic representation of the surface Raman experiment

5.1 THE THEORY OF RAMAN SCATTERING

The quantum mechanical description of the scattering process known as the Raman effect invokes a 'pseudo absorption' process in which the incident radiation is absorbed into a virtual electronic state of the molecule, followed by emission back to the first excited vibrational state, Case 1 in Figure 7.27.

The energy difference between the incident and emitted radiation is thus equal to one quantum of vibrational energy and the emitted photons are termed Stokes photons. An alternative situation is described in Case 2, where the molecular vibration is already described by $v = 1$ and, upon re-emission, photons of

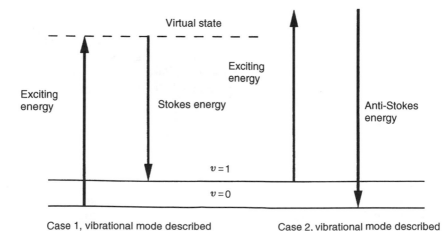

Figure 7.27. A description of the vibrational Raman effect based upon an energy level approach

higher energy than the exciting energy result. These are known as the anti-Stokes photons. The virtual state is not generally a true electronic state of the molecule but a composite function involving all possible states, rotational, vibrational and electronic. Where the energy of the incident radiation does coincide with a true molecular state the cross-section for Raman scattering increases enormously. This situation is referred to as the resonance Raman effect, for obvious reasons, and may result in an increase in scattering levels of a factor of 10^6 as compared to conventional Raman scattering.

Classically the effect is described in terms of the induction of a dipole moment into the molecule of magnitude proportional to the polarisability, which is itself a time-dependent function, the time dependence arising from the distortion to the molecule that accompanies a vibration. The combination of the frequency of the incident radiation v_i and the superimposed vibrational frequency v_{vib} gives rise to inelastic scattering at frequencies $v_i + v_{vib}$ and $v_i - v_{vib}$, which are known as the anti-Stokes and Stokes components respectively. This classical description is conceptually easier to understand than the quantum mechanical description but cannot explain the relative intensities of the Stokes and anti-Stokes components which are of course determined by Boltzmann factors.

5.2 THE STUDY OF COLLECTIVE SURFACE VIBRATIONS (PHONONS)

As with IR spectroscopy, there are a wide variety of applications of laser Raman spectroscopy to the study of surface processes. An application which illustrates some of the advantages of the technique is depicted in Figure 7.28.

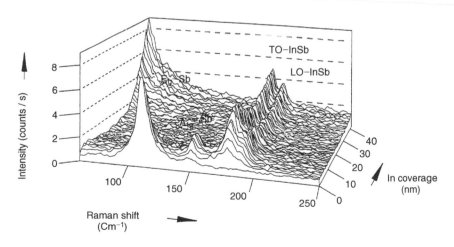

Figure 7.28. Raman spectra recorded from an Sb (111) surface as a function of varying In coverage, after Zahn [27]

Figure 7.28 illustrates immediately that via the use of multiple monochromators/notch filters and sophisticated detection systems, it is possible to detect extremely low frequency vibrations, i.e. at absolute frequencies close to that of the exciting laser frequency, which could only be accessed in the IR by using far-IR optics and sources. The particular vibrations in question are as a result of the collective oscillations of atoms in the surface layers. Such vibrations are known as *phonons*. The frequency of a surface phonon is a marked function of the structure of the surface layer and the extent of any surface symmetry. Thus it is noted that as the coverage of In increases, the phonon bands of InSb appear, indicating the reactive nature of the In overlayer with the Sb substrate. The width of the phonon bands is a measure of the degree of surface order.

The high level of sensitivity displayed in this example serves to illustrate that fact that Raman spectroscopy has many direct surface applications in the electronics industry where oriented wafers are used as substrates for the construction of devices based upon layer structures. However, the sensitivity obtained from these systems is somewhat atypical due to the careful choice of excitation wavelength. The materials under examination are direct bandgap semiconductors, capable of absorbing or emitting photons with reasonably high efficiency. The excitation wavelength is chosen so as to coincide with a particular electronic transition within either substrate or the deposited film. Under these conditions, the virtual levels displayed in Figure 7.27 become real levels and the process of Raman scattering becomes resonant. Resonance Raman systems similar to this are well known in other areas of chemistry and physics where the enhancement of the Raman scattering efficiency that occurs when the resonance condition is satisfied can reach up to 10^6 times that observed off-resonance.

5.3 RAMAN SPECTROSCOPY FROM METAL SURFACES

The most widely known application of Raman spectroscopy to metal surface chemistry involves a phenomenon known as surface enhanced Raman spectroscopy or SERS. This was first demonstrated by Fleischmann and co-workers [28, 29] in 1974 who were attempting to record vibrational spectra from silver electrode surfaces immersed in an aqueous solution containing KCl as a supporting electrolyte and pyridine as the specific adsorbate. The pyridine/silver system was chosen because pyridine has a relatively high Raman scattering cross-section, particularly for the ring breathing modes and was known from other measurements to adsorb readily at silver electrode surfaces. In order to increase the sensitivity of the technique, the silver electrode was repeatedly oxidised and reduced *in situ*, resulting in a surface with a grey spongy appearance with a surface area about ten times that of a corresponding flat, polished surface. The resulting Raman spectra were intense, exhibiting signal/noise ratios in excess of 100:1 and were highly sensitive to changes in electrode potential, Figure 7.29.

Since the electrode potential is effectively felt by only the first one or two molecular layers, this was a remarkable display of sensitivity. It was soon realised

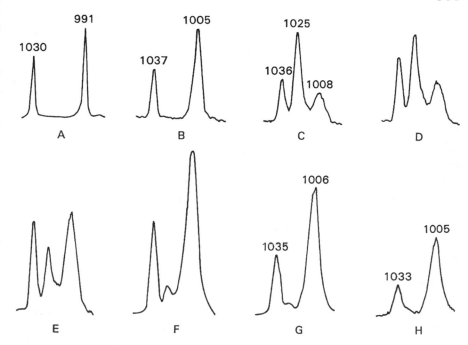

Figure 7.29. Raman spectra recorded from the surface of a polycrystalline Ag electrode immersed in an aqueous KCl/pyridine solution as a function of applied electrochemical potential, measured with respect to the potential of a saturated calomel reference electrode, after Fleischmann *et al.* [28]. (A) liquid pyridine; (B) 0.05 M aqueous pyridine; (C) silver electrode 0 V (S.C.E.); (D) −0.2 V (E) −0.4 V; (F) −0.6 V (G) −0.8 V; (H) −1.0 V

that the spectra were too intense to be accounted for in the normal way and that the Raman scattering cross-section for the adsorbed pyridine species was between 10^4 and 10^6 times greater than that for liquid pyridine! [30, 31]. Much subsequent study of this and related systems revealed that the surface enhancement could only be observed from silver, copper and gold surfaces although the effect was not restricted to electrochemical systems and similar observations were made from metal surfaces under high vacuum and even from colloidal metal particles in solution. A large number of mechanisms were postulated in order to explain the enhancement process including resonance phenomena similar to the resonance Raman process discussed in the previous section, but involving novel surface complexes and surface plasmons activated by the electrochemical oxidation/reduction cycle. The 'conventional' resonance Raman process was discounted since molecules that exhibited resonance Raman scattering (i.e. those capable of absorbing the incident laser energy) gave rise to a further enhancement when adsorbed at a silver electrode. It is now felt that it is unlikely that any one mechanism will be sufficient to explain the phenomena observed from a large number of different surface chemical systems and that the limitation of the technique in terms of the 'active' metals precludes its use as a general method of surface analysis. However, for certain analytical applications of Raman spectroscopy, sensors made of silver, copper and gold have been employed in order to obtain increased sensitivity [32–34].

5.4   SPATIAL RESOLUTION IN RAMAN SPECTROSCOPY

As a photon-based technique, the resolution is typically limited to the best focus possible for a given exciting laser wavelength. This is not limited by the focusing optics but by self-diffraction of the light, which results in a beam 'waist' of fixed dimensions for a given wavelength, that cannot be reduced. Using short wavelengths such as the 488 nm blue line available from a continuous-wave argon ion laser, it is possible to achieve a spot size of around 20 microns.

5.5   FOURIER TRANSFORM RAMAN TECHNIQUES

The use of a visible laser source to stimulate Raman scattering from a sample, while convenient, is not without difficulties. Until recently, by far and above the major difficulty in this area was the stimulation of fluorescence from impurity species. Early studies of species adsorbed at supported metal catalyst surfaces were constrained by the appearance of intense fluorescence emission resulting from electronic excitation of impurity species such as vacuum greases. This fluorescence manifested itself as a broad spectral envelope on the Stokes side of the exciting laser line and was often too intense to enable the weak Raman bands to be observed. To overcome this, lasers were chosen such that the excitation energy was far below any possible electronic excitations, i.e. in the IR region of the spectrum. The difficulty associated with this was that spectral discrimination in the IR region was necessary and thus all of the advantages of working with visible radiation were lost. The advent of Fourier transform IR techniques overcame this difficulty. With some modification, interferometers were developed that could perform both conventional IR absorption measurements and IR-laser Raman experiments. This combination of techniques is proving particularly powerful for surface analysis.

6   INELASTIC NEUTRON SCATTERING (INS)

6.1   INTRODUCTION TO INS

This is a particularly powerful form of vibrational spectroscopy since both IR and Raman active modes, as well as those vibrations which are neither IR or Raman active, may appear in an inelastic neutron scattering experiment. Like some of the other methods described in this chapter, INS is applicable to a wide range of samples, including surfaces and adsorbed species. It is not the intention to provide a comprehensive overview of this method here, but rather to concentrate on a particular surface chemical study and direct the interested reader to the review by Parker [35] where further references may be found.

The experiment is clearly not trivial since a supply of neutrons is required! Two methods are commonly employed to create a flux of neutrons: spallation and fission. The first of these, spallation, involves the shattering of nuclei using very high energy protons. Generating high energy protons is also non-trivial. Typically in the UK this is achieved using a proton accelerator or proton synchrotron such as the one based at the Rutherford Appleton Laboratories. The second process of nuclear fission requires a nuclear reactor to operate. Thus it may be seen that the generation of a neutron flux takes some considerable effort and expense. The expense is justified by the large number of experiments which may then be performed, of which INS is only one type. The directional nature of the neutron flux coupled with the property of being able to penetrate most matter easily, is used to great effect in INS, where the time-of-flight of the neutrons around the system is used to determine the energy transfer to the sample directly. Thus in terms of the other methods described in this section, the energy transfer is perhaps similar to that in Raman spectroscopy, while the mode of sampling is perhaps most similar to that of transmission IR methods.

6.2 THE INS SPECTRUM

There are no selection rules in INS. Thus, as has been stated earlier, modes that are IR-active, Raman-active or neither IR nor Raman-active appear in the INS spectrum. However, an INS spectrum does not usually appear as a combination of IR and Raman spectra together with new bands, because of the nature of the neutron scattering cross-section — the fundamental property which determines band 'intensity'. This does not depend upon such factors as dynamic dipole or polarisability but rather on a more basic 'billiard ball' approach to momentum transfer. This approach predicts that momentum transfer occurs most efficiently between two particles of comparable mass. Thus the process is most efficient for modes involving hydrogen. In practice it is found that the inelastic cross-section for hydrogen is at least an order of magnitude larger than that of other elements, including deuterium, which makes possible the use of isotopic substitution for the immediate identification of hydrogenic modes. Thus a typical INS spectrum is dominated by bands arising from the motion of hydrogen atoms. While there are obviously many applications for a technique of this nature, it may also be exploited by the surface chemist. An example of which is presented in the next section.

6.3 INS SPECTRA OF HYDRODESULFURISATION CATALYSTS

Since crude oil contains a number of contaminants including sulphur, nitrogen and trace metals, part of the refining process involves hydrogenation to remove these contaminants as volatile hydrides. This process also serves to reduce the unsaturated fractions of the crude. Of the contaminants, perhaps sulphur is the one where environmental factors place the most stringent limits on the levels of emission of substances such as SO_2, which may be formed if a sulphur-containing oil is combusted.

The present generation of hydrodesulfurisation catalysts, as they are known, are unlikely to be able to reduce levels of sulfur in crude sufficiently, given the particular economic factors that are involved and thus potential new catalysts are currently being sought. One possible new catalyst is the metal sulfide, Ru_2S. This material is capable of adsorbing hydrogen which is then chemically activated towards the hydrodesulfurisation process. The activity of the material has been directly correlated with the concentration of RuH species although until the application of INS, such metal-hydride species had not been formally identified by vibrational spectroscopy. Jobic *et al.* [36] have conclusively identified Ru–H species on partially desulfurised Ru_2S exposed to hydrogen, Figure 7.30.

These spectra also show bands assigned to overtones and combination modes — a feature which is typical of an INS spectrum. The cells used in INS are simply stainless steel or aluminium 'tubes', these metals being basically transparent to the neutron flux, and thus large amounts of material may be employed, e.g. up to 100 g. Under these circumstances it is fairly certain that the spectrum measured is indeed a true representation of the catalyst material. By comparison with other forms of surface vibrational spectroscopy described here, the following features are noteworthy:

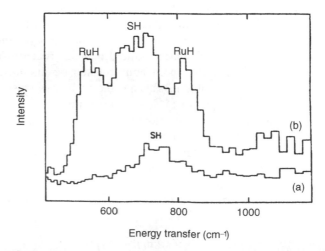

Figure 7.30. INS spectrum recorded from a partially desulfurised Ru_2S catalyst exposed to hydrogen, showing the features assigned to the formation of Ru–H species at energy transfer values of 540 and 821 cm^{-1}, after Jobic *et al.* [36]

(i) INS is extremely sensitive to hydrogenic modes, but as atomic number increases so its sensitivity decreases dramatically.

(ii) Neutron sources are comparatively weak, such that best results are obtained where there is a relatively large amount of sample, usually in the form of fine particulates.

(iii) Reliable neutron sources are not always readily available!

7 SUM-FREQUENCY GENERATION METHODS

The reader will appreciate that many of the methods described here cannot be applied to surfaces in contact with liquids or significant overpressures of gases. For others, such as IR methods, gas-phase interferences require the use of complex normalisation/background subtraction routines. In the mid-1980's researchers began to study the non-linear optical properties of surfaces, and one resulting technique, sum frequency generation (SFG), has been developed which in principle may be used to study certain substrate/adsorbate systems, irrespective of the medium over the surface. The technique is based upon the interaction of two photons at a surface, induced via wave-mixing, such that, upon reflection from the surface, a single photon emerges. This process conserves energy (and also frequency) which is where the name sum-frequency generation originates. The full theory of non-linear optics required for a thorough understanding of this method is beyond the scope of this text. The reader is referred to the book by Shen for an in-depth review of the topic [37].

Three-wave mixing processes of this type rely upon the inherent dipole resulting either from the interface between two media, e.g. solid–liquid, solid–gas, liquid–liquid (for two immiscible liquids), or dipoles present in the bulk, found in ordered materials under circumstances where the unit cell lacks inversion symmetry. Thus if we consider a substrate such as a single crystal surface of metals like Ag, Cu, Ni, Pt etc., the non-linear optical process will not arise from photons that penetrate into the bulk since the bulk face-centred cubic structure possessed by these metals has inversion symmetry across the unit cell. However, the interface between the crystal surface and the medium in contact will be 'active' towards sum-frequency generation.

The simplest process to envisage is where both photons possess the same energy $E(\omega)$ (frequency ω), such that when the three-wave (two in, one out) process occurs, the outgoing photon has energy $2E(\omega)$ and frequency 2ω. This process is known as second harmonic generation.

The efficiency (intensity) of this process depends upon a variety of factors but where either $E(\omega)$ or $2E(\omega)$ correspond to allowed transitions of some description, e.g. electronic or vibrational transitions, then the process is said to become resonant, with the result that the efficiency of the process, and hence the intensity at 2ω is increased. Sum-frequency generation thus exploits this possibility by ensuring that one incident photon is generated from a source capable of producing radiation that may be tuned in energy across a suitable transition.

Thus as a form of surface vibrational spectroscopy, SFG operates as follows: one incident photon is usually produced by a tunable laser operating in the IR spectral region, producing radiation that may be absorbed directly by the species adsorbed at the surface. Another pump photon, which may be visible, or IR, is used to provide the intensity necessary to observe weak second-order effects such as this. If the pump photon has frequency ω_p and the tunable photon has frequency ω_t, then when ω_t corresponds to the frequency required to excite the surface vibration, there results a change in the intensity of the sum photon, frequency $\omega_p + \omega_t$. Due to the difficulty in providing laser systems with tunable output in the useful part of the IR spectral region this method is currently of limited application. However, as advances in laser technology are made it is likely to become more widely used.

An example of the application of SFG to a surface chemical problem is provided here. Bain and co-workers at Oxford have studied the adsorption of surfactant molecules at the liquid/air interface [38, 39] Figure 7.31.

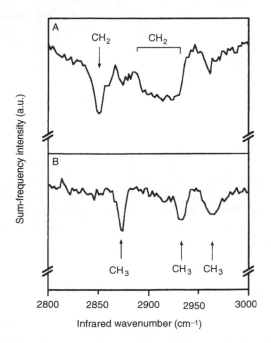

Figure 7.31. Sum-frequency generation spectrum recorded from a hydrophobic surface in contact with the anionic surfactant SDS (sodium dodecyl sulphate) and the analogous spectrum recorded in the presence of deuterated dodecanol, after Bain [38]

Figure 7.31 shows a typical SFG spectrum in which the intensity of the sum-frequency component varies dramatically as the tunable laser sweeps over the frequency corresponding to the vibrations in the surfactant species. In this example, the observed resonances were interpreted in terms of a model in which, in the absence of the deuterated alcohol, the surfactant species were oriented such that the methylene groups were observed. While in the presence of the alcohol, the structure of the overlayer changed dramatically revealing sharp methyl resonances. The principal advantage of sum-frequency generation arises from its selectivity. For substrates in which the bulk unit cell possesses inversion symmetry bulk SFG, is disallowed. Furthermore, a gas or a liquid represents an isotropic medium to the incident laser beams. Isotropic media also behave as if they possess inversion symmetry. Thus there can be no contribution from either a gas or a liquid in contact with the substrate. Therefore, the only SFG component that may be observed from such a system, arises from the interface itself. In this respect given the correct choice of substrate, SFG is the ultimate pressure independent form of surface vibrational spectroscopy. The principle disadvantage of the method is that it is based upon an inherently weak phenomenon such that intense, usually pulsed laser sources are required in order to provide sufficient photons in a given time interval (typically 10 ns or less), in order to observe SFG. There is thus the possibility that surface damage may occur due to interaction of the intense laser pulses with the surface. As implied earlier, the method is also technically difficult at the present time due to the limitations of current tunable laser technology, although as laser technology improves, these methods will become increasingly used.

REFERENCES

[1]   C.N. Banwell and E.M. McCash, *Fundamentals of Molecular Spectroscopy*, Fourth Edition, McGraw-Hill 93–96, (1994).

[2]   J.R. Ferraro and L.J. Basilo, *Fourier Transform Infrared Spectroscopy: Application to Chemical Systems*, Academic Press, vol 1 (1978) vol 2 (1979).

[3]   P.R. Griffiths and J.A. de Hareth, *Fourier Transform Infrared Spectroscopy*, John Wiley, (1986).

[4]   R.P. Eischens, W.A. Pliskin and S.A. Francis *J. Chem. Phys.*, **22**, 1786 (1954).

[5]   R.P. Eischens, S.A. Francis and W.A. Pliskin, *J. Phys, Chem.*, **60**, 194 (1956).

[6]   C. Delacruz and N. Sheppard, *J. Chem. Soc. Chem Comm.*, **24**, 1854 (1987).

[7]   M.A. Chesters, C. Delacruz, P. Gardner, E.M. McCash, P. Pudney, G. Shahid and N. Sheppard, *J. Chem. Soc. Faraday Trans.*, **86**, 2757 (1990).

[8]   P.D. Holmes, G.S. McDougall, I.C. Wilcock and K.C. Waugh, *Catalysis Today*, **9**, 15 (1991).

[9]   R.G. Greenler, *J. Chem. Phys.*, **44**, 310 (1966) and *J. Vac. Sci. Tech.*, **12**, 1410 (1975).

[10] M.A. Chesters, *J. Electr. Spectrosc. Rel. Phenom.*, **38**, 123 (1986).

[11] P. Hollins, *Surface Science*, **107**, 75 (1981).

[12] P. Hollins and J. Pritchard, *J. Chem. Soc. Chem. Commun.*, **21**, 1225 (1982).

[13] M.A. Chesters, S.F. Parker and R. Raval, *J. Electr. Spectrosc Rel. Phenom.*, **39**, 155 (1986).

[14] A. Bewick and B.S. Pons in 'Advances in Infrared and Raman Spectroscopy', vol 12, (ed) R.J.H. Clark and R.E. Hester, Wiley Heyden, (1985).

[15] B. Beden and A. Bewick, *Electrochimica Acta*, **33**, 1695 (1988).

[16] H. Froitzheim, H. Ibach and S. Lehwald, *Rev. Sci. Instr.*, **46**, 1325 (1975).

[17] H. Ibach, H. Hopster and B. Sexton, *Appl. of Surface Science*, **1**, 1 (1977).

[18] P.A. Thiry *J. Electr. Spectrosc. Rel. Phenom.*, **39**, 273 (1986).

[19] M.A. Chesters, G.S. McDougall, M.E. Pemble and N. Sheppard, *Applied Surface Science*, 22/23, 369 (1985).

[20] H. Ibach and D.L. Mills, *Electron Energy Loss Spectroscopy and Surface Vibrations*, Academic Press, (1982).

[21] N. Sheppard, *J. Electr. Spectrosc. Rel. Phenom.*, **38**, 175 (1986).

[22] B.J. Bandy, M.A. Chesters, D.I. James, G.S. McDougall, M.E. Pemble and N. Sheppard, *Phil. Trans. R. Soc. Lond. Ser. A*, **318**, 141 (1986).

[23] M.J. Grogan and K. Nakamoto, *J. A.m. Chem. Soc.*, **88**, 5454.

[24] See for example '*Group Theory and Chemistry*', D.M. Bishop, Clarendon Press, Oxford (1973).

[25] A. Smekal, *Naturwiss*, **11**, 873 (1923).

[26] C.V. Raman and K.S. Krishnan, *Nature* 121 (1928) 501.

[27] D.R.T. Zahn, *Phys. Stat. Solidi, a* **152**, 179 (1995).

[28] M. Fleischmann, P.J. Hendra and A.J. McQuillan, *Chem. Phys. Lett.*, **26**, 163 (1974).

[29] M. Fleischmann, P.J. Hendra, A.J. McQuillan, R.L. Paul and E.S. Reid, *J. Raman Spec.*, **4**, 269 (1976).

[30] R.P. Van Duyne and D.L. Jeanmaire, *J. Electroanalytical Chem.*, **84**, 1 (1977).

[31] M.G. Albrecht and J.A. Creighton, *J. Am. Chem. Soc.*, **99**, 5215 (1977).

[32] W. Hill, B. Wehling, V. Fallourd and D. Klockow, *Spectroscopy Europe*, **7**, 20 (1995).

[33] K. Mullen and K. Carron, *Anal. Chem.*, **66**, 478 (1994).

[34] K. Carron, L. Pietersen and M. Lewis, *Env. Sci. Technol.*, **26**, 1950.

[35] S.F. Parker, *Spectroscopy Europe*, **6**, 14 (1994).

[36] H. Jobic, G. Clugnet, M. Lacroix, S.B. Yuan, C. Mirodatos and M. Breysse, *J. Am. Chem. Soc.*, **115**, 3654 (1993).

[37] Y.R. Shen, *The Principles of Non-Linear Optics*, John Wiley, New York, (1984).

[38] D.C. Duffy, P.B. Davies and C.D. Bain, *J. Phys. Chem.*, **99**, 15241 (1995).

[39] C.D. Bain, *Biosensors and Bioelectronics*, **10**, 917 (1995).

QUESTIONS

1. Using group theory, show that for ethene adsorbed at a flat, featureless surface with the plane of the ethene molecule parallel to the surface plane, we predict a maximum of four possible bands in the RAIRS or dipolar EELS spectrum.

2. For the example of the formation of ethylidyne on Pt (111), predict the number of possible bands in the RAIRS or dipolar EELS spectrum.

3. Given that a surface vibration is observed at 3000 cm^{-1} in a RAIRS experiment, calculate the energy loss in meV and the absolute energy in eV where

it may appear in an EELS experiment involving the use of an electron beam of energy 5.0 eV.

4. The same band as in question 3 in found to have some Raman activity. Calculate the wavelength and absolute energy that this band would appear at in a Raman experiment using as excitation source an argon ion laser operating at 488.0 nm.

5. The molecule HCl is found to adsorb on a Pt (110) surface with the H−Cl bond parallel to the surface plane and aligned along the rows of atoms. Using the selection rules for impact scattering deduce the required orientation of the EELS experiment such that (a) H−Cl vibration is observed and (b) the H−Cl vibration is not observed.

6. In a RAIRS experiment, an infra-red beam of diameter 10 mm strikes a pt (111) single crystal surface at an angle of incidence of 88°. Determine the area of the surface sampled in the experiment.

7. Calculate the real and imaginary parts of the bulk dielectric function of a simple metal in the infra-red region of the spectrum given that the real imaginary parts of the refractive index of the metal in this spectral region are 3.0 and 30.0 respectively.

8. By means of a simple diagram, distinguish between a and p-polarised light in the context of a surface experiment and further distinguish between the p-perpendicular and p-parallel components.

9. Via the use of a microscopic dipole model, show how the so called surface selection rule arises for the detection of the vibrations of a species adsorbed at a metal surface.

10. Determine the critical angle for total reflection from the boundary between vacuum and a hypothetical solid medium with refractive index $n = 2.7$, $k = 0$ for infra-red photons of energy equivalent to 2000 cm^{-1}, determine the depth of penetration of the radiation into the solid at an angle of incidence of 60°.

11. Calculate the energy and frequency of the second harmonic generation (SHG) response produced via the interaction of a Nd;YAG laser operating at 1064 nm with a surface of an fcc metal.

12. Account for the fact that both SHG and SFG may be observed from the (001) surface of Pt and Si, but not from the (001) surface of GaAs.

13. An SFG experiment aims to measure the C−H vibrations in an adsorbed hydrocarbon species. Given that the pump laser is operating at a wavelength of 532 nm and that in a typical IR spectrum the C−H vibrations for the particular adsorbed species lie in the range 2800–3000 cm^{-1}, calculate the wavelength range that should be measured by the detector in the SFG experiment.

CHAPTER 8

Surface Structure Determination by Interference Techniques

WENDY R. FLAVELL

Department of Physics, UMIST, Manchester

1 INTRODUCTION

In the earlier chapters of this book, a range of techniques have been introduced which allow us to determine the chemical composition of a surface. Just as important as this is the way in which the atoms making up the surface are arranged relative to one another, as the geometry of the surface determines the way in which new molecules adsorb at the surface, and hence influences the reactivity of the surface. The problem of determining surface structure may be divided into two parts. First, we would like to know the *symmetry* of the surface atomic arrangement, in other words where a surface has some *long range order* we would like to know the size and shape of the repeat unit on the surface (the surface unit cell). Secondly, we would like to know the precise details of the atomic positions themselves, in other words the number of neighbours surrounding a particular atomic site, and the distances and direction vectors to each surrounding atom (in other words the *short range order* of the surface).

Not all solid surfaces will possess the first type of order, long range order, this will only normally be a property of the surfaces of single crystal materials. If we wanted to investigate the long range order in the *bulk* of a single crystal material, we would use the diffraction pattern obtained when a beam of X-rays, electrons or neutrons is incident on the sample. By carefully measuring the intensities of the diffracted beams, and comparing them to those calculated for various models of the bulk structure, we could also obtain the details of the atomic coordination, in other words, information about short range order in the bulk of the crystal. We shall see that diffraction techniques, notably electron diffraction techniques can be adapted to give information about the *surface*, rather than the bulk of the material, and that these techniques give analogous information to bulk diffraction

Surface Analysis — The Principal Techniques. Edited by John C. Vickerman
© 1997 John Wiley & Sons Ltd

techniques, i.e. details of symmetry *and* atomic coordination. The first part of this chapter deals with these electron diffraction techniques.

There are many important classes of material, such as polycrystalline and amorphous materials, glasses and gels, where no long range order exists, and diffraction techniques cannot be used. In these cases one of the most important techniques currently available is based on creation of an electron wave at a particular atomic site (by absorption of a photon), followed by backscattering of the wave from neighbouring atoms. Such *absorption/scattering* techniques which give information about the *short range order* of the surface or bulk of a material are discussed in the second half of this chapter.

The techniques available for analysing surfaces are still developing at a rapid rate, and surface structure determination is no exception. In the last few years, there have been a number of important developments in the field, and a very wide range of interrelated techniques are now available. Some of the most important of these new techniques are discussed at the end of the chapter.

1.1   BASIC THEORY OF DIFFRACTION – THREE DIMENSIONS

Before considering the diffraction processes which may occur at surfaces, it is useful to remind ourselves of some of the important ideas underlying the treatment of diffraction from the bulk of a solid, as we will see that most of these general concepts may be readily extended to the treatment of surface diffraction.

The most widely used technique for studying bulk structure is X-ray diffraction. X-rays are scattered by the electron distribution around atom cores in the solid. As the X-rays are uncharged, this interaction is rather weak, with the result that the X-rays penetrate deeply into the solid and act as a bulk probe (although we shall see in Section 4 that it is possible to use X-rays to probe surface diffraction).

Diffraction effects will only be observed from crystalline solids. A crystal is distinguished from a glass or an amorphous material by possessing long range order. The crystal structure itself may be regarding as being made up of two parts, the *lattice* and the *basis*. The crystal lattice is a three-dimensional array of points which repeats itself periodically in all three dimensions, so providing the framework for the crystal structure. In fact we find that for this infinite repetition to occur, only 14 unique lattices are possible in three dimensions, known as the 14 Bravais Lattices. The basis is the arrangement of atomic positions around each lattice point. When this arrangement is superimposed on the crystal lattice, the full crystal structure results,

$$\text{lattice} + \text{basis} = \text{crystal structure.}$$

The *unit cell* for a particular structure is usually the simplest possible unit of the structure which contains everything which is unique about that structure, so that if the unit cell is repeated infinitely in space in three dimensions, the macroscopic crystal structure results. The unit cell lengths (the cell parameters) are usually denoted a, b, and c (and the corresponding vectors in these directions a, b, and c), and the cell angles (which need not be 90°) α, β, γ. In the simplest possible case of a cubic unit cell (e.g. NaCl), we have $a = b = c$ and $\alpha = \beta = \gamma = 90°$.

Any plane of atoms within a crystal structure must be defined by three coordinates, and we find it convenient to label planes using *Miller Indices*, (h, k, l). The Miller indices of a plane are obtained by calculating the intercepts of the plane with the a, b and c axes, as a fraction of a, b, and c, then taking the reciprocals of these numbers, and expressing the result in the form of the smallest whole numbers. An example is shown in Figure 8.1. A bar above a figure is used where the intercept with the a, b or c axis is negative — for example the six faces of a cube of length a are denoted (100), (010), (001), ($\bar{1}$00), (0$\bar{1}$0) and (00$\bar{1}$). This illustrates an important and useful property of Miller indices — wherever an intercept is infinite, the corresponding index is zero. We know, for example, that the (100) plane is parallel to the b and c axes.

The Miller indices of a plane bear a reciprocal relationship to the real intercepts of the plane with the axes; in the same way the wavelength of the X-ray beam (a real distance) bears a reciprocal relationship to the wavevector, k, of the beam, whose magnitude is defined as:

$$k = 2\pi/\lambda. \tag{1}$$

The wavevector is an important quantity, as it is a measure of the momentum of the incident and diffracted beams* — the change in wavevector of a beam on

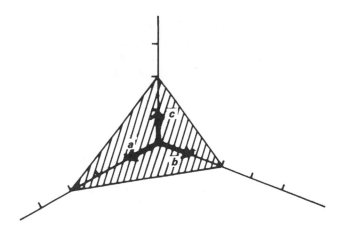

Figure 8.1. An example of the derivation of Miller indices for a plane. The plane has intercepts with the axes at 3a, 2b, 2c. The reciprocals of these numbers are 1/3, 1/2, 1/2. The smallest whole integers having the same ratio are 2, 3, 3, so the Miller indices for this plane are (233)

---

* Remember that the inverse relationship between the wavelength and the momentum of a photon comes from the deBroglie equation, $\lambda = h/p$, where p is momentum and h is Planck's constant — hence $k = p/\hbar$.

scattering from a plane of atoms will determine the direction of any emergent diffracted beams. Thus, the diffraction pattern obtained is a reflection of changes in wavevector. It is thus very useful when treating diffraction to work not in 'real' space, but in reciprocal space, which bears an inverse relationship to real space. Instead of the real crystal lattice, where diffraction is occurring, we can create a reciprocal lattice, where the distances between points are inversely proportional to the corresponding distances in the real lattice, but are a direct measure of k (hence the alternative name of 'k-space'). A diffracted beam of X-rays will emerge from the crystal whenever constructive interference occurs between successive planes of atoms in the real lattice. The advantage of the reciprocal lattice is that we may easily determine when this will be, by applying the law of conservation of momentum in our reciprocal lattice, where distances are directly proportional to momentum. The geometrical construction which we employ to do this is known as the *Ewald sphere construction*. This is shown in Figure 8.2, for a cubic lattice. The lattice shown here is the reciprocal lattice, where the distance between adjacent lattice points is $2\pi/d$, with d the distance between points in the real lattice. Clearly, the full reciprocal lattice is three-dimensional, as is the real lattice, but it can be shown on the page in only two dimensions. A vector, k_0, is drawn to scale on this diagram, with its tip pointing towards the origin of reciprocal space, (000), representing the wavevector of the incident X-ray beam. A circle is then drawn with radius $|k_0|$, and centre the origin of the vector, point P. This is the Ewald sphere. Its significance is that it maps the magnitude of k_0 onto the reciprocal lattice — obviously, by conservation of momentum, no diffraction events can occur outside this sphere. If any of the reciprocal lattice points are intersected by the Ewald sphere, then the condition for elastic scattering is satisfied (i.e. there is a change in momentum of the beam, but not its energy), with the scattered beam having a wavevector k'. By conservation of momentum,

$$k_0 = k' + g, \tag{2}$$

where the change in momentum on scattering is represented by the reciprocal lattice vector, g. Note that for elastic scattering, $|k_0| = |k'|$, i.e. there is a change in *direction* of the beam, but not in the *magnitude* of the momentum of the beam, and this is represented by the radius of the Ewald sphere. In Figure 8.2, the diffraction angle is shown as 2θ, in accordance with convention[†]. It can be seen that

$$\sin\theta = \frac{|k_0|}{|g|/2} \tag{3}$$

Pythagoras' theorem (in three dimensions) gives us the result that

$$|g| = (h^2 + k^2 + l^2)^{1/2} 2\pi/d \tag{4}$$

[†] Note that in the treatment of diffraction in two dimensions (Sections 1.2.2 and 2.2.3.1), the diffraction angle is often denoted as θ, not 2θ; this arises in particular from the special case where a beam is incident along the surface normal (Section 2.2.3.1).

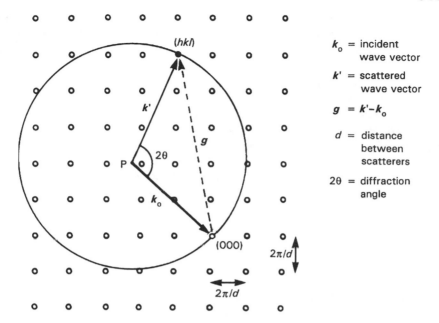

Figure 8.2. The Ewald sphere construction for a cubic lattice, (hkl) are the Miller indices of the point on the Ewald sphere. Reproduced by kind permission from ref. [1]

and in addition, we know that $|k_0| = 2\pi/\lambda$. Hence we obtain

$$\sin \theta = (h^2 + l^2 + k^2)^{1/2}\lambda/2d, \tag{5}$$

or the familiar Bragg condition,

$$n\lambda = 2d \sin \theta, \tag{6}$$

where $n = (h^2 + l^2 + k^2)^{1/2}$, known as the 'order of diffraction'.

The Bragg condition leads us to expect that diffraction will occur for all possible values of $(h^2 + k^2 + l^2)$. However, in many crystal systems, this is not the case, as in some cases the diffracted beams from one sub-set of planes may be exactly cancelled by diffracted beams from another sub-set having the same amplitude but exactly opposite phase. This leads to the absence of a diffracted beam at some θ values where one is expected, i.e. to *systematic absences*. As an example, in the body centred cubic system, only beams with $(h + k + l)$ even are found. These patterns of systematic absences are particularly useful in determining the symmetry of the lattice. By relating the interplanar spacing, d, in the Bragg equation to the unit cell lengths, a, b and c, the angles at which diffraction occurs may be used to calculate the unit cell size.

We can thus obtain information about the crystal lattice size and symmetry, by an analysis of the positions of the diffracted beams. However, this does not give us information about the arrangement of atoms within the lattice, i.e. the crystal basis. The exact atomic positions can only be obtained from an analysis of the intensities, rather than the positions of the diffracted beams. Each atom within a lattice scatters X-rays to an extent which is dependent on the charge distribution around the atom (i.e. on the number of electrons surrounding the atomic core), so that each atom i has an atomic scattering factor, f_i associated with it. The scattering produced by the crystal as a whole will depend on the number of atoms in each unit cell, and on their positions relative to one another (as scattering from atom i, say, may interfere constructively or destructively with scattering from atom j). This is summarised in the *structure factor*. F_{hkl}, which expresses the scattering amplitude expected for each diffracted beam, and is obtained by summing the individual atomic scattering factors over all the atoms of the unit cell, taking into account their phase differences:

$$F_{hkl} = \sum_i f_i \cos[2\pi(hx + ky + lz)], \tag{7}$$

for a centrosymmetric lattice. In this expression, the atomic scattering factors are summed over i atoms in the unit cell, the cos term representing the correction for their phase differences. In the case of a non-centrosymmetric lattice, this term is complex. x, y and z are the fractional coordinates of the ith atom in the cell. Unfortunately, the intensity of a wave is the square of its amplitude, so that the intensity, I_{hkl}, of a spot with Miller indices (hkl) is related to F_{hkl} through

$$I_{hkl} \propto |F_{hkl}|^2 \tag{8}$$

This means that a measurement of the intensity of a diffracted beam tells us only the magnitude of the structure factor F_{hkl}, and not its sign (phase), so important information which we need to determine the arrangement of atoms within the lattice is lost. This difficulty is known as the *phase problem*. It creates some complexity in the determination of atomic positions from diffraction data, but in general the problem is not insuperable. A number of methods have been developed to allow atomic positions to be extracted from the experimental data. It is beyond the scope of this book to discuss these methods in detail. However, in general, the methods rely on making an initial guess at the atomic positions, consistent with the symmetry of the lattice determined from the positions of the diffracted beams. Results from other experiments, such as spectroscopic data, and general chemical knowledge of the expected coordination of particular atoms/ions may feed into this initial guess. The intensities of the expected diffracted beams are then calculated for this arrangement, as a function of a change in the diffraction conditions, usually a change in the angle of incidence, or azimuthal rotation of the sample around the surface normal. This calculation is compared with a detailed set of experimental measurements of a large set of diffracted beams, and the guessed structure is refined as a result. A new set of intensities is calculated, and the procedure is repeated until satisfactory agreement is obtained between theory and experiment.

The interaction of the X-ray beam with the solid is quite weak, as we have already noted. This means that in the calculation of diffracted beam intensities, we may assume that each scattered beam undergoes only one scattering event before emerging from the solid. The theory of X-ray diffraction is said to be *kinematic*, in other words, multiple scattering effects are ignored, which considerably simplifies the theoretical treatment. We shall see that we can no longer make this assumption when dealing with an electron beam, which interacts strongly with the solid.

1.2 EXTENSION TO SURFACES – TWO DIMENSIONS

We have seen that in bulk structure measurements, X-ray diffraction is the most commonly used method. X-rays are essentially scattered by the charge distribution around atoms; this scattering is very weak, so that X-rays penetrate materials very deeply, and bulk structure can be probed. Neutrons are even more weakly scattered by solids. These properties of X-rays and neutrons, while useful for bulk measurements, mean that these beams are not the first choice for surface sensitive measurements. (This does not mean that they cannot be used in such experiments — see Section 4.) Electrons, on the other hand, as charged particles, interact very strongly with matter; we have already seen in previous chapters that the mean free path length of low energy electrons (say <500 eV) is very short, of the order of a few tens of Å. In addition, the wavelength of such electrons is of the order of an Ångstrom, slightly smaller than a typical interatomic spacing, and hence suitable for diffraction experiments. These properties, combined with the ease of producing a monochromatic electron beam, make electrons the primary tools for surface diffraction measurements.

The two most important 'surface diffraction' techniques currently being used are Low Energy Electron Diffraction (LEED) and Reflection High Energy Electron Diffraction (RHEED). These are discussed in Section 2. First, however, we will discuss the main ways in which our theoretical treatment of diffraction is adapted to apply to surfaces.

Firstly, if we are using a very surface sensitive probe, such as a beam of low energy electrons to make the measurement, it may be reasonable to suppose that we will be observing diffraction from the topmost layer of ordered atoms, i.e. a *two-dimensional surface unit mesh*, rather than a three-dimensional unit cell, as with a bulk measurement. This reduction in the number of dimensions means that instead of the 14 Bravais lattices which are possible in three dimensions, there are only five possible unit meshes, or *surface nets*, which may be repeated infinitely in two dimensions to build up the planar, periodic net of the surface. In this net, every lattice point can be reached from the origin by translation vectors,

$$T = ma_s + nb_s, \tag{9}$$

where m and n are integers, and the vectors a_s and b_s define the unit mesh, with the subscript s denoting the surface. The five surface nets are illustrated in Figure 8.3 [1]. As in the case of three dimensions, generation of the complete surface structure requires us to attach the basis atoms to the unit mesh consistent with certain symmetry restrictions. In fact, we find that only 17 two-dimensional space groups are possible.

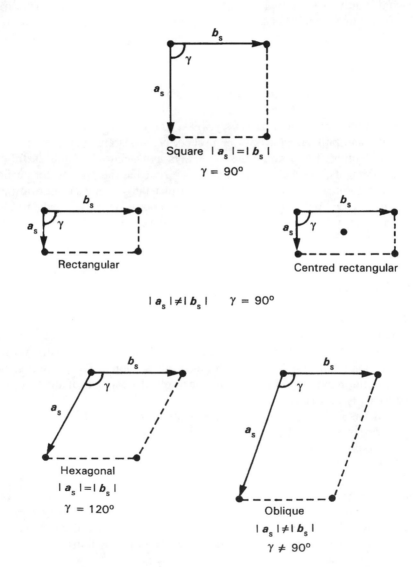

Figure 8.3. The five surface nets. Reproduced by kind permission from ref. [1]

1.2.1 Notation for Surface Structures

An ideal surface may be identified easily by reference to the bulk plane of termination, e.g Pt(100), Ni(110), NaCl(100). However, it is very common for the atoms in the topmost surface layer to rearrange themselves into a new net which is not a simple termination of the bulk; this is known as *surface reconstruction*, and occurs essentially because the minimum energy conformation for the atoms at a newly created surface, which have a reduced number of nearest neighbours, may not be the same as for the same atoms in the bulk of the material. We need a form of notation which describes the orientation of the new net of the reconstructed surface on the bulk, and which can also be used to describe the orientation of adsorbed overlayers on any surface. If the periodicity and orientation of the surface net is the same as the underlying bulk lattice, the surface is designated (1×1) (i.e. unreconstructed). However, it is more general for the translation vectors of the surface net to differ from those of the underlying lattice, so that

$$a_s = Ma, \quad b_s = Nb, \tag{10}$$

where a and b are the translation vectors of the ideal, unreconstructed surface. The nomenclature for this structure is $(M \times N)$. If, in addition, the surface net is rotated with respect to the underlying lattice by an angle ϕ degrees, the notation becomes $(M \times N)R\phi$. If the surface net is best described using a centred, rather than a primitive net (i.e. one with a surface lattice point at the centre), this is indicated as $c(M \times N)$. If the overlayer consists of an adsorbate, rather than simply reconstructed substrate atoms, this is also usually indicated. These points are illustrated in Figure 8.4.

1.2.2 The Ewald Sphere Construction in Two Dimensions

In Section 1.3, some basic ideas about reciprocal space were discussed, and the Ewald sphere construction for three-dimensional diffraction was introduced. This construction is also helpful to us when considering two-dimensional diffraction from surfaces, and is shown in Figure 8.5. It can be seen that instead of showing reciprocal lattice points (as in the case of diffraction in three dimensions, Figure 8.2), the diagram shows reciprocal lattice *rods*. This arises because, if the surface forms a completely two-dimensional net, the periodic repeat distance normal to the surface is infinite. As we saw earlier, the distance between adjacent points in a reciprocal lattice is inversely proportional to the corresponding distance in the 'real' lattice. This means that the reciprocal lattice 'points' along the surface normal are infinitely dense, forming rods (Figure 8.5). The diffraction condition is satisfied for every beam that emerges in a direction along which the sphere intersects a reciprocal rod. By using a construction similar to that used in Figure 8.2) (for three dimensions), we can show that this corresponds to a Bragg relationship (Section 2.2.3.1), although this condition is often expressed slightly

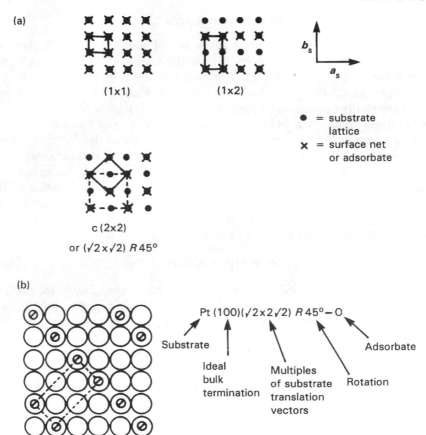

Figure 8.4. Surface net nomenclature: (a) some examples of surface nets, with their corresponding notation; (b) an example of full notation for a real surface (Pt (100) exposed to oxygen)

differently to the condition used in three dimensions. The diffracted beams, which each produce a spot in a LEED pattern, are indexed according to the reciprocal lattice vector which produces the diffraction. Because of the loss of periodicity in one dimension, only two Miller indices, h and k are needed to label a reciprocal lattice rod.

One important consequence of the loss of periodicity normal to the surface is that the conditions for observation of a diffraction pattern are relaxed relative to those for bulk diffraction. This is because observation of a diffraction pattern from the bulk rests on the constructive interference of the outgoing waves in this direction, which can no longer occur when the structure is not periodic in this dimension. This relaxation has the result that diffracted beams may occur at all

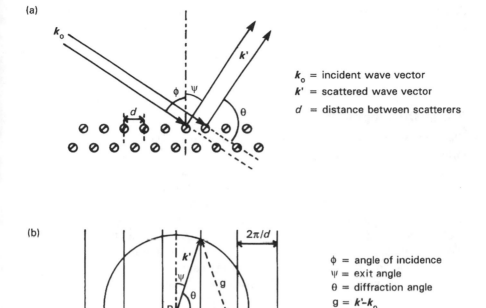

Figure 8.5. A schematic diagram of a diffraction process occurring at a surface in real space (a), with the corresponding Ewald sphere construction in reciprocal space shown in (b) (see text)

energies, and not just at certain discrete energy values, provided the corresponding rod lies within the Ewald sphere.

2 ELECTRON DIFFRACTION TECHNIQUES

2.1 GENERAL INTRODUCTION

An electron beam of energy around 150 eV has a wavelength of around 1 Å, making it suitable for electron diffraction experiments. However, this energy is roughly at the minimum in the universal pathlength curve, giving these electrons optimum surface sensitivity. The elastic backscattering of low energy electrons incident normally on a crystal surface forms the basis of the technique of Low Energy Electron Diffraction (LEED). An alternative is to use high energy electrons incident at a grazing angle on the crystal surface. In this case the penetration depth of the electron beam into the surface is also very small, as the component of the incident electron momentum normal to the surface is very small. This forms the basis of Reflection High Energy Electron Diffraction (RHEED). LEED and RHEED are discussed in greater detail in Sections 2.2 and 2.3.

2.2 LOW ENERGY ELECTRON DIFFRACTION (LEED)

2.2.1 Development

The first experiments which showed that electrons could be diffracted by crystalline solids, in the same way as X-rays, were conducted almost simultaneously by Davisson and Germer and by Thomson and Reid, in the late 1920's. The latter observed diffraction of a beam of electrons transmitted through a thin metal foil. However, it was Davisson and Germer who might be said to have performed the first LEED experiment, when they observed diffraction effects in the electrons backscattered from a single crystal of nickel [2]. At the time, these experiments provided evidence of the wave properties of electrons, and the wavelength of the electrons was found to be consistent with the value of h/mv (Planck's constant divided by the momentum of the electron) predicted by the then new theories of wave mechanics.

However, despite these initial experiments, the technique was not developed further until 1960, when further work by Germer and co-workers led to the development of the modern LEED display system [3]. It is only since the development of UHV technology that LEED has been widely studied and used.

2.2.2 Experimental Arrangement

A typical experimental arrangement used in a LEED experiment is shown in Figure 8.6. An electron beam of variable energy is produced by an electron gun, and is incident on the sample. The electrons are then backscattered from the sample surface onto a system of grids surrounding the electron gun. The backscattered electrons are of two types; elastically scattered electrons forming a set of diffracted beams which create the LEED pattern, and inelastically scattered electrons, which may make up 99% of the total flux, but which are not required. After reaching the first grid, G1, which is earthed, the elastically scattered electrons are accelerated towards the fluorescent screen, S, which carries a high positive potential (of the order of 5 kV). This provides the electrons in the diffracted beams with enough energy to excite fluorescence in the screen, so that a pattern of bright LEED spots is seen. The grids G2 and G3 are held at an adjustable negative potential, and are used to reject the majority of the electron flux, which is made up of inelastically scattered electrons, and which otherwise contribute to a bright, diffuse background across the whole of the LEED screen. The potential on these grids is adjusted to minimise the diffuse background to the LEED pattern.

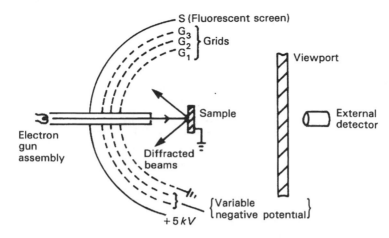

Figure 8.6 Schematic diagram of conventional RFA-type LEED optics

The LEED pattern which is observed may be recorded using a still or video camera mounted onto a chamber window placed directly opposite the LEED screen. This of course has the disadvantage that parts of the pattern will be obscured by the experimental arrangement around the sample, and possibly by any sources or detectors used for other techniques and mounted on the same UHV chamber. This arrangement is known as 'front view LEED', and has remained very popular as the screen and grid arrangement is essentially a retarding field analyser, (RFA), and may also be used for Auger spectroscopy. The problem of obscuring of the LEED pattern can be alleviated to a large extent by the use of 'reverse' or 'rear-view LEED', where the pattern may be viewed from a window placed on the rear side of the screen system (Figure 8.7), but this results in technical difficulties in designing a grid system which will function as a high resolution Auger system. These problems have only recently been overcome, and the first combined rear-view LEED and Auger systems are now commercially available.

In many laboratories, the type of LEED system described above is used almost entirely for measuring spot positions on the screen. As we shall see, just as important as this are the intensities of individual spots, and their widths (particularly in applications where surface phase transformations are being measured). Spot intensities can be measured using a conventional system, but accurate analysis of spot profiles can be difficult. For this purpose, the spot profile analysis LEED system (SPA-LEED) has recently been developed by U Scheithauer and co-workers [4]. A schematic representation of this system is shown in Figure 8.8. An electron beam passes through a series of deflection plates, which provide two octopole fields, and strikes the crystal sample. The diffracted beam is detected by a channeltron. The angle between the electron gun and the channeltron is fixed; a spot profile is obtained by scanning the potential on the electrostatic deflection plates.

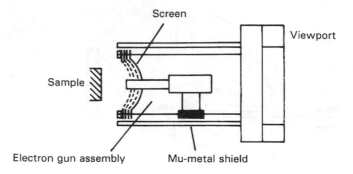

Figure 8.7. Schematic diagram of a reverse view LEED/Auger system (courtesy of Omicron)

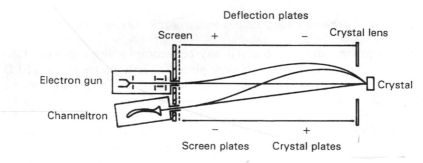

Figure 8.8. Schematic diagram of the SPA-LEED system, showing the beam paths with and without deflection voltages applied. Reproduced by kind permission of Elsevier Science – NL from ref. [4]

As shown in the digram, this has the effect of changing the angle of incidence of the beam. At constant beam energy, an arc is described through reciprocal space. In this mode, a small current of the order of 0.1–50 nA is used, to scan individual spots. As no mechanical movement is involved, and the deflection plate voltages are under computer control, spot profiles can be recorded with speed and accuracy. An overview of the diffraction pattern may be obtained using larger gun currents of up to 10 µA; the resulting pattern is then observed on a glass phosphor screen, which is viewed from the electron gun side (Figure 8.8). In this mode, the system has higher spatial resolution than RFA systems, as the screen–sample distance is greater than in these conventional systems.

2.2.3 Elements of the Theory of LEED

In the LEED experiment, information may be obtained from analysis of both the spot positions and their intensity profiles. The former is much more straightforward than the latter, and this is where we begin our discussion.

2.2.3.1 Analysis of Spot Positions

The simple production of a LEED photograph, without an analysis of the intensities of the individual spots, is by far the most widespread use of LEED. In most surface science laboratories around the world, LEED is often routinely used to check the cleanliness and order of surfaces being prepared for other experiments. Because of the sensitivity of LEED to surface contamination and surface roughness, the appearance of a LEED pattern with bright, sharp spots is widely regarded by surface scientists as evidence of a completely clean, ordered surface. (In fact, our assumptions here are beginning to be called into question, as we can now image these surfaces directly using scanning tunnelling microscopy (STM) or atomic force microscopy (AFM) (Chapter 9). It appears that surfaces which show sharp LEED patterns can sometimes look like the surface of the moon in STM/AFM! [5].)

In addition to its routine use in sample preparation in UHV, the pattern of LEED spots can be used to obtain information about surface symmetry or surface reconstruction, or about imperfections in the surface, such as steps or islands. It can also be used to determine whether any molecules on the surface are adsorbed in an ordered or random way. If an overlayer is ordered, its surface unit mesh size can be determined, and if the layer is adsorbed commensurately with the substrate, its orientation relative to the underlying substrate may be determined.

We have seen in Section 1.3.2 that in LEED, diffracted beams may occur at all energies, provided the corresponding rod lies within the Ewald sphere. Changing the incident beam energy will change the radius of the sphere, $|k_o|$, and so the number and directions of the scattered beams will vary (Figure 8.5). We find that the LEED pattern contracts towards the specularly reflected beam as the incident electron beam energy is increased. Usually, the incident electron beam is normal to the surface, in which case a symmetrical LEED pattern is obtained, which converges towards the (0,0) specular beam (i.e. the centre of the pattern) as the beam energy is increased. (In fact, of course we do not observe a (0,0) spot, as the centre of the fluorescent screen is occupied by the electron gun.) This is because the energy of the incident electron beam, E, is given by

$$E = (\hbar^2/2m)k^2 \tag{11}$$

where

$$k = 2\pi/\lambda.$$

Thus the incident wavevector k_0 increases with increase in the electron beam energy, so that the size of the Ewald sphere increases, cutting more and more rods. This means that, as the beam energy increases, more and more spots will appear on our LEED screen, and the spacing between spots will progressively decrease, i.e., the pattern converges towards the centre of the screen.

The discussion above reminds us that the LEED pattern is a reflection of reciprocal space — the distance between adjacent points in the LEED pattern

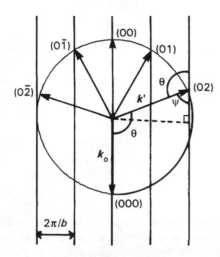

k_o = incident wave vector
k' = scattered wave vector
θ = diffraction angle
ψ = exit angle w.r.t. surface
   normal

Figure 8.9. The Ewald sphere construction for an electron beam incident normal to the surface (i.e. along the (0,0) direction, compared with Figure 8.5, ϕ is now zero.) Here we show only diffracted beams where the Miller index, $h = 0$

(reflecting the reciprocal lattice) is inversely proportional to the distance between points in the corresponding direction of the real surface unit mesh (the direct lattice). Figure 8.9 shows a Ewald sphere construction for an electron beam at normal incidence, i.e. along the (0,0) direction. (Comparing this with Figure 8.5, ϕ is now zero.) Note that each Ewald sphere construction can show only the diffraction conditions which are satisfied in one azimuth, rather than all the diffracted beams which are emerging in all directions from the crystal. This is because only that part of reciprocal space in the plane of the paper can be drawn. In Figure 8.9 we have chosen to keep the Miller Index h constant (and equal to zero), and vary the index k, corresponding in real space to probing diffraction from a row of atoms along the b-direction of the surface unit mesh.) The angle of diffraction between the incident wavevector and the backscattered beam having (h,k) equal to (0,2) is shown as θ. From the construction in the diagram, we can see that, for the (0,2) beam,

$$\sin\theta = \sin\theta' = \frac{2(2\pi/b)}{|k_0|} \tag{12}$$

or, as $|k'| = |k_0| = 2\pi/\lambda$, where λ is the wavelength of the electron beam,

$$\sin\theta = 2\lambda/b \tag{13}$$

for this reflection, or more generally for this azimuth,

$$b\sin\theta = k\lambda. \tag{14}$$

The integer k is sometimes known as the *order of diffraction*. The rows giving rise to the (0,2) beam will, by definition of Miller indices, be spaced a distance $b/2$ apart, (and will be parallel to a). Thus we arrive at the condition

$$\lambda = d_{02} \sin \theta. \tag{15}$$

Using a Ewald sphere for the perpendicular direction $((0,0)\ldots(h,0))$ we can arrive at the corresponding condition for diffraction from atoms along the a axis, i.e.

$$a \sin \theta = h\lambda. \tag{16}$$

We have thus identified the conditions which must be satisfied in order to observe diffraction from two perpendicular sets of rows of atoms. In order to observe diffraction from our two-dimensional surface, both these conditions must be satisfied simultaneously. In the case of a simple cubic lattice $(a = b)$, we find the more general result

$$a \sin \theta = (h^2 + k^2)^{1/2}\lambda \tag{17}$$

where $(h^2 + k^2)^{1/2}$ is the order of diffraction, sometimes written as n:

$$a \sin \theta = n\lambda. \tag{18}$$

equivalent to the Bragg condition for diffraction in three dimensions, but expressed slightly differently*.

In the case above of a simple cubic mesh, the mesh side 'a' may be easily obtained from a plot of $\sin \theta$ versus λ for any spot (h,k). Obviously, the sample–screen distance and the size of the screen must be known in order to calculate the diffraction angle. More generally, the observed LEED pattern will reflect the symmetry of the surface under study, and the surface unit mesh size may generally be easily obtained. Some examples of the possible types of two-dimensional lattice, and their corresponding LEED patterns are shown in Figure 8.10. In the case of a surface reconstruction of an ideal surface, or overlayer adsorption, new spots will usually appear in the LEED pattern. Figure 8.11 shows some examples of overlayer structures due to adsorbates or reconstructions, with their corresponding LEED patterns. Note the reciprocal relationship between the real overlayer structure on the surface and the spot density in the LEED pattern. It is important to remember that the surface atom arrangement can have *at most* the symmetry indicated by the LEED pattern; the true symmetry could be lower than that indicated by the LEED pattern. An example is shown in Figure 8.11, where we show the LEED pattern expected from a surface covered by domains of a (1×2) and a (2×1) overlayer. The

---

* NB: note that in our treatment, θ is the angle between the incident wavevector, and the wavevector of the backscattered beam; some treatments use 2θ, and the corresponding angle in the three-dimensional treatment is usually denoted as 2θ. From Figure 8.5(a), the general pathlength difference between successive diffracted beams is $d(\sin\theta + \sin\chi)$. Usually in LEED, $\theta = 0°$, and so the condition for constructive interference becomes $n\lambda = d\sin\chi = d\sin\theta$, as θ is $180° - \chi$ when $\theta = 0°$.

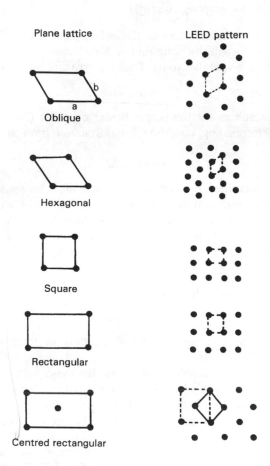

Figure 8.10. The five plane lattice types(see Figure 8.3) and their corresponding LEED patterns. Note the reciprocal relationship. In high-symmetry cases, systematic absences can occur — for example, compare the rectangular pattern with the simple rectangular pattern

LEED pattern which results is a composite of the individual patterns of both overlayers, achieved by averaging over the physical size of the LEED beam. The resultant pattern has four-fold symmetry, even though neither of the overlayers has this property.

From the above discussion, it would seem likely that there is a finite distance over which LEED can detect features such as disorder occurring. In fact, because of the energy spread of the incident beam, and its angular divergence, the electrons have a limited *coherence length* at the surface, typically $\approx 50\text{--}100$ Å, depending on beam energy. Features which occur at larger scales than this will not be detected. For example, it will generally be difficult to distinguish an adsorbate forming large islands at the surface from one forming a uniform overlayer.

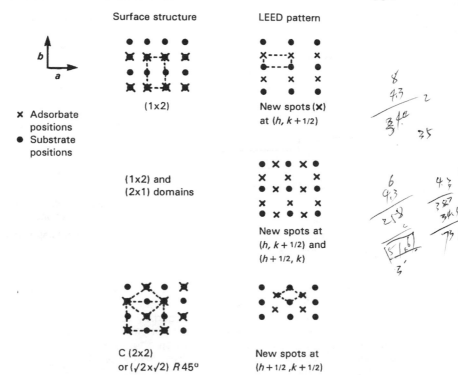

Figure 8.11. Some possible overlayer structures caused by absorbates or reconstructions, with their corresponding LEED patterns

Features such as surface steps give rise to identifiable features in LEED. In the case of a regular array of steps, the effect is to increase the repeat unit of the surface mesh in the direction perpendicular to the steps, as illustrated in Figure 8.12. This gives rise to a corresponding decrease in spot spacing in the corresponding direction in the LEED pattern as illustrated. Irregular steps give rise to streaking of spots in the direction of disorder (Figure 8.12). Facetted surfaces may be distinguished in LEED by changing the beam energy. We have seen that for a perfect surface, the LEED pattern converges to the (0,0) specular beam as the beam energy is increased. In the case of a facetted surface, there are different (0,0) beams corresponding to different facets, and this should be apparent on increasing the beam energy.

2.2.3.2 Analysis of Spot Intensity Profiles

We have seen that the positions of the spots appearing in the LEED pattern may be used to determine the size and symmetry of the surface unit mesh. This is analogous to the situation in X-ray diffraction in three dimensions, where the

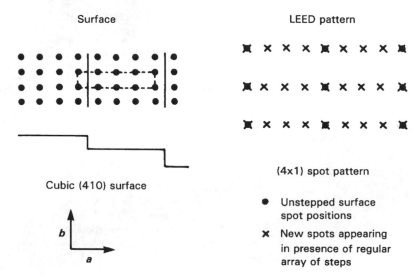

Figure 8.12. The effect on a LEED pattern of steps on a surface, where the distance between steps is less than the coherence length of the technique. In the example, the surface repeat unit is enlarged by four times in the horizontal direction, leading to the appearance of a LEED pattern which is roughly (4 × 1) (but not exactly, as the real repeat length is from step to step edge, which is $(17)^{1/2}$). The presence of irregular steps on a surface will lead to streaking of the LEED spots in the direction of the disorder

positions of diffracted beams may be used to determine the size and symmetry of the bulk lattice. In X-ray diffraction, although as we have seen, there are complications due to the phase factor, it is generally possible to determine precise coordinates for the atoms inside each unit cell (forming the basis) from the intensities of the diffracted beams. We might therefore hope that it would be possible to determine the positions of the atoms within each surface unit mesh by measuring the intensities of the diffracted beams in the LEED pattern. Indeed, it is possible to measure the intensities of particular LEED spots as the diffraction conditions are changed, for example as a function of beam energy (known as $I(V)$ curves), or as a function of azimuthal rotation ($I(\psi)$).

Unfortunately, the theoretical interpretation of such data is extremely difficult. Paradoxically, this arises because of the extremely strong interaction between electrons and atoms, precisely the property which makes LEED useful to the surface scientist. Because the cross-section for elastic scattering (as well as inelastic scattering) of low energy electrons by atoms is high, it is possible for diffracted beams to be elastically scattered several times at the surface, and still emerge with a measurable intensity. This multiple scattering complicates the analysis of the intensities of the resultant beams. In contrast, as we have seen (Section 1.2), in X-ray diffraction, the interaction between X-rays and the charge distribution around atoms is very weak, and as a result, the probability of such multiple scattering effects may be regarded as negligible; each photon is backscattered after a single encounter with an atom. This type of scattering is known as *kinematic* and forms the basis of the theoretical treatment used to interpret X-ray diffraction. In LEED, however, multiple scattering effects cannot be neglected, and the intensity of a particular spot may only be obtained by adding together all the waves scattered into a particular direction from many different scattering sequences, taking into account their amplitude and phase differences. This is known as a dynamical theory, and is essential in the treatment of LEED.

Much of the development of a dynamical theory of LEED is due to Pendry [6]. The theoretical difficulty associated with computing the multiple scattering intensity is eased slightly by the fact that, although the cross-section for elastic scattering is high in LEED, the corresponding cross-section for inelastic scattering is high also. This means that the mean-free-pathlength of electrons in this energy range within the solid is of the order of a few tens of Å. There is thus a limit on the number of elastic scattering events which can occur before inelastic scattering destroys the coherence of the diffracted beam. Thus reasonable agreement between theory and experiment may be achieved using a limited number of multiply-scattered beams.

The normal procedure in this type of analysis is to determine the variation in intensity of the LEED beam intensities, as a function of some change in the diffraction parameters, such as azimuthal rotation, or variation in beam energy. A purely kinematic theory of LEED would predict that there would be no variation in spot intensity as a function of azimuthal rotation of the sample, as this does not vary the angle of incidence ϕ (Figure 8.5). In fact, such plots tend to show strong intensity minima at certain angles where strong multiple scattering can occur in some direction other than that being measured. Similarly, a kinematic theory would predict that an $I(V)$ curve would only show strong maxima whenever the incident electron wavelength satisfies the diffraction conditions discussed in Section 2.2.1 (known as 'Bragg peaks'). Usually, we are able to see a number of secondary peaks, caused by multiple scattering in such curves. An example is shown in Section 2.2.4.

An iterative procedure has been developed by Pendry to determine the geometrical arrangement of surface atoms within the surface unit mesh from an experimentally determined set of $I(V)$ curves. The starting point for the calculation is an initial guess at the arrangement of atoms on the surface, which is chosen to be consistent with the symmetry of the LEED pattern. The intensity of a number of the diffracted beams expected for this arrangement is then calculated as a function of electron beam energy. This is done by solving the Schrödinger equation for the electron wavefunction in the first few atomic layers of the solid. The resulting calculated $I(V)$ curves are compared to the experimental result, and the guessed atomic arrangement is adjusted, and a new set of curves is calculated.

The process is repeated until satisfactory agreement is obtained. In practice, this procedure is very difficult and requires enormous computational effort.

One major drawback of this type of treatment is that the amount of computer time necessary to solve a particular structure scales exponentially with the size of the problem. For example, if we are interested in just three atoms (say CO on a Cu atom), nine coordinates in space are involved. If we need 10 trials for each coordinate to obtain a good fit to the data, we need to run 10^9 trials for the system as a whole! One treatment aimed at cutting down the computational effort involved has been developed in the last few years by Rous and Pendry [7]. This is known as 'Tensor-LEED'. The procedure involves guessing an initial trial structure as close as possible to the expected structure. Perturbation theory is then used to move the *individual* atoms about by small amounts, until a good fit is obtained. Now for the example above, treating the three atoms independently of one another, we need 10^3 trials for each atom, and so only 3×1000 for the whole system. This means that if the contribution to the LEED data from each atom can be picked out independently (easiest for heavy atoms which scatter strongly), the computer time required now scales linearly with the complexity of the problem. In its original form, Tensor-LEED was restricted to structural models within 0.1 Å of the guessed reference structure (as first-order perturbation theory was assumed valid, so this limited the size of the pertubation possible). More recently, the model has been developed further by Oed, Rous and Pendry, into a second-order Tensor-LEED approximation, which appears to be valid for displacements up to ≈ 0.2 Å perpendicular and 0.4 Å parallel to the surface [8].

In general, in the analysis of LEED data, the agreement between experiment and theory is not always good, and there may sometimes be more than one computed structure which fits equally well with the data. This can lead to arbitrary and subjective assignments. An attempt to overcome this is the use of reliability factors (R-factors), which attempt to provide objective criteria for the quantitative evaluation of the closeness of curve-fitting. There are various ways of calculating R-factors, but in general these are designed to emphasis features of the LEED data which are very sensitive to structural details, such as peak positions and shapes. Usually, a fit to LEED data will have an R-factor associated with it; the lower the value, the better the fit. (As an example, Somorjai *et al.* use Tensor-LEED to obtain a structural analysis of the (2×1) surface reconstruction of β-SiC, and obtain a quite a good R-factor of 0.27 [9].)

Using the types of procedure described above, atom positions may now be obtained optimally to an accuracy of ± 0.01 Å, very close to the accuracy of bulk diffraction techniques.

2.2.4 Applications of LEED

LEED remains one of the most widely used tools for surface structure determination. Here we discuss just one or two of the many applications of LEED which

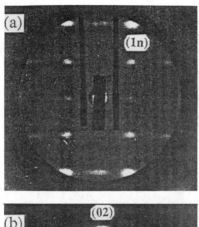

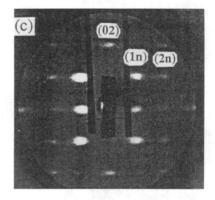

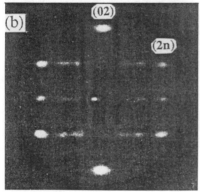

Figure 8.13. Photographs of LEED patterns at: (a) 94 eV; (b) 123 eV; and (c) 171 eV from the Be (1$\bar{1}$20) surface. The orientation of the surface corresponds to Figure 8.14. Reproduced by kind permission of Elsevier Science – NL from ref. [13]

may be found in the literature. Some other recent representative examples may be found by consulting the reference list [10–12].

Our first example concerns the surface reconstruction of a metal surface, Be (1$\bar{1}$20) [13]. Figure 8.13 shows the LEED patterns obtained from this surface at three different beam energies, and a temperature of −160°C. In Figure 8.13(a) recorded with a beam energy of 94 eV, the $\{1, n\}$ spots are evident. As the beam energy is increased (Figure 8.13(b)), these spots disappear, and the intensity is split into pairs of spots at the third-order positions $\{2/3, n\}$ and $\{4/3, n\}$, indicating a (1×3) reconstruction of the surface. At higher beam energy still (Figure 8.13(c)) the original integer-order, unsplit spots are regained. These observations may be explained by a surface reconstruction in which every third row of atoms on the surface is missing. Two possible ways in which this might occur were considered (Figure 8.14), corresponding to a simple missing row

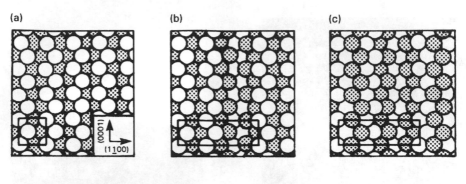

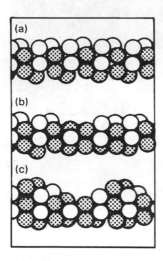

Figure 8.14. Top and side views of: (a) the bulk termination of Be (1$\bar{1}$20); (b) proposed surface structure based on the removal of every third surface chain; (c) the facetted surface produced by moving every third surface chain to the bridge site of the remaining two surface chains. Some atoms are shown hatched, merely to distinguish them. Reproduced by kind permission of Elsevier Science – NL from ref. [13]

(Figure 8.14(b)) and a situation where the removed row sits on top of the bridge sites between the remaining two rows (a facetted surface, Figure 8.14(c)). A simple kinematic analysis was initially applied to the data. For a simple stepped surface (Figure 8.14(b)) this predicts that when the scattering from the up-steps and down-steps is in phase, integral order diffraction spots will appear unsplit, whilst when the scattering is completely out-of-phase, the spot will be split in a manner related to the terrace width. The model therefore predicts that the split-spot intensity will oscillate completely out-of-phase with the integer-order spot intensity, as is observed. For this model, a value of the step height of

1.62 Å, 40% larger than the interplanar spacing in the bulk can be obtained. For the facetted surface (Figure 8.14(c)) the dependence of spot intensity on beam energy is more complex, and in particular the intensity of integral order spots does not vanish. The kinematic analysis therefore supports the missing row model (Figure 8.14(b)) However, as this treatment ignores multiple scattering effects, the analysis cannot be taken as definitive proof. Total energy calculations showed that in fact the facetted surface is likely to be more stable than the missing row structure, so a full dynamical calculation is necessary to resolve this problem.

Our next example is one where a full multiple-scattering calculation has been carried out, and concerns the geometry of overlayers of Sb on GaAs (110) [14]. Here the calculation was undertaken to distinguish between two alternative models proposed for the surface structure. Sb forms a simple overlayer on GaAs (110), denoted GaAs(110)p(1 × 1)-Sb. The two proposed structures for this overlayer are shown in Figure 8.15). In the experiments, normal incidence LEED $I(V)$ curves were obtained at 2 eV intervals in the range ≈50 eV–300 eV. Figure 8.16 shows some of the resulting $I(V)$ curves, for the most intense diffracted beams, compared with the best fit to the data which can be obtained from the multiple scattering calculation. A number of geometries were tested, including those shown in Figure 8.15. Very satisfactory R-values (of around

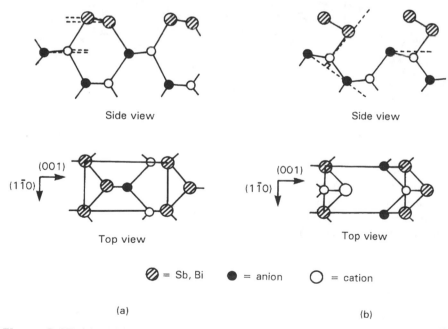

Figure 8.15. (a) and (b): two proposed models for the GaAs(110)p(1 × 1)–Sb system. Reproduced by kind permission of the American Institute of Physics from ref. [14]

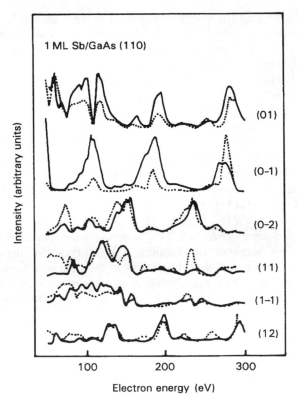

1 ML Sb/GaAs (110)

Intensity (arbitrary units)

(01)

(0-1)

(0-2)

(11)

(1-1)

(12)

100   200   300

Electron energy (eV)

Figure 8.16. $I(V)$ curves for the most intense diffracted beams for the GaAs(110)p(1 × 1)–Sb system (solid lines), compared to the best fit to the data which can be obtained from multiple scattering calculations. A good fit is only obtained for the model shown in Figure 8.15(a). Reproduced by kind permission of the American Institute of Physics from ref. [14]

0.2) were obtained for the model shown in Figure 8.15(a), but for the model in Figure 8.15(b), and for a disordered model, R-values were unsatisfactory (>0.3). It was therefore concluded that the geometry shown in Figure 8.15(a) is the most probable atomic geometry for Sb/GaAs (110).

One potential future development of LEED is in LEED holography [15]. The principle behind holography is to produce interference between two coherent waves that have travelled along different spatial regions. One of them (the reference wave) must be as simple as possible, and known, whilst the other (the object wave) is the unknown, to be found. In principle it is possible to obtain the amplitude and phase of the object wave by illuminating with the time-reversal of the reference wave. It could be possible to obtain three-dimensional holographic images of the surface. The development of holographic techniques using electrons has been hampered relative to light holography until recent years by problems with the coherence of the electron beam sources. However, it is clear that most of the techniques discussed in this chapter, which are based on the interference between electron beams, could lend themselves to holographic applications. Indeed, such experiments are now becoming possible for LEED, DLEED (diffuse LEED, produced by the secondary electron background, rather than the Bragg reflected beams), SEXAFS (surface extended X-ray absorption fine structure, Section 3.3) and PD (photoelectron diffraction, Section 4.3). The interested reader is referred to references [15] and [16], and references therein; photoelectron holography is discussed further in Section 4.3.

2.3 REFLECTION HIGH ENERGY ELECTRON DIFFRACTION (RHEED)

2.3.1 Introduction

An alternative electron diffraction technique for the determination of surface structure is Reflection High Energy Electron Diffraction (RHEED). Here, a relatively high energy electron beam (5–100 keV, electron mean free pathlength 20–100 Å) is used, but the electron beam is directed towards the sample at a very grazing angle of incidence. Surface sensitivity is obtained because the component of the incident electron momentum normal to the surface is very small (even though the electron mean free pathlength is longer than for the low energy electrons used in LEED), and the penetration of the electron beam is small. The high energy electrons are scattered through small angles, and sample only the first one or two atomic layers of the target material under these conditions. RHEED has been developed during the last thirty or so years, alongside LEED, and in parallel with the development of UHV technology. As we shall see, there are some disadvantages to the technique when compared with LEED, which have resulted in RHEED being less widely and generally used than LEED. However, in recent years, specific applications where the RHEED technique offers unique advantages have emerged; one of the most significant of these is discussed in Section 2.3.4. The grazing incidence arrangement means that RHEED is more sensitive to surface roughness on an atomic scale than LEED (where roughly normal angles of incidence are usually used); this property is exploited in the applications of the technique.

2.3.2 Experimental Arrangement

The experimental geometry employed in the technique is illustrated in Figure 8.17. A high energy (5–100 keV) fine parallel beam of electrons is incident on the surface at a very grazing angle of incidence (near $\phi = 90°$). A RHEED pattern may then be collected in a similar manner to LEED, for example with a retarding field analyser (RFA), so that electrons which have lost energy of

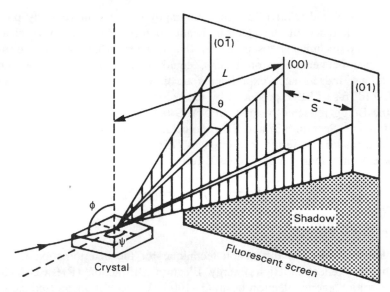

Figure 8.17. A schematic diagram of the experimental geometry of a LEED experiment, ϕ is the angle of incidence (which is close to 90°), ψ is the azimuthal angle,
and θ is the diffraction angle. Reproduced by kind permission from ref. [1]

more than a few electron volts are removed. The elastically diffracted electrons
may then be detected by a fluorescent screen and photomultiplier. Alternatively,
the intensity of specific diffracted beams may be monitored, for example by using
an optical fibre with a well-collimated entrance aperture and a photomultiplier.
Scanning High Energy Electron Diffraction (SHEED) attachments are available,
which enable the diffraction pattern to be scanned in a raster way in order to
build up a map of intensity of the RHEED spots along various directions in the
diffraction pattern.

2.3.3 Elements of the Theory of RHEED

The major difference in appearance between RHEED and LEED patterns is
caused by the higher energy (small wavelength, large $|\mathbf{k}_o|$, equations (1,11)) of
the incident beam in RHEED. The Ewald sphere (Figure 8.5) is now very large
compared to the reciprocal lattice vectors. This means that the Ewald sphere
cuts the (0 0) rod almost along its length (Figure 8.18) In the resulting RHEED
pattern, the (0 0) rod will give rise to a long streak, rather than a spot. Other reciprocal lattice rods which intersect the Ewald sphere will also give rise to streaks,
but in practice the sphere is so large that these are very few. In order to explore
the arrangement of reciprocal lattice rods in three dimensions, it is necessary to
change the angle of incidence, ϕ, so that additional diffraction conditions may

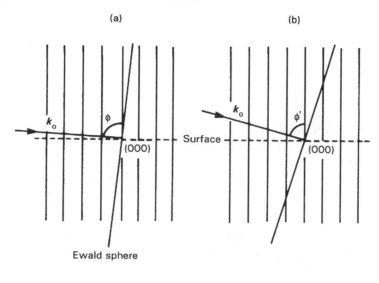

k_o = incident wave vector
ϕ, ϕ' = angles of incidence

Figure 8.18. The Ewald sphere construction of RHEED, at two slightly different angles of incidence, ϕ(a) and ϕ'(b). Because the sphere is very large, the sector shown is almost linear, and the angle of incidence relative to the surface normal must be reduced (as in (b)) to cut reciprocal rods away from (0,0)

be satisfied, as the Ewald sphere cuts other reciprocal lattice rods (Figure 8.18). For most general purposes, this requirement to change the diffraction geometry is something of a disadvantage when compared with LEED, where a large number of reciprocal lattice rods may be probed simultaneously, allowing a surface unit mesh size and arrangement to be quickly and easily obtained. A change in ϕ is usually obtained by rocking the sample about an axis in its surface, although this has the disadvantage that it changes the component of the incident electron beam normal to the surface, and thus changes the surface sensitivity of the technique during an experiment. As an alternative, the sample may simply be rotated about its surface normal, in which case, well-defined streak patterns will be produced from a single crystal surface when the incident electron beam lies along high symmetry directions of the surface.

The RHEED pattern may be used to obtain the size of the surface unit mesh. Taking s as the separation between the streaks in the pattern, we can see from Figure 8.17 that the diffraction angle, θ, is given by

$$\tan \theta = s/L, \tag{19}$$

where L is the distance between the sample and the screen. Because θ is very small in RHEED, it may be difficult to measure s. For this reason, RHEED cameras are constructed with L as long as possible, as this increases the streak separation on the screen (equation (19)). Using the diffraction condition obtained in Section 2.2.3.1 (equation (17)) for a square surface unit mesh, and remembering that in RHEED, θ is very small, so that $\sin\theta \approx \tan\theta$, we find

$$a = (h^2 + k^2)^{1/2}\lambda(L/s), \tag{20}$$

allowing a to be determined.

Due to the finite energy spread of the incident beam, and its angular divergence, the RHEED beam has a limited coherence length at the surface, analogous to that in LEED. This is longer than the corresponding length in LEED, and is typically of the order of 2000 Å. Disorder occurring on the surface on larger scales than this may be difficult to detect. Because of the grazing incidence used in the experiment, RHEED is extremely sensitive to surface roughness on scales smaller than the coherence length. It may be very difficult to obtain a surface which is sufficiently flat to give a clear pattern of streaks. In some cases, the high energy RHEED beam may pass through any protruberances on the surface, giving rise to bulk diffraction, and hence spots, rather than streaks in the RHEED pattern.

As in the case of LEED, information may be obtained about the arrangement of atoms within the surface unit mesh from an analysis of the variation in intensity of the diffraction features as a function of change in the diffraction conditions. As we have seen, the usual variables in this case are the angle of incidence, ϕ, or the azimuthal angle, ψ (Figure 8.17). Intensity data is usually in the form of 'rocking curves', $I(\phi)$, as the sample is rocked about an axis in its surface, changing the angle of incidence, or rotation diagrams, $I(\psi)$, produced as the sample is rotated around its surface normal. The cross-section for elastic scattering at the high beam energies used in RHEED is smaller than in LEED, which leads to less multiple scattering in the diffraction process. Unfortunately, however, the cross-section for inelastic scattering is also smaller than in LEED, which, as we have already seen, leads to a longer electron mean free pathlength. This means that diffracted beams may travel a longer distance through the solid before losing their coherence by inelastic scattering than they can in a LEED experiment. The net effect is that any accurate description of the intensities of the diffracted beams must include multiple scattering effects, and as for LEED, a *dynamical* theoretical treatment is essential.

2.3.4 Applications of RHEED: Temporal intensity variations in RHEED patterns during film growth of semiconductors by Molecular Beam Epitaxy (MBE)

Because of its sensitivity to surface roughness, RHEED has been extensively used in the study of thin surface coatings, surface disorder and processes such as surface passivation and hardening of metals. One of the most important applications of recent years has been its use to monitor the layer-by-layer growth necessary for the production for semiconductors (e.g. GaAs) by MBE. During MBE deposition, gated sources or effusion (Knudsen) cells of the elements are used to lay down successive alternating single atomic layers of the components (e.g. Ga and As in the case of GaAs), in a UHV environment. The ability to deposit material in this layer-by-layer (Frank–Van der Merwe or FV) manner is essential to the process, but is understandably a difficult process to control.

RHEED has been shown to be a particularly reliable monitor of FV growth. In this experiment, the intensity of the specular diffracted beam is simply monitored as a function of time during the layer growth process [17, 18]. The intensity exhibits very regular oscillations as a function of time (Figure 8.19). In the example shown (the growth of GaAs (001)) control of the Ga beam in the presence of a continuous flow of As is crucial, as the rate determining step is the sticking of As in the MBE growth process. The period of the oscillations corresponds exactly to the growth rate of one layer of GaAs in the [001] direction (i.e. one layer of Ga plus one layer of As). The maxima in the reflectivity correspond to atomically smooth surfaces — i.e. before deposition of a layer, and when deposition is complete ($\theta \approx 0$ and $\theta \approx 1$ in Figure 8.19). The reflectivity minima correspond to the most disordered surfaces, i.e. $\theta \approx 0.5$. As can be seen in Figure 8.19, the intensity of the oscillations progressively decreases as the overall surface roughness gradually increases as more and more layers are deposited. The RHEED technique for monitoring MBE growth is simple, practical and accurate. It is particularly well suited to the geometry of the MBE process — as the Knudsen effusion cells producing the molecular beams of the elements (e.g. Ga and As) generally occupy positions fairly directly in front of the substrate, LEED cannot be used to monitor the growth. However, the grazing geometry of RHEED means that it is one of very few techniques which can be used inside the MBE machine without interfering with the growth process (Figure 8.20).

As we have already noted, RHEED is particularly sensitive to surface imperfections. RHEED may therefore be used very effectively to study the presence of any imperfections on the resulting MBE-grown semiconductor wafer. Regular and random steps on GaAs (001) have recently been studied by Toyoshima et al. [19]. We saw in Section 2.3.3 that the presence of protruberances on the surface may lead to the appearance of diffraction spots, rather than streaks. This is well-illustrated in the experiments of Saiki et al. [20], where NaCl is grown heteroepitaxially on GaAs (111) and (001) surfaces. Deposition on the (001) surface results in a streaked RHEED pattern, suggesting that the NaCl film grown on this surface has a rather flat surface. However, deposition onto the GaAs (111)A surface, results in a spot pattern, suggesting a rough surface (Figure 8.21). A closer examination of the pattern reveals off-angle streaks, in every {001} direction, suggesting that the roughness takes the form of triangular pyramids with three exposed {001} facets.

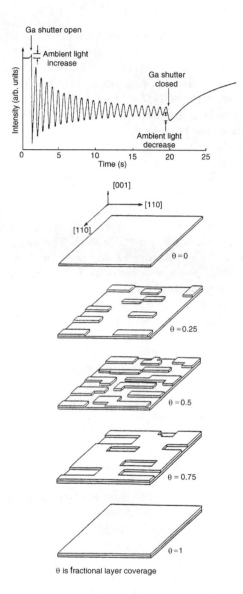

θ is fractional layer coverage

Figure 8.19. Intensity oscillations of the specular beam in the RHEED pattern from a GaAs(001) (2 × 4) reconstructed surface, during semiconductor growth by MBE. Intensity maxima correspond to atomically smooth surfaces ($\theta \approx 0$ and $\theta \approx 1$), while minima correspond to completely disordered surfaces ($\theta \approx 0.5$). The period of the oscillations then exactly corresponds to the growth rate of a single Ga + As layer. Inflections at the beginning and end of growth result from ambient light change as the effusion cell shutters are opened and closed. Reproduced by kind permission of Elsevier Science – NL from ref. [18]

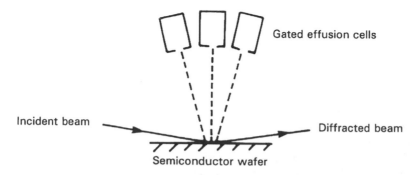

Figure 8.20. Schematic diagram illustrating the advantages of the RHEED geometry over LEED during semiconductor growth by MBE. Effusion cells are usually mounted directly in front of the sample, so that LEED cannot in general be accommodated. The grazing incidence geometry of RHEED means that it can be used to probe surface order during growth

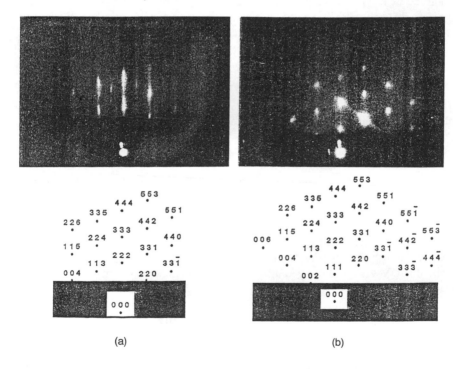

Figure 8.21. RHEED patterns taken before (a), and after (b) the growth of NaCl on a GaAs (111)A surface, showing the change between a streaked and a spot pattern. The beam is parallel to the [$\bar{1}$10] azimuth of the substrate, and the beam energy is 15 keV. Reproduced by kind permission of Elsevier Science – NL from ref. [20]

3 ABSORPTION/SCATTERING TECHNIQUES

3.1 INTRODUCTION

So far, our discussion has centred on surface diffraction techniques which give us information about the long-range order of a single crystal surface, or an adsorbate on the surface. However, we will see that there are many classes of material which do not possess long-range order, but where it is important to determine the details of local coordination, both in the bulk and at the surface. One of the most important techniques currently available which allows us to do this is Extended X-ray Absorption Fine Structure (EXAFS), and its surface modification, surface-EXAFS (SEXAFS). These techniques rely on two sequential processes. The first is the absorption of X-ray radiation by an atom in a condensed medium (solid or liquid), causing excitation of a core electron. The departing photoelectron wave is then backscattered by surrounding atoms in the medium, resulting in interference effects which modulate the X-ray absorption observed. This second step can be regarded as a type of 'local' diffraction, or scattering. As EXAFS and SEXAFS involve absorption, followed by scattering, we will consider the initial absorption step, before going on to consider the scattering which gives rise to the EXAFS itself.

3.1.1 X-ray Absorption in Solids

The parameter used to describe the absorption of photons by a medium is the absorption coefficient α, defined according to Beer's law:

$$I = I_0 \exp(-\alpha 1) \tag{21}$$

i.e. a photon beam having intensity I_0 at $l = 0$ is attenuated by absorption on travelling through a homogeneous medium; after travelling a pathlength l, its intensity is reduced to a value I. The extent of absorption, and thus the absorption coefficient α depend very strongly on the photon energy; the photon beam may be transmitted right through a medium unless it is of the correct energy to cause an excitation within the medium.

In the UV/visible energy range, photons are generally sufficiently energetic to cause excitation of the outermost valence electrons of a species in solution or an atom in a solid. Examples are the $d-d$ and charge–transfer transitions of transition metal ions in solution, which generally give rise to the colours of these solutions. In metallic solids, valence electrons may be excited into the empty density of states above the Fermi level. In semiconductors and insulating solds, photon absorption will begin when the UV/visible photons are sufficiently energetic to cause excitation of electrons from the filled valence levels across the forbidden band gap to the unfilled conduction levels. If higher energy UV photons are supplied, these electrons may be excited into an unbound continuum state above the vacuum level, i.e., ionised (Figure 8.22).

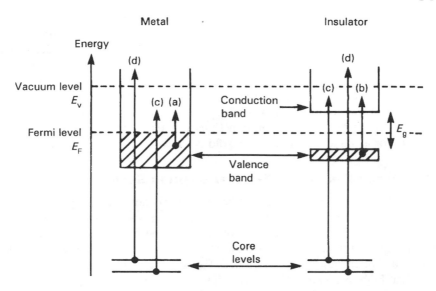

Figure 8.22. Optical absorption processes in solids for metals and insulators or semiconductors. E_g is the forbidden band gap of the semi-conductor. (a) Low-energy transition from the filled to empty band states of a metal. (b) Interband transition from valence to conduction band in an insulator or semiconductor, typically observed in UV/visible absorption. (c) Higher energy transitions from filled core levels to empty bound states below the vacuum level (d) Transitions to continuum states above the vacuum level, E_v corresponding to ionisation

In contrast, X-ray photons are sufficiently energetic to excite core electrons, either into empty valence states, or at higher energies into the ionisation continuum (Figure 8.22). The general form of the X-ray Absorption Spectrum (XAS) for an atom (such as, say, Xe) will consist of a series of core level excitation thresholds (each of which occurs close to the ionisation potential of the core level), corresponding to excitation from $n = 1, n = 2, n = 3 \ldots$ levels (i.e. K, L and M-edges) of the atom. In fact, although the K-edge is a single edge, higher edges are a collection of edges close together. This is because while the K-edge corresponds to excitation of electrons having $n = 1$ and orbital angular momentum $l = 0$, the L-edge corresponds to $n = 2$ ($l = 0, 1$) and states 2s, $2p_{1/2}$ and $2p_{3/2}$. The L-edge thus consists of three separate components known as L_1, L_2 and L_3. Similarly, the M-edge gives rise to M_1, M_2, M_3, M_4 and M_5 components, corresponding to states 3s, $3p_{1/2}$, $3p_{3/2}$, $3d_{3/2}$ and $3d_{5/2}$.

The probability, P_{if} of an optical transition occurring between an initial state $|\psi_i\rangle$ and a final state $|\psi_f\rangle$ is given by

$$P_{if} \propto |\langle \psi_i | \mu | \psi_f \rangle|^2 \tag{22}$$

where μ is the dipole operator. The contribution of each optical transition to the total optical absorption coefficient is proportional to the corresponding transition probability P_{if}. The study of the absorption coefficient as a function of photon energy, $\alpha(h\nu)$, provides important information about the initial and final states involved in the transitions. The magnitude of α is clearly strongly dependent on the overlap between the initial core level wavefunction and the final state wavefunction. After crossing the core level threshold (at the ionisation energy), α decreases, with decreasing overlap of the initial state wavefunction with the photoelectron wavefunction. However, absorption into the continuum will generally continue far above the absorption threshold. The general form of an XAS spectrum for a single absorption threshold is shown in Figure 8.23.

3.1.2 Fine Structure in X-ray Absorption Spectra

Figure 8.23 shows only an idealised form of an X-ray absorption edge. In general, in the case of species which are not isolated atoms, we find two types of fine structure superimposed on the X-ray absorption edge. The first of these, which gives structure up to around 50 eV above the absorption edge is known as Near Edge X-ray Absorption Fine Structure (NEXAFS), or alternatively as X-ray Absorption Near Edge Structure (XANES) (Figure 8.24). This near-edge structure is determined by the details of the final density of states, the transition probability, and resonance and many-body effects. As a result, the analysis of NEXAFS may be very complex, requiring in some cases theoretical background not yet available. This is an expanding area of research, and recent work has shown that it is possible to obtain information about surface structure from NEXAFS. For this reason, we will return to the subject of near-edge structure in Section 3.3.3.

In addition to the near-edge structure, the optical absorption coefficients of molecules and condensed media show fine structure which extends from about 50 eV above the absorption threshold for several hundred eV. This second type of fine structure is known as Extended X-ray Absorption Fine Structure, or EXAFS (Figure 8.24).

3.2 EXTENDED X-RAY ABSORPTION FINE STRUCTURE (EXAFS)

This structure is due to interference effects in the wavefunction of the excited electron. After absorption of a photon, this wavefunction propagates away from the atomic core where excitation occurred, and is partially backscattered by the surrounding atoms in the medium. Interference between the outgoing wave and the backscattered wave produces the extended oscillations observed above the absorption edge (Figure 8.24).

As the structure arises due to the presence of atoms around the core which absorbs a photon, the effect is not observed in the case of isolated atoms. We shall see that in the case of molecules and condensed media, it is possible to extract information about local coordination number and coordination distances from the amplitude and period of the EXAFS oscillations. At above 50 eV beyond the absorption threshold, the final excited electron state can be regarded as a nearly

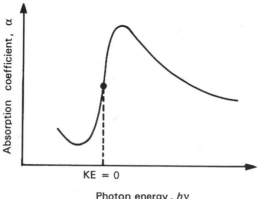

Figure 8.23. The general form of an X-ray absorption spectrum of an isolated atom for a single core-level threshold. In the treatment of EXAFS data (see section 3.2.2, equation (27)), we need to know the photon energy, $h\nu$, at which the kinetic energy of the photoelectron created is zero. This is usually chosen to be halfway up the absorption edge

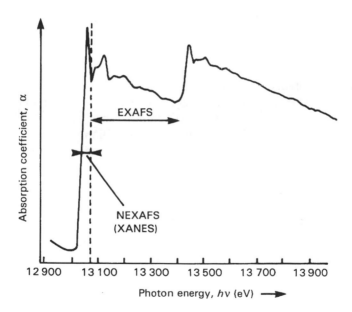

Figure 8.24. A typical X-ray absorption spectrum of a condensed medium. The sample is $BaPb_{1-x}Bi_xO_3$, and the figure shows the L_{III} absorption edge of Pb, with the NEXAFS and EXAFS regions indicated. Absorption above around 13 400 eV is due to the L_{III} edge of Bi, which is close in energy to the Pb edge

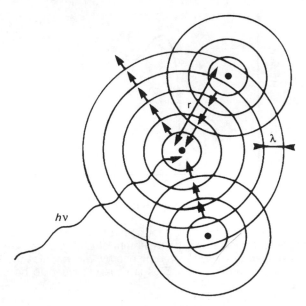

Figure 8.25. In the EXAFS process, the excited electron wavefunction (wavelength λ) propagates away from the atom at which photon absorption occurs, and is backscattered from surrounding atoms. The interatomic distance is *r*

free electron wavefunction — i.e. a nearly spherical wave centred on the emitting atom (like the ripples caused by throwing a stone into a pond) (Figure 8.25). The backscattering of this wave, and the resulting interference effects which give rise to the EXAFS can be described to a first approximation using atomic quantities which are independent of the chemical environment of the atom, and may be calculated or determined experimentally. The interpretation of EXAFS is thus much easier than interpretation of NEXAFS data.

3.2.1 The Importance of EXAFS

From the preceeding sections of this chapter, you will be familiar with the idea that the mean free pathlength of low energy (say 50–500 eV) electrons in solids is extremely small — of the order of a few atomic spacings in some cases. In addition, the amplitude of the outgoing spherical electron wavefunction is inversely proportional to the radius. This means that EXAFS arises from backscattering only from atoms very close to the emitting atom, with the nearest neighbours playing the most important role. By careful analysis, some information may be obtained about the second nearest neighbour coordination shell, or sometimes about more distant shells. EXAFS is thus a very important probe of local coordination and interatomic distances. This is in contrast to the diffraction techniques which we discussed in the first half of the chapter. These collect information simultaneously on a large number of atoms of the system, and give us information about long range order, for example in single crystals. In contrast, in EXAFS, if we select an absorption edge belonging to a particular atomic species of interest, we may probe the structure specifically around that chemical species. There are many applications where such information is very important, such as metal atom coordination in supported catalysts or in biomolecules such as haemoglobin. As the EXAFS oscillations are produced only by atoms close to the emitting atom, EXAFS studies are not limited to systems having long range order, like single crystals. Systems which have only a well defined coordination around a central atom may be studied. These include polycrystalline and amorphous materials, glasses, gels and solutions; examples include amorphous semiconductors, supported catalysts and biological systems where single crystal samples may be very difficult and expensive to grow.

3.2.2 Basic Theory of EXAFS

We have seen in a very qualitative way how the EXAFS modulations arise. Whether the interference between the outgoing and backscattered waves is constructive or destructive will depend on the phase shift between the two waves, in turn determined by the difference in pathlengths. If we assume, for the time being, that there is no phase shift due to the backscattering process itself, and that backscattering occurs only from the shell of nearest neighbours itself, then constructive interference will occur when the pathlength of the backscattered wave, $2r$ (Figure 8.25) is equal to a whole number of wavelengths,

$$n\lambda = 2r, \tag{23}$$

where λ is the wavelength of the excited electron, and n is an integer. If, on the other hand,

$$n\lambda/2 = 2r, \tag{24}$$

destructive interference will occur. For other values, some intermediate interference will occur. This interference modulates the amplitude of the final state wavefunction, which in turn implies modulation of the transition probability P_{if} (equation (22)) and thus the absorption coefficient α. Normally, our EXAFS spectrum will be recorded as a plot of the variation of α with photon energy, $\alpha(h\nu)$. Thus, we need to be able to relate the photon energy to the electron wavelength, λ. The wavelength λ can be related to the electron kinetic energy, E, by noting that

$$E = p^2/2m, \tag{25}$$

where p is the momentum, and m the mass of the electron, and remembering the de Broglie equation,

$$p = h/\lambda, \tag{26}$$

where h is Planck's constant. The excited electron only possesses kinetic energy if it is created above the ionisation threshold — and at the ionisation threshold itself E is zero.
Thus

$$E = h\nu - h\nu_{E_0} \tag{27}$$

where $h\nu_{E_0}$ is the photon energy at which the electron kinetic energy is zero. This is usually chosen as half-way up the absorption edge (Figure 8.23); however, there are limits on the validity of this assumption, and in practice, E_0 is used as an adjustable parameter.

By combining equations (25) (26) and (27) we arrive at the result

$$\lambda = h[2m(h\nu - h\nu_{E_0}]^{-1/2}, \tag{28}$$

in other words, the excited electron wavelength decreases as the photon energy $h\nu$ increases above the absorption threshold. Combining with equation (23) we find that constructive interference occurs whenever

$$(2r/h)[2m(h\nu - h\nu_{E_0})]^{1/2} = n, \text{ an integer.} \tag{29}$$

This means that if we acquire an EXAFS spectrum as $\alpha(h\nu)$, the period of the EXAFS oscillations increases with $h\nu$ (an example is seen in Figure 8.27). The analysis of the EXAFS is made easier if the absorption coefficient, α, is plotted not as a function of $h\nu$, but as a function of the electron wavevector, $k = 2\pi/\lambda$. We then obtain

$$k = (1/\hbar)[2m(h\nu - h\nu_{E_0})]^{1/2} \tag{30}$$

$$\text{or } k = 0.5123(h\nu - h\nu_{E_0})^{1/2}, \tag{31}$$

where k is in Å^{-1}, and $h\nu$ is in eV. To a first approximation, the oscillations in a plot of $\alpha(k)$ have a constant, rather than an increasing period (Figure 8.26 [21]). It follows that the amplitude of the interference oscillations will depend on the number of nearest neighbours, with higher coordination numbers characterised by larger amplitudes. In addition, examination of equation (29) suggests that shorter bonds (smaller r) will give rise to larger spaced oscillations (Figure 8.27 [22]).

In order to be able to work out absolute values of the bond distance and coordination number, we must look at the physics of the absorption/scattering process in more detail. Mathematically, the EXAFS oscillations are often expressed as the 'EXAFS modulation', χ, corresponding to the difference between the EXAFS-modulated part and the unmodulated part of the absorption coefficient, normalised to the background due to absorption by core levels at lower IP, i.e.

$$\chi(h\nu) = (\alpha - \alpha^* - \alpha_0)/\alpha_0 \tag{32}$$

Here $\chi(h\nu)$ is the EXAFS modulation per absorbing atom corresponding to ionisation of a particular core level, and α^* is the background absorption coefficient representing background due to absorption by core levels with ionisation

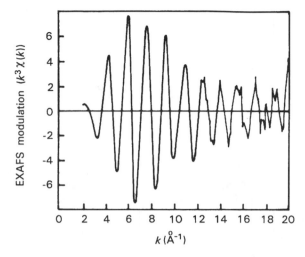

Figure 8.26. The EXAFS modulation expressed as a function of k, rather than $h\nu$, showing roughly regularly-spaced oscillations. The sample is a light-exposed As_2S_3 film. Data is taken at the As K-edge, and the primary modulation is caused by backscattering from S nearest neighbours [21]. The quantity plotted on the y axis is $k^3\chi(k)$, and is a measure of the EXAFS modulation (see Section 3.2.4)

potentials at lower photon energy; α_0 is the unmodulated part of the absorption coefficient, known as the 'atomic' absorption coefficient (referring to the fact that there is no EXAFS for isolated atoms).

Assuming that the scattering can be treated within a plane wave approximation, then the EXAFS modulation for core level absorption can be written as a function of wavevector in an expression of the type

$$\chi(k) = \sum_i A_i(k) \sin[2kr_i + \Phi_s^i(k) + \Phi_d^i(k)]. \tag{33}$$

Here we are no longer assuming that backscattering occurs only from the shell of nearest neighbours. r_i is the interatomic distance to the ith shell of neighbouring atoms, and contributions to $\chi(k)$ from shells other than that nearest to the emitting atom ($i = 1$, Figure 8.28) are included by summing over a range of i values. However, as pointed out in Section 3.2.1, the major contribution is from the nearest neighbours of the emitting atom.

Equation (33) has two basic components; the second, sine, term determines the frequency of the oscillations, while the prefactor A determines their amplitude.

The phase shift difference between outgoing and backscattered waves is

$$\Delta\Phi = 2\pi \cdot 2r_i/\lambda = 2kr_i, \tag{34}$$

(Figure 8.25), which appears in the argument of the sine term. The basic EXAFS modulation then corresponds to a sine function for each value of r_i. This phase

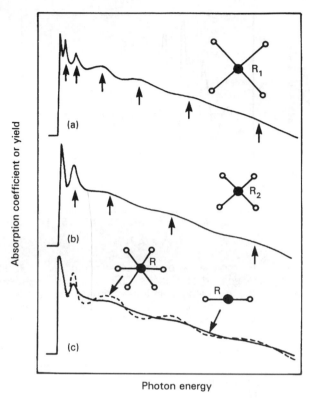

Figure 8.27. The effect of coordination number and nearest-neighbour distance on the form of the EXAFS spectrum. Shorter bonds give rise to larger spaced oscillations (compare R_1(a) and R_2(b)), whilst an increase in coordination number gives a larger amplitude (compare the traces for a two-coordinate and a six-coordinate species in (c)) [22]. Shorter bonds are characterised by larger-spaced EXAFS oscillations, and the EXAFS amplitude increases with the number of nearest-neighbour atoms

shift is adjusted by further angles, Φ_s and Φ_d. Φ_s reflects the fact that our initial assumption that there was no phase shift due to the backscattering process itself was not strictly correct; it is the sum of the phase shifts induced for the outgoing and backscattered electron wavefunctions by the core atomic potentials of the emitting atom and the backscattering atoms. The appearance of this second angle, which is itself a function of k means that $\chi(k)$ does not in fact show oscillations of exactly constant period. A further correction to the argument of the sine function, $\Phi_d^i(k)$ may be necessary to account for deviations of the actual positions of the atoms from their ideal positions, which may arise due either to thermal motion or to structural disorder. In the case of thermal vibrations, the phase shift correction $\Phi_d^i(k)$ is zero, although as we will see, thermal motions modulate the amplitude term.

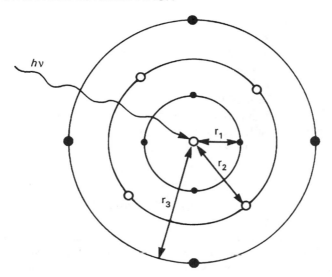

Figure 8.28. Summation over successive shells ($i = 1, 2, 3$) of neighbouring atoms in the EXAFS equation (equation (33)). Nearest neighbours correspond to $i = 1$, and are predominantly responsible for the EXAFS

The amplitude term, $A_i(k)$ consists of a number of components:

$$A_i(k) = (\pi m_0/h^2 k)(N_i^*/r_i^2) F_i(k) \exp(-2\sigma_i^2 k^2) \exp(-2r_i/\lambda(k)). \quad (35)$$

$F_i(k)$, the atomic backscattering amplitude, is an important term, as it describes the amplitude of the EXAFS oscillations. It is dependent on the atomic number, z, of the backscattering species, so its variation with k may be used to recognise the species around the emitting atom.

$A_i(k)$ also contains two exponential terms. The first of these, $\exp(-2\sigma_i^2 k^2)$, is the Debye–Waller factor, and describes the amplitude correction due to thermal displacement of the backscatterer relative to the emitting atom. σ_i is the mean square deviation from the ideal distance, r_i. A more complex correction may be necessary in cases where deviations from ideal distances arise due to structural disorder.

The second exponential term, the attenuation factor, describes the fact that, as we have discussed, the excited photoelectron has only a short mean-free-pathlength, λ in the solid. This causes an $\exp(-2r_i/\lambda(k))$ attenuation in the intensity of the electron wave over the path $2r_i$ (in accordance with Beer's Law). N_i^* is the effective coordination number, given by

$$N_i^* = 3 \sum_{j=1}^{N_i} \cos^2 \theta_j. \quad (36)$$

The summation extends over all j neighbours in the ith shell. The angle θ_j is the angle between the electric vector of the light, and the vector connecting the absorber and the backscatterer. Electron emission is strongest in the electric vector direction. This means that if the electric vector is pointing along the axis between the emitter and the backscatterer, the backscattering amplitude is at a maximum. This may be used to great effect in studies of ordered materials, such as single crystals, or their surfaces. However, the effect is averaged out in media having no long range order, such as liquids, gels and amorphous samples. In this case, N_i^* is simply equal to N_i, the true coordination number.

3.2.3 Experimental Techniques

The basic method for detecting EXAFS is to perform an optical absorption measurement. In principle, the absorption coefficient can then be measured (via equation (21)) by measuring the intensities of the incident and emergent beams in parallel, if the thickness of the sample is known. X-ray absorption measurements themselves are not new; as early as the 1930's measurements were being carried out using a continuum source of X-rays, a dispersive spectrometer and a film detector. However, the technique progressed very little until the advent of synchrotron radiation sources. These sources produce X-ray radiation which is of the order of 10^6 times more intense than that from conventional sources, and is in addition, continuously tunable.

3.2.3.1 Synchrotron Radiation

There are now over 20 sites worldwide where synchrotron radiation research is carried out, including a site at Daresbury, near Manchester, UK. The synchrotron radiation is produced from a high energy electron storage ring. At Daresbury, an electron beam is produced by a 12 MeV linear accelerator. Pulses of these electrons are injected tangentially into a small booster synchrotron, where they are accelerated by 500 MHz radiation. This causes the electrons to travel in bunches, with the separation between bunches dictated by the radio frequency (i.e. the electrons travel in *synchronous* orbit). The electrons are steered round the ring by dipole magnets. When the electron energy is 600 MeV, the electrons are injected tangentially into the main storage ring (around 32 m in diameter), where RF power (from klystron cavities) is again used to raise the electron energy to 2 GeV (Figure 8.29). In this energy region, the electrons are travelling at velocities close to the speed of light, and emit electromagnetic radiation with a continuous distribution from the infra-red to the hard X-ray part of the spectrum. The radiation is predominantly plane polarised, and emerges tangentially in pulses of length 0.17 ns and spacing 2.0 ns (in normal operation).

The energy lost to synchrotron radiation emission is replaced by RF power. However, even though the storage ring itself is under high vacuum, electrons are removed by collisions with gas molecules, so that the beam current decays

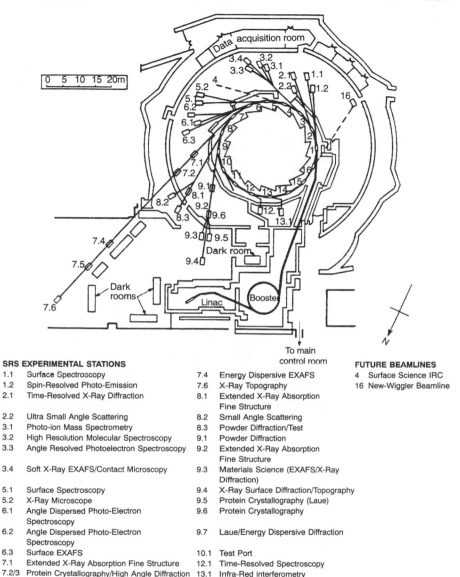

SRS EXPERIMENTAL STATIONS

1.1	Surface Spectroscopy	7.4	Energy Dispersive EXAFS
1.2	Spin-Resolved Photo-Emission	7.6	X-Ray Topography
2.1	Time-Resolved X-Ray Diffraction	8.1	Extended X-Ray Absorption Fine Structure
2.2	Ultra Small Angle Scattering	8.2	Small Angle Scattering
3.1	Photo-ion Mass Spectrometry	8.3	Powder Diffraction/Test
3.2	High Resolution Molecular Spectroscopy	9.1	Powder Diffraction
3.3	Angle Resolved Photoelectron Spectroscopy	9.2	Extended X-Ray Absorption Fine Structure
3.4	Soft X-Ray EXAFS/Contact Microscopy	9.3	Materials Science (EXAFS/X-Ray Diffraction)
5.1	Surface Spectroscopy	9.4	X-Ray Surface Diffraction/Topography
5.2	X-Ray Microscope	9.5	Protein Crystallography (Laue)
6.1	Angle Dispersed Photo-Electron Spectroscopy	9.6	Protein Crystallography
6.2	Angle Dispersed Photo-Electron Spectroscopy	9.7	Laue/Energy Dispersive Diffraction
6.3	Surface EXAFS	10.1	Test Port
7.1	Extended X-Ray Absorption Fine Structure	12.1	Time-Resolved Spectroscopy
7.2/3	Protein Crystallography/High Angle Diffraction	13.1	Infra-Red interferometry

FUTURE BEAMLINES

4 Surface Science IRC
16 New-Wiggler Beamline

Figure 8.29. Plan of the synchrotron radiation source at the CCLRC Daresbury Laboratory, UK. Pulses from the electron beam produced in the linear accelerator (linac) are injected tangentially into the small booster synchrotron, where they are accelerated to 600 MeV energy, before being injected into the main storage ring. Here, their energy is raised to 2 GeV. Numbers 1–16 around the ring indicate dipole magnets which steer the beam around the ring. Synchrotron radiation emerging at tangents to the ring travels down assorted beam lines to the numbered experimental stations (courtesy CCLRC Daresbury Laboratory)

appreciably over a 24 hour period. At Daresbury, a new beam is normally injected once in every 24 hours.

The radiation frequency required is selected out using a monochromator. Up to photon energies of around 1500 eV, a ruled Bragg diffraction grating can be used. At higher energies, the wavelengths are so small that cut single crystals (e.g. Si(111), Ge(111)) may be used to diffract out the required component. The broad range of frequencies available means that synchrotron radiation may be used for a wide variety of different techniques, ranging from photoemission to X-ray diffraction, in addition to XAS and EXAFS.

Almost all EXAFS experiments currently use a synchrotron light source. There have been attempts to build laboratory EXAFS sources, the main design problem being how to produce enough X-ray flux to be able to detect EXAFS oscillations, especially for light atoms. However, laboratory sources are now becoming more competitive, with improvements both in the design of the sources, which are rotating anode X-ray sources and the type of detection (which usually employs a solid state detector) [23, 24].

3.2.3.2 Detection Techniques (EXAFS)

The most usual method of measurement of the intensities of the incident and transmitted light uses gas-filled ion chambers (Figure 8.30). The incident synchrotron radiation ionises the gas in ion chamber 1 before passing through the sample (the ion count provides a measure of I_0). The transmitted intensity, I, is measured by ion chamber 2. It is very important that I_0 and I are measured simultaneously, allowing for cancellation of sudden changes in intensity of the incident photon beam. Even using a synchrotron radiation source, EXAFS data may take some time to accumulate; in general, the data is best acquired in the form of several short scans of the entire spectral range, rather than a single slow scan, as this minimises the effects of a sudden step loss in the beam current in the storage ring.

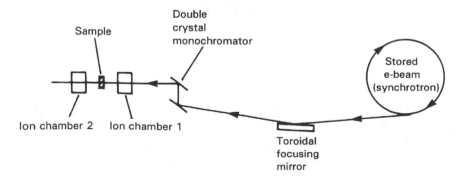

Figure 8.30. Schematic experimental arrangement for transmission EXAFS

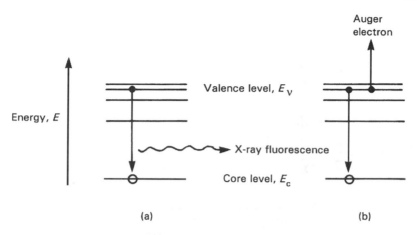

Figure 8.31. (a) The core hole created in XAS may be filled by an electron from a higher level, with the excess energy emitted as X-ray photons. (b) Alternatively, the excess energy may be dissipated by the emission of a second, Auger electron

In a transmission EXAFS experiment, the absorption coefficient of the entire system is measured — as α at a given photon energy is affected by all the absorption processes whose thresholds occur at lower photon energies. This can make it very difficult to measure the EXAFS of a dilute atomic species, as the EXAFS for that species is a very small modulation of the total absorption coefficient. An important alternative is fluorescence detection. This relies on detection of the X-ray fluorescence which occurs when the core hole created in the initial X-ray absorption process is filled by electron decay from an upper, filled level (Figure 8.31(a)). The energy of the photon emitted is dependent on $(E_v - E_c)$, and is thus characteristic of the element in question. It is thus possible, using an energy dispersive X-ray detection system, to single out the absorption coefficient of the species of interest by selecting one of the characteristic emission frequencies for that species. For small values of the absorption coefficient, α_i, for the species, the X-ray fluorescence is proportional to α_i. Using the enhanced sensitivity which fluorescence detection gives, it is possible to detect isolated impurity atoms with concentrations of 10^{19} cm^{-3} or less.

3.2.4 Data Handling

We have seen that the EXAFS modulation, $\chi(k)$ is a superposition of sine waves (equation (33)), the period and amplitude of which give us information about coordination distances and coordination numbers. The deconvolution of the contributions to $\chi(k)$ requires the use of Fourier transformation techniques.

Initial manipulation of the data involves subtraction of contributions from other lower lying absorption edges, and subtraction of the non-modulated part of α (the

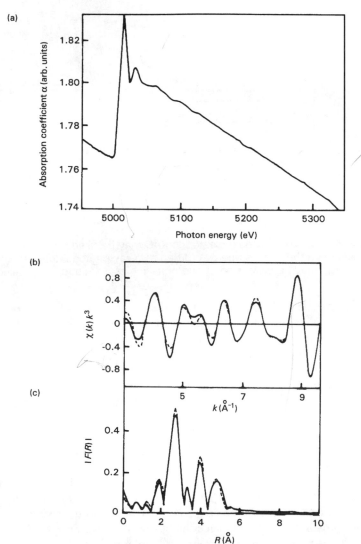

Figure 8.32. Stages in the manipulation of EXAFS data. The spectra shown are from surface EXAFS (SEXAFS) experiments using various adsorbates on Ag(111) [25]. (a) Raw EXAFS data as a function of photon energy, at the Cs L_{III}-edge for Cs on Ag(111). (b) $k^3\chi(k)$ plot for Cl on Ag(111). The data has been Fourier filtered to include real space components up to 5 Å from the absorbing atom, then back-transformed into k space (as described in the text). The data is compared with a theoretical calculation (broken line). (c) The experimental and theoretical Fourier transforms of $k^3\chi(k)$. The major peak gives a nearest neighbour Ag–Cl distance of 2.70 ± 0.01 Å. Reproduced by kind permission from ref. [25]

atomic absorption coefficient), in accordance with equation (32). The data is then transformed onto a momentum (k) scale (via equation (31)), using an emprical estimate of the threshold frequency, $h\nu_{E_0}$. The data is often displayed as $k^3\chi(k)$ or $k^2\chi(k)$ vs k (Figure 8.32 [25]) — as the atomic backscattering amplitude $F_i(k)$, which forms a part of the amplitude term of $\chi(k)$ (equations (33) and (35)) is proportional to k^{-2}, and the amplitude term $A_i(k)$ is additionally dependent on k^{-1} (equation (35)). The imaginary part of the Fourier transform of $k^2\chi(k)$ or $k^3\chi(k)$ has one peak for each value of r which corresponds to one of the interatomic distances r_i.

In practice, calculation of the Fourier transform of χ requires knowledge of the phase shifts $\phi_s^i(k)$ induced by backscattering by the ith shell of atoms, as the Fourier transform of $\chi(k)$,

$$F(r) = \int_0^\infty \chi(k)\exp\{-i(2kr + \phi_s^i(k))\}\,dk. \tag{37}$$

These shifts can be calculated, but it is more usual to derive them experimentally from the EXAFS of the same combination of emitting and scattering atoms in a reference material. The data may be initially transformed neglecting the phase shift term, i.e.

$$F(r) = \int_0^\infty \chi(k)\exp(-2ikr)\,dk; \tag{38}$$

the peaks in the transformed data then give a first approximation to the interatomic distances. The major peak in the transform is then 'Fourier filtered' — a window is introduced along the r axis around the feature, and the data is back transformed to k-space only in that window. This gives a good approximation to the EXAFS modulation due only to the single interatomic distance in question. From this back-transformed EXAFS modulation, the phase shift for the shell of backscattering atoms in question and their backscattering amplitude can be obtained. The phase shift so obtained is generally different to that of the reference material. This difference is reduced by re-adjusting the position of the threshold frequency $h\nu_{E_0}$ until the phase shift between the two materials becomes zero at $k = 0$. Using this procedure, it is possible to obtain interatomic distances to an accuracy of 0.01 Å. Using the backscattering amplitudes obtained, the number of atoms in each shell around the emitting atom may be obtained. This is generally done by calculating the amplitude for a range of model systems, and identifying the calculation which corresponds most closely with the experimentally determined amplitude.

3.2.5 Applications of EXAFS

Because EXAFS does not rely on long range order within a sample, it has become one of the work-horses of materials science, and the range of current applications is vast (see for example [26]). These include determination of metal ion coordination in biomolecules, such as ferritin and haemoglobin, studies of glasses and gels, determination of metal cluster size in supported metal catalysts, studies of impurity sites in semiconductors and in naturally occurring minerals, and studies of local coordination in complex materials, such as high temperature superconductors. As the purpose of this chapter is to discuss surface structure determination, we will not discuss these in detail, but simply show one or two examples below.

One of the advantages of EXAFS in the study of catalyst materials is that samples may be relatively easily studied under real catalytic environments (usually high temperature and pressure). If there is sufficient photon flux to allow data to be accumulated rapidly, then catalytic processes may be observed in a time-resolved way. The feasibility of this type of study has recently been investigated by Catlow *et al.* [27], who studied the high temperature reduction of NiO supported on α-alumina. This is a reforming catalyst, and can also be used in the oxidation of methane, which involves reduction of the catalyst at high temperatures. In order to simulate this, the material was ramped to a temperature of 500°C, under a hydrogen/nitrogen atmosphere.

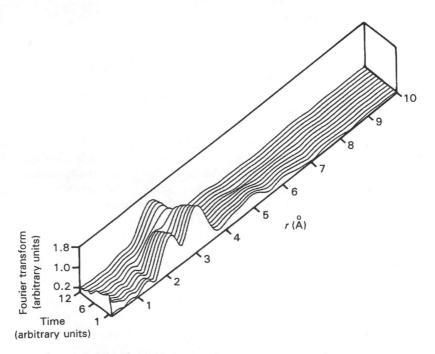

Figure 8.33. Fourier transforms of the Ni K-edge EXAFS data from NiO on α-Al$_2$O$_3$ as a function of time, during the high-temperature reduction of the catalyst. Reproduced by kind permission from ref. [27]

EXAFS spectra at the Ni K-edge were recorded every three minutes. The resulting data showed that reduction of the NiO occurs over around 15 minutes at a temperature of 400–500°C. The phase-corrected Fourier transforms of the data for times around the transition point are shown in Figure 8.33. The two large peaks in the first spectrum are due to Ni atoms surrounded by six oxygen atoms in the first coordination shell, and twelve Ni atoms in the second, i.e. crystalline NiO. In the last spectrum, only one broad peak is seen, due to the Ni neighbours surrounding the central nickel atom in Ni metal. The major problem in accumulating EXAFS in a time-resolved way is of course one of signal; as new generation synchrotrons become available, studies of this type will become increasingly feasible.

One way in which the surface sensitivity of EXAFS may be improved, without moving into a UHV environment as with SEXAFS (discussed in the next section), is simply to use a glancing angle of incidence. The EXAFS oscillations may then be picked up using either fluorescence detection, or, if the angle of incidence is small enough, in the externally reflected beam (rather than the transmitted beam, as with conventional EXAFS), as shown in Figure 8.34. By varying the angle of incidence, the depth of the surface sampled may be varied from nm's to mm's. This technique is known as glancing angle XAFS. Greaves et al. [28] have applied this technique to the study of the initial stages of corrosion of polished stainless steel surfaces when exposed to air at 1000°C, by taking reflectivity measurements at the Fe K-edge. These were compared with standard spectra from Fe and its oxides. The angle of incidence was chosen so that the sampling depth was of the order of 30 Å. The resulting Fourier transforms of the data are shown in Figure 8.35). The unoxidised surface shows a spectrum very similar to that of fcc Fe, but after four minutes' oxidation, an oxide coating with a local structure similar to that of Fe in Fe_2O_3 has developed. The magnitude of the absorption threshold increased during oxidation, indicating that Fe was migrating to the surface of the steel during the process, and being oxidised there.

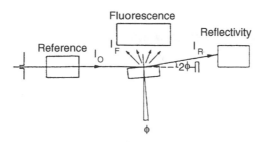

Figure 8.34. Schematic arrangement for glancing angle XAFS. Reproduced by kind permission from ref. [28]

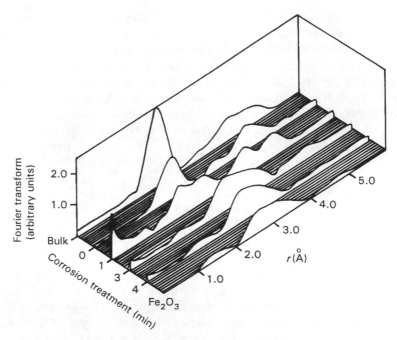

Figure 8.35. Fourier transforms of the Fe K-edge EXAFS data from a glancing XAFS investigation of polished stainless steel as a function of corrosion time, compared with bulk steel and Fe_2O_3. Reproduced by kind permission from ref. [28]

3.3   EXTENSION TO SURFACES – SURFACE EXAFS (SEXAFS)

SEXAFS is the surface-sensitive modification of EXAFS, and may be used for the determination of local surface geometry and coordination, for example of molecules chemisorbed at a surface. As we shall see, the technique can be made surface specific, rather than simply surface sensitive (like glancing angle XAFS), and as such provides detailed information on the nature of adsorption sites in the topmost plane of the surface.

3.3.1   Experimental Techniques

The surface of a material represents a very small part of the solid (for a single crystal, around 1 part per million for a 1 mm thick crystal). The main problem to be overcome in obtaining SEXAFS is thus one of obtaining sufficient sensitivity to the surface species. The usual transmission technique is not suitable, as the bulk absorption coefficient generally swamps that of the surface.

A number of alternative techniques are available which measure the absorption coefficient indirectly, and which give enhanced sensitivity in SEXAFS. All the methods require the use of ultra-high vacuum instrumentation. Several methods rely on the decay of the core hole created in the absorption event. These include fluorescence detection (as used in EXAFS and described in Section 3.2.3.2), Auger yield and total electron yield detection. The core hole created in the initial absorption process may decay with the emission of a photon as we have seen (X-ray fluorescence). Alternatively, the hole may decay via the ionisation of another, Auger electron (as described in Chapter 4, see also Figure 8.31(b)). The relative proportions of the two decay pathways depend on the atomic number of the species in question, with Auger electron production being favoured for the lighter elements (with $z \leqslant z_{Si}$). The energy of the released Auger electron is dependent on the energy difference ($E_v - E_c$), and thus Auger emission occurs at characteristic energies for each element. The yield of Auger electrons is proportional to the number of core holes produced, and hence to the absorption coefficient of the species, $\alpha_i(h\nu)$.

As discussed in previous chapters, and in Section 3.2.1, the mean free pathlength of low energy electrons in solids is of the order of a few atomic spacings. Thus, the Auger electrons being detected come mainly from the top 10 Å of a solid, giving this method its surface sensitive characteristics. The Auger electrons are detected in an electron energy analyser similar to those used in XPS experiments described in Chapters 3 and 4.

A related detection technique is total electron yield detection, in which *all* electrons emitted from the sample are detected. In the main, these are electrons which were emitted as Auger electrons, but which have lost varying amounts of energy in inelastic scattering processes in the solid. The total electron yield is thus also proportional to the absorption coefficient. One disadvantage of the technique compared with Auger detection is that it is not element specific, as all absorption processes contribute to the low energy tail of 'secondary' electron emission. The technique is surface sensitive for the same reasons as Auger detection. However, most of the electrons collected have very low energies, and have a longer mean free pathlength in the solid (≈ 50 Å) than the primary Auger electrons (remember the universal pathlength curve for electrons in solids, Chapter 3). In addition, the mean free pathlength changes rapidly at low energies, so that the surface sensitivity of detection can be varied by changing the energy range of the detection window. A further advantage of total electron yield detection is that it requires only a biased metal plate to collect the electrons, coupled to a picoammeter.

Although both Auger yield and total electron detection are surface, *sensitive*, neither is specific *only* to surfaces. The signal may be very difficult to disentangle from bulk contributions. Complete surface specificity is only obtained by ensuring that the adsorbate species studied is present only at the surface. The signal in SEXAFS experiments may be very low, necessitating long data accumulation times (several hours). As discussed in earlier chapters, surface-sensitive techniques of this type require the use of ultra-high vacuum techniques, but even with pressures in the 10^{-10}–10^{-11} mbar range, it may be difficult to avoid unwanted surface contamination for the duration of data accumulation.

One further technique which may be used is ion yield SEXAFS. Figure 8.31(b) shows that following emission of an Auger electron, the emitting atom is left in a very unstable doubly positively charged state. This state may be produced via either inter- or intra-atomic process, and the positive ions produced at the surface may spontaneously desorb, and can be detected using a time-of-flight mass spectrometer. Due to the high probability of re-neutralisation, and the very low escape depth for ions, only ions produced on the surface can escape into the detector, giving the technique inherent surface specificity. However, the low yield for the desorption process means that it cannot be used to study dilute surface species, and competition with fluorescence decay for heavier elements means that the ion yield method is only applicable to the lighter elements (up to, say, Si).

3.3.2 Applications of SEXAFS

Although in principle, SEXAFS probes local structure, and so can be used to study amorphous and polycrystalline materials, in practice these types of surface give rise to a large amount of random electron scatter, so a high photon flux is necessary to carry out these experiments. The vast majority of SEXAFS work to date has used single crystal substrates. In the case of a single crystal material, we saw from equations (35) and (36) that the backscattering amplitude from a particular atom is at a maximum if the electric vector of the light is pointing along the axis between the emitter and the backscatterer. Thus by rotating the crystal sample to vary the angle between the electric vector and the surface, and observing the effect on the SEXAFS, the geometry of, say, an adsorption site can often be determined. This approach has been quite widely adopted, for example in the determination of the coordination site of sulfur on Ni(110)c(2×2)-S [29], and the coordination site of O on Ni(111)p(2×2)-O [26]. Figure 8.36 shows the SEXAFS data of Haase *et al.* [30], for the latter system, taken at the oxygen K-edge, and recorded in partial electron yield mode, together with the corresponding Fourier transforms. Both normal incidence ($\theta = 90°$) and grazing incidence ($\theta = 20°$) have been used. The plots of $\chi(k)$ vs k show nearly sinusoidal forms, and the corresponding Fourier transforms are dominated by a single peak, corresponding to the nearest-neighbour O–Ni distance of around 1.85 ± 1.03 Å. In the normal incidence data, there is a small second peak (labelled R_2), which can be tentatively assigned to the next-nearest neighbour O–Ni distance, corresponding to around 3.1 Å. Having determined the O–Ni nearest-neighbour distance, the amplitudes of the SEXAFS oscillations expected in both polarisations for different coordination sites with this nearest-neighbour distance may be calculated. The ratio of the amplitudes in the polarisations is then compared with the experimentally determined value. No satisfactory agreement can be obtained for oxygen in atop or bridged sites, and the most convincing agreement by far is obtained when oxygen is in a three-fold site. Further information about whether the surface is reconstructed around the three-fold site can be obtained from an analysis of

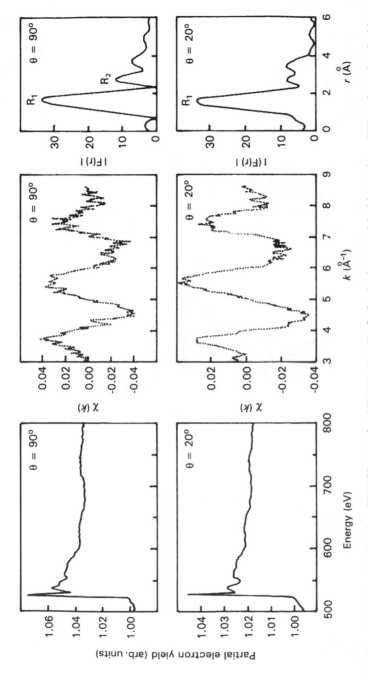

Figure 8.36. Oxygen K-edge SEXAFS data for Ni(111)p(2 × 2)–O, for both normal incidence ($\theta = 90°$) and grazing incidence ($\theta = 20°$). Raw data as a function of photon energy is shown (left), together with $\chi(k)$ vs k plots (centre) and finally the Fourier transforms of the data (right). Reproduced by kind permission of Elsevier Science – NL from ref. [30]

the coordination shells outside the first. Because of the polarisation dependence, Ni atoms in the second layer of the sample do not contribute to the normal incidence SEXAFS. However, the grazing incidence data are affected by the separation between the first and second Ni layers, and so should be capable of distinguishing between fcc and hcp three-fold coordinated sites. Best agreement is obtained for the fcc site, shown in Figure 8.37.

As we saw above, the vast majority of SEXAFS measurements to date have been carried out on single crystals. The first study to obtain the structure of an adsorbate on a polycrystalline surface was work by Norman *et al.* on the surface structure of BaO/W matrix thermionic emission cathodes [31]. The work studied the adsorption geometry of BaO on either a W matrix or a W/Os alloy, using real dispenser cathodes, in an attempt to determine why the cathodes are more effective when the W cathode surface is coated with a W/Os alloy. Total electron yield detection was used to monitor the absorption coefficient above the Ba L_3-edge. SEXAFS data from the uncoated cathode revealed that Ba is attached to oxygen, and this is bonded to W. The Ba-O distance is unchanged on the alloyed surface, but on the uncoated surface, Ba is bonded to one O atom, whereas on the

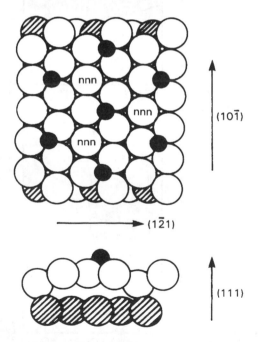

Figure 8.37. Top and side views of the Ni(111)p(2 × 2)-O system, with oxygen atoms (the small filled circles) in three-fold coordinated fcc sites, suggested by the SEXAFS data of Figure 8.36. Next-nearest-neighbours are marked 'nnn'. Reproduced by kind permission of Elsevier Science − NL from ref. [30]

alloy-coated surface it is bonded to two O atoms. This was seen as enhancing the surface dipole (producing a larger effective positive charge on Ba, which helps electrons to escape the material), and hence lowering the work function and increasing the emission of the cathode.

3.3.3 NEXAFS (XANES) in Surface Structure Determination

In Section 3.1.2, we noted that NEXAFS may be used to yield structural information, but that the analysis of NEXAFS data may be complex. In the EXAFS part of the spectrum, the interaction between the photoelectron wave and the backscatterer is relatively weak, so single scattering predominates (as we saw, for example in X-ray diffraction). However, in the NEXAFS region, the interaction is much stronger, and multiple scattering of the outgoing photoelectron must be considered. As the photoelectron wavefunction is strongly dependent on the form of the potential around the absorber, in principle NEXAFS can be used to investigate the oxidation state of the absorber, and its coordination geometry. Investigation of the oxidation state is relatively straightforward, as the position of the absorption edge changes as the charge on the absorber changes (i.e., there is a chemical shift in the edge position). For small molecules adsorbed on surfaces, the NEXAFS at an adsorbate core level threshold is dominated by resonances within the adsorbed molecule, which means that the technique may sometimes be very powerful in this type of study. However, for atomic adsorbates, particularly those on metallic substrates, long range multiple scattering dominates, so that modelling the spectra then requires intensive multiple scattering calculations. As an example, in a study of Ni(110)c(2×2)-S [32], five shells of Ni atoms surrounding a surface S atom (corresponding to 42 atoms) needed to be included in the calculation in order to obtain good agreement with experiment, indicating significant multiple scattering within this cluster. In the corresponding SEXAFS measurement from the same system, only one shell of Ni atoms (four atoms) contributed significantly.

The polarisation dependence of NEXAFS can be used in the same way as that in SEXAFS, for example to determine the orientation of a small molecule on a surface. As an example, we can take the adsorption of CO on Ni(100) [33]. In this study, NEXAFS at the C K-edge is observed, using C-KVV Auger emission detection. The NEXAFS in this case is dominated by intramolecular excitations, within the CO molecule. When the photons are incident at 10° to the (100) plane, two peaks are observed (labelled A and B in Figure 8.38). As the angle of incidence is increased to 90°, peak B disappears. The two peaks are due to transitions from the C 1s level to the $2\pi^*$(A) and σ^*(B) molecular orbitals of CO. (Peak B is much broader than peak A. This is because the $2\pi^*$ state is pulled below the vacuum level on ionisation of the C 1s electron, so peak A is a bound state resonance. In contrast, peak B is what is known as a continuum or shape resonance, and arises because an electron excited into the

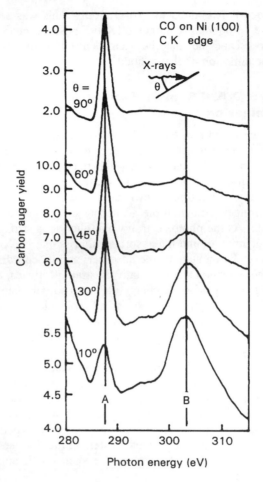

Figure 8.38. NEXAFS data above the C K-edge for CO on Ni(100) at 180 K, as a function of incidence angle. Reproduced by kind permission of the American Institute of Physics from ref. [33]

σ^* orbital remains above the vacuum level and will escape from the solid. The σ^* wavefunction has a large amplitude near the C atom, but tails off rapidly to a free-electron wavefunction.) Using the dipole selection rule, it is easy to show that the C 1s (σ) $\rightarrow$ $2\pi^*$ transition is only allowed if the electric vector of the light is perpendicular to the CO bond axis, whilst the C 1s (σ) $\rightarrow$ σ^* transition is allowed if the vector is parallel to the CO bond axis. This means that the intensity ratio between peaks A and B can be used to determine the orientation of the CO molecule on the surface, by comparing calculated ratios for different CO orientations with the experimentally obtained ones. This produces the result that the CO molecule is perpendicular to the Ni(100) plane, in agreement with photoemission results.

Much work in developing NEXAFS as a structural probe has been undertaken by Thornton and co-workers (for examples see references [34] and [35]). Recently, in work on Si(111)(7 × 7)-Cl and Si(100)(2 × 1)-Cl [36], this group have demonstrated that for atomic adsorbates on semiconductors, rather than metals, the NEXAFS displays 'pseudo-intramolecular' behaviour, arising from the local nature of the adsorbate–semiconductor bonding. This means that NEXAFS from these substrates may be used to give some indication of the adsorbate–substrate bondlength and geometry, in the same way as for molecular adsorbates. In this case, the polarisation dependence of the Si-Cl σ^* resonance is used to infer that the Si-Cl bond angle is $<10°$ away from the surface normal in Si(100)(2 × 1)-Cl, by comparison with the NEXAFS observed for the Si(111)(7 × 7)-Cl surface.

4 ADDITIONAL TECHNIQUES FOR SURFACE STRUCTURE DETERMINATION

Electron diffraction (LEED and RHEED) and surface-EXAFS are the techniques currently most widely used in surface structure determination. However, the last few years have seen very rapid progress in this area, particularly with the advent of scanning tunnelling microscopy (STM) and atomic force microscopy (AFM); both these techniques are discussed, along with other forms of microscopy, in Chapter 9.

Techniques based on the scattering of small atoms such as He, or ions have also been developed; ion scattering spectroscopy (ISS) has been discussed in Chapter 7. In addition, with the increasing availability of synchrotron radiation, the possible role of X-rays in determining surface structure has been re-examined, leading to the development of several new techniques. Some of these techniques are described briefly below.

4.1 TWO-DIMENSIONAL X-RAY DIFFRACTION (SURFACE X-RAY DIFFRACTION)

Our initial negative assessment of X-rays as a probe of surface structure was based on the weak interaction between X-rays and solids, with the result that X-rays penetrate very deeply into solids. However, this weak interaction also means that multiple scattering effects can be ignored, and bulk X-ray diffraction can be treated by a kinematic theory. From this point of view, a technique which uses X-rays, rather than electrons, to probe surface structure should have considerable advantages over LEED and RHEED, as the theoretical description of the scattering will be much simpler, and the arrangement of atoms within the surface unit mesh may be obtained in a relatively straightforward way.

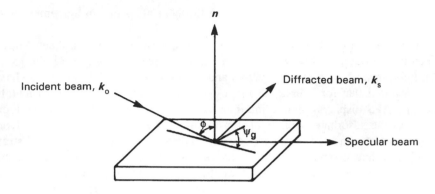

Figure 8.39. The grazing angle geometry used in surface X-ray diffraction. An incident beam of wavevector k_0 is incident at a grazing angle to the surface, such that the polar angle, ϕ, with the surface normal n is typically $\approx 89.5°$. After diffraction through an azimuthal angle ψ_B the beam leaves the surface at a grazing angle similar to the angle of incidence

The index of refraction of solids at X-ray wavelengths is only very slightly less than unity. From Snell's Law, we can therefore deduce that at very grazing angles of incidence, we should obtain total external reflection of an incoming X-ray beam (Figure 8.39). For example, for a typical X-ray wavelength of around 1.5 Å, we calculate that the critical angle from the surface below which we should obtain total external reflection is around $0.2°$–$0.6°$, depending on the sample. Under these conditions however, there is a finite penetration depth of the X-ray beam into the solid, caused by absorption of X-rays by the sample, but for angles of incidence below the critical angle, this is only of the order of 10–50 Å. This means that scattering is essentially from the reciprocal lattice vectors of the surface. The two-dimensional treatment which we considered for LEED and RHEED is again appropriate, so that diffracted beams emerge wherever the Ewald sphere intersects the surface reciprocal lattice *rods* (see, e.g. Figure 8.5). However, we now have the additional advantage that multiple scattering effects are negligible, so that the diffraction process may be treated by a *kinematic* rather than a dynamical theory. The two-dimensional X-ray diffraction experiment will, of course, still be subject to the phase problem, but using refinement techniques developed from conventional X-ray diffraction, atom positions can be estimated to within ± 0.01–0.03 Å.

In practice, the scattered light intensity is low, (as the flux density of the incident radiation on the crystal at a grazing angle is low) and an intense source of X-ray photons, such as a synchrotron, is necessary to produce acceptable signal-to-noise levels. As with LEED and RHEED, the experiment must be carried out in UHV, but the need to control the experimental angles to very fine tolerances means that the experimental arrangement can be rather complex. At the

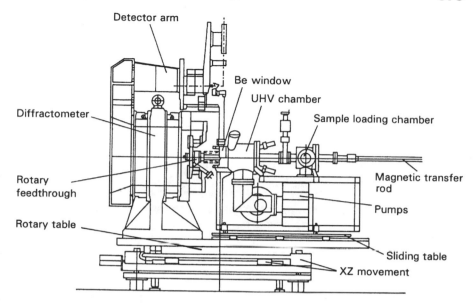

Figure 8.40. Schematic diagram of the large four-circle diffractometer and UHV chamber dedicated to surface X-ray diffraction, at the Daresbury synchrotron radiation source, UK. The UHV chamber is mounted on a sliding table, and both chamber and diffractometer are placed on a rotary table. Access to the X-ray beam is provided by a large Be window. Reproduced by kind permission of Elsevier Science – NL from ref. [37]

Daresbury synchrotron source in the UK, a beamline dedicated to this experiment is available, which terminates in a large purpose-built four-circle diffractometer, incorporating a UHV chamber and its associated equipment (Figure 8.40). The monochromated X-ray beam enters the UHV chamber through a large Be window.

As we would expect from the grazing incidence geometry, the technique (like RHEED) shows extreme sensitivity to surface roughness. It can sometimes be difficult to find a suitably flat area of the surface to obtain a pattern. The coherence length of the technique is of the order of 1000–2000 Å, again similar to RHEED, and disorder occurring on a scale larger than this will not be detected.

A further difficulty results from the absorption of a proportion of the incident X-ray flux at the surface (which gives rise to the finite penetration depth). The scattered intensity is a sum of the surface and near-surface components, which can lead to difficulties in background subtraction, if the contributions from the surface plane are to be separated from other scattering events. A surface reconstruction, for example, will give rise to new reciprocal rods at fractional order positions. In order to observe these rods, it is necessary to change the diffraction geometry, as in the RHEED experiment. In the case of RHEED (Section 2.3.3), we have seen that the experimental data is usually recorded as $I(\phi)$, i.e. the

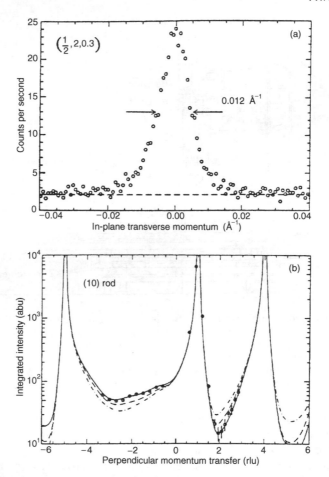

Figure 8.41. Examples of an azimuthal scan (a) and a rod scan (b), from the Ge(111)(2 × 1)-Sb surface, used to determine the surface structure shown in Figure 8.42. In (b), the open circles represent experimental data, while the dashed and full curves represent the fits to the data obtained for different models for the heights of the Sb atoms above the surface. The best agreement is found for the full curve, which corresponds to the model shown in Figure 8.42. Reproduced by kind permission of Elsevier Science − NL from ref. [41]

angle of incidence is varied, or as $I(\psi)$, i.e. the sample is rotated about its surface normal (Figure 8.39). Unfortunately, the first of these is not practical for two-dimensional X-ray diffraction. This is because the angle of incidence is usually kept fixed exactly at the critical angle for total external reflection. In this geometry, constructive interference occurs between the reflected and refracted waves, resulting in an improvement in the signal-to-background ratio. The data is thus generally accumulated as plots of intensity of a particular reflection versus azimuthal rotation, $I(\psi)$. A structure factor analysis of the resulting plots (adapted from the kinematic theory of bulk diffraction) can then be used, for example to determine the displacements of the atoms on a reconstructed surface relative to those of the underlying bulk, in the plane of the surface (the 'in-plane' displacements).

It is also possible to obtain some information about the displacements of any surface atoms out of the plane of the paper (the 'out-of-plane displacements'); in reciprocal space this corresponds to probing a reciprocal rod at different points along its length (see Section 1.3.2). This is essentially achieved by changing the radius of the Ewald sphere (Figure 8.5), which is most conveniently done in this case by varying the exit angle with respect to the surface plane. In the case of a completely flat surface, this 'rod scan' would be featureless. In practice of course, this is not the case, and the variation in intensity of a diffraction feature along a reciprocal rod gives us information about any displacements of the surface atoms normal to the surface.

The technique is potentially extremely powerful in the study of reconstructed and relaxed surfaces, and in studies of surface roughening, melting and interface formation. In practice it is currently limited to studies of heavier atoms (say $z \geqslant$ Si), as the intensity of X-ray scattering depends on z^2. Representative examples include studies of reconstructions on Pt(111) [38], epitaxial growth of Ag(111) [39], the TiO_2 (100)(1 × 3) reconstruction [12], and structure determinations of the InSb (111)(2 × 2) surface [40] and the Ge (111)(2 × 1)-Sb surface [41].

As an example, we show some data from the latter determination in Figure 8.41. Here azimuthal scans (made by rotating the crystal around the surface normal) were used to determine the in-plane structure of the Sb layer, and its registry with the Ge (111) substrate, whilst rod scans were used to determine out-of-plane displacements. Figure 8.41 shows one of the azimuthal scans around a fractional order peak (0.5,2,0.3), due to the adsorbate. A structure factor analysis of this data gives the in-plane registry of the Sb on the substrate. This showed that the Sb atoms are arranged on the surface in zig-zag chains (Figure 8.42), with a nearest-neighbour distance very close to the bulk Sb–Sb bond length. The Sb atoms are slightly out of registry with the first layer of underlying Ge atoms, which implies some tilting of the downward Sb–Ge bonds. Rod-scans were used to investigate this out-of-plane structure. An example of a rod scan for the (1 0) rod is shown in Figure 8.41. Several different calculated rod scans were determined, for different models of the heights of the Sb atoms above the Ge surface. The best fit corresponds to the situation shown in Figure 8.42, where it is found that the Sb atoms are at two different heights, with half lying 0.2 Å above the other half. This buckling continues into the Ge substrate.

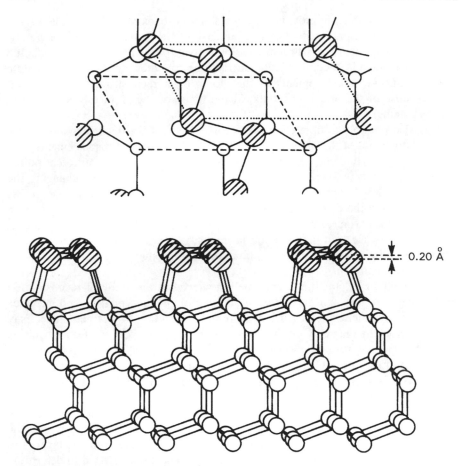

0.20 Å

Figure 8.42. Top and side views of the best fit model to the surface XRD data of Figure 8.41, for the Ge(111)(2 × 1)–Sb surface. Half the Sb atoms lie slightly above the other half. Larger hatched atoms are Sb. Reproduced by kind permission of Elsevier Science – NL from ref. [41]

4.2   *X-RAY STANDING WAVES (XSW) OR STANDING X-RAY WAVEFIELD ABSORPTION (SXW)*

XSW is a powerful technique for determining surface structure, and is closely related to 2-D X-ray diffraction. When an X-ray beam is incident on a surface, interference between coherent incident and diffracted beams will result in the generation of an *X-ray Standing Wave (XSW)* above (and below) the surface. This means that the electric field intensity associated with the X-ray photons will vary in a periodic way above the surface. An example of an X-ray standing wave is shown in Figure 8.43. The XSW periodicity, D is related to θ, the angle of

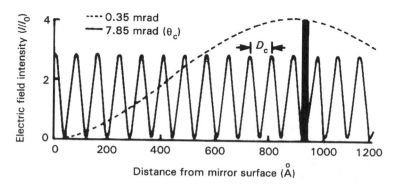

Figure 8.43. An example of an X-ray standing wave. The diagram shows the varia-tion of the normalised electric field intensity generated during specular reflection of a 9.8 keV X-ray plane wave from the surface of a gold mirror. Dashed and solid lines correspond to the field profile with the beam incident at 0.02° (0.35 mrad) and 0.449° (7.85 mrad, the critical angle) to the gold surface. The shaded rectangle shows the position of Zn atoms 925 Å above the surface, when the gold film is coated with a lipid multilayer (Figure 8.44). Reproduced by kind permission from ref. [47]

incidence (relative to the surface) by a Bragg relationship in the following way:

$$D = \lambda/(2\sin\theta). \tag{39}$$

Thus the period of the standing wave may be altered either by changing the energy of the incident radiation, and hence λ, or by changing its angle of inci-dence at the surface. Atoms, particularly heavy atoms lying within the XSW may absorb X-ray radiation of a suitable frequency, with a number of possible results, including photoemission and X-ray fluorescence. As we saw in Sections 3.2.3.2 and 3.3.1, for dilute species, the fluorescence yield, total electron yield or Auger emission intensity following X-ray absorption may be shown to be directly related to the probability of initial absorption. In the case of absorption from the XSW, this is in turn directly related to the intensity of the sinusoidally varying electric field above the surface. For a particular set of experimental conditions, this can be related to the distance of the atom from the surface. The basic XSW experiment therefore involves measuring the X-ray fluorescence or photoemission observed from an atom of interest as the phase of the standing wave is varied.

Clearly sensitivity is improved if the absorption cross-section of the atom of interest is quite large; the method was originally developed in the 1960's as a method for determining heavy atom positions in and on silicon and germanium crystals. In general, the technique is particularly useful in cases where heavy atom impurities are present at a surface, or where an overlayer of a relatively heavy species is adsorbed to a surface made up of other elements. Such impurity atoms may generally be located precisely with respect to the bulk lattice. The problems of obtaining a sufficiently good signal-to-noise ratio are similar to those in SEXAFS (Section 3.3.1). In some cases, measurement of the Auger electron yield may not be a good method as the signal may be superimposed on a background arising from electron emission from the substrate. For reliable measurements of the SW profiles from adsorbates, fluorescence yield detection is often more sensitive than the Auger yield method.

A variety of photon energy ranges and geometries have been used experimentally. Most measurements have been carried out in the 'Bragg geometry'. This utilises the XSW associated with a particular diffracted beam, for example, in the case of a Cu (111) surface, the (111) reflection normal to the surface has been used [42]. Measurement of the X-ray fluorescence or photoelectron yield as the phase of the XSW is varied at this Bragg condition permits the location of impurity or adatoms along the direction of the reciprocal lattice vector associated with the diffracted beam. Some experiments using this geometry have used a highly collimated beam of hard X-rays as a source of photons, providing both surface and bulk information, sometimes about surfaces exposed to high pressure liquids or gases.

The Bragg relationship above (equation (39)) indicates that by lowering the energy of the X-ray beam, and hence increasing its wavelength, the angle of incidence relative to the surface plane may be increased. In fact, by using soft X-rays (with energies say in the range 800 eV–5 keV), it is possible to make the incident beam normal to the lattice plane for most metal and semiconductor surfaces. This forms the basis of normal incidence standing X-ray wavefield absorption (NISXW), first applied by Woodruff *et al.* to Cl/Cu(111) [43, 44]. The method offers a number of advantages over the hard X-ray method. The width of a Bragg reflection becomes narrower as the angle of incidence is increased towards 90°, and this allows the method to be applied to less perfectly crystalline surfaces, such as those made up of a mosaic of small crystallites of differing orientations. Precise collimation of the soft X-ray beam is in general not required, which means that synchrotron beamlines used for example for SEXAFS measurements can be used without modification, as both techniques require a UHV chamber, and similar methods of detection, involving for example electron yield or fluorescence yield measurement. The two techniques may be powerfully employed together to give a complete surface structure determination; this approach has been adopted, for example, in the case of c(2×2)Cl/Ni(100) [45]. In this case, the Cl K fluorescence yield spectrum in NISXW revealed that Cl atoms are located 1.80 ± 0.03 Å above the Ni (200) lattice plane. On the other hand, SEXAFS results showed the Cl–Ni layer spacing to be 1.60 ± 0.02 Å. These two results taken in combination must mean that the surface nickel layer undergoes a substantial relaxation outwards, by 0.2 Å ± 0.05 Å, representing 11% of the bulk spacing [45].

If higher energy X-rays are used, problems of sensitivity may be overcome to some extent by using very grazing angles of incidence, so that total external reflection of the X-ray beam occurs, in exactly the same way as for 2-D X-ray diffraction (Section 4.1). The resulting standing waves are sometimes known as

GAXSW (grazing-angle X-ray standing waves). Unfortunately, possible advantages of this geometry are somewhat offset by the low flux density onto the surface if the beam is incident at a grazing angle. At grazing incidence, the coherently related X-ray beams that generate the SW correspond to the incident X-ray beam, and the beam totally reflected at the mirror surface below its critical angle θ_c. We saw in Section 4.1 that this angle is, typically, fractions of a degree at hard X-ray frequencies. In this geometry, XSW should provide some information about the positions of atoms *parallel* to a surface ('in-plane') as well as *perpendicular* to it ('out-of-plane'). (Compare 2-D X-ray diffraction (Section 4.1), which utilises this geometry, and where both types of information may be obtained.) A combination of XSW in Bragg and grazing angle geometries may be particularly powerful [46]. Equation (39) shows that at zero angle when the incident beam and the surface are parallel, the period of the XSW, D, is infinite, and the phase of the standing wave is such that the surface coincides with a node in the SW (Figure 8.43). As θ is increased, the period of the XSW decreases, and the first maximum in XSW field intensity approaches the surface, and is roughly coincident with the surface at θ_c. (Remember that in Section 4.1, we saw that at θ_c, constructive interference occurs between reflected and refracted waves, resulting in an improvement in signal-to-noise-ratio.) The XSW may be regarded as 'collapsing in like a bellows' as θ is increased towards θ_c [47].

XSW is one of few techniques which may be used to probe surface structure at rather large distances above a surface, as the XSW has a long range nature. The distance above a surface is limited by the distance over which the XSW remains coherent. To date, it is still rather unclear what this limit is, but structural studies of membranes up to 1000 Å thick have recently been reported [47], making the technique particularly important in studies of large biomolecules at surfaces. Figures 8.43–8.45 illustrate the result of such a study on a lipid multilayer (Figure 8.44), where the metal atoms in a zinc arachidate bilayer were located at Å resolution at a distance of almost 1000 Å above the surface of a gold mirror. Figure 8.45 shows the Zn fluorescence yield and the reflectivity of the surface as θ is increased form zero to θ_c. From Figure 8.43, it can be seen that for the heavy atom layer at around 925 Å above the surface, 12 maxima in the XSW will have passed through the heavy atom end of the bilayer during this process, giving the oscillations in fluorescence intensity observed. This information is used to calculate the position of the heavy atom layer relative to the surface, using the fact that the fluorescence yield is proportional to the modulated part of the absorption coefficient, χ (as for SEXAFS); thus χ^2 is calculated for a number of possible configurations of the heavy atom layer, and the best fit to the data over a range of θ values is taken as the solution to the problem.

4.3 PHOTOELECTRON DIFFRACTION (PD)

In Section 3.3, the basic principles of SEXAFS were introduced, and we noted that information may be obtained about the local coordination of a particular

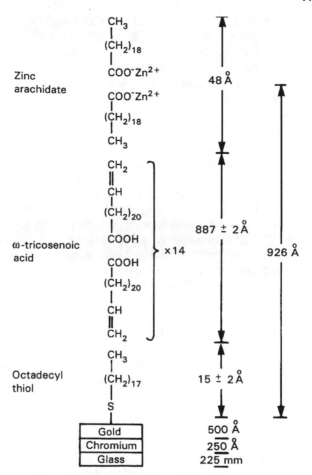

Figure 8.44. Schematic drawing of lipid multilayer LB films deposited on a gold surface. The Zn atoms at 925 Å from the surface gave rise to the XSW data shown in Figure 8.45. Reproduced by kind permission from ref. [47]

surface species by selecting the incident X-ray wavelength to correspond with an absorption edge of that species, and by analysing the SEXAFS modulations at that edge. At the absorption edge, deep lying core electrons are excited into empty valence states or at higher energies into the ionisation continuum (Figure 8.22). The SEXAFS modulations in the absorption signal above the absorption edge then arise due to interference between the outgoing excited electron wavefunction and the waves backscattered by the surrounding atoms in the medium. In practice in SEXAFS, we usually measure the Auger or fluorescence yield following deep core hole production, as these reflect the absorption coefficient (Section 3.3.1).

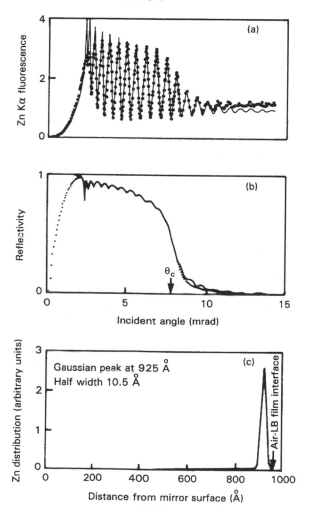

Figure 8.45. The experimental (dots) and theoretical (lines) angular dependence of the Zn Kα fluorescence yield (a) and specular reflectivity (b) at 9.8 keV photon energy for the lipid multilayer shown in Figure 8.44. The resulting calculated zinc distribution above the surface is shown in (c). Reproduced by kind permission from ref. [47]

A related technique, which also involves the production of deep-lying core holes is X-ray Photoelectron Diffraction (XPD). The principles of photoelectron spectroscopy were introduced in Chapter 3. In Photoelectron Diffraction (PD), adsorbate-specific surface structural information (similar to that obtained from SEXAFS) may be gained by measuring the direct photoelectron current following creation of a core hole.

As we saw in Section 3.1, the probability, P_{if} of a transition occurring between an initial state $|\psi_i >$ and a final state $|\psi_f >$ is given by

$$P_{if} \propto |\langle \chi_i | \mu | \chi_f \rangle|^2 \qquad (40)$$

where μ is the dipole operator. After creation of a photoelectron by X-ray absorption, the excited photoelectron wavefunction propagates away from the absorbing atom, with wavevector k. Some beams may leave the solid without undergoing a scattering event, but others may propagate first to a neighbouring atom, and undergo elastic scattering into the k direction (Figure 8.46). Other contributions may arise from beams scattered by several other atoms, or from beams backscattered to the emitting atom (the latter beams being those which give rise to EXAFS modulations). Analysis of the interference between the outgoing and scattered beams will provide bond length information.

The first theoretical model of PD was developed by Liebsch [48]. For the photoelectrons created in this process with an energy greater than ≈ 200 eV, some simplifications are introduced into the analysis of the scattering. This is because the cross-section for elastic scattering of high energy electrons is lower than that for low energy electrons (as we noted when comparing LEED with RHEED),

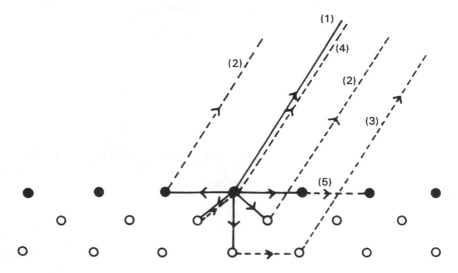

Figure 8.46. Photoelectron scattering events occurring close to a surface, following X-ray absorption. The photoelectron wavefunction propagates away from the absorbing atom (solid arrows). The direct beam (1) propagates freely away from the central absorption site. Some beams may propagate first to a neighbouring atom, and undergo elastic scattering into the k direction (2). Other contributions to this k direction may arise from beams scattered by several atoms (3), or from beams backscattered to the emitting atom, (4) (these last give rise to EXAFS). However, most scattering is in the forward direction (e.g. (5))

and so *multiple* scattering of the photoelectron beams is unlikely. The intensities of diffracted beams may then be analysed simply in terms of the interference between the outgoing wave and waves which have undergone one scattering event (Figure 8.46). This scattering event will usually be at the nearest neighbouring shell of atoms, as in EXAFS, because of the relatively short mean-free-pathlength, λ, of the photoelectron wave due to inelastic scattering (Section 3.2.2). In photoelectron diffraction, therefore, the outgoing photoelectron signal is modulated by interference from scattered beams, and carries information about the local coordination of the emitting atom. Liebsch estimated the modulation to be of the order of 40%.

The PD experiment requires a photon source of the type generally used in a conventional photoemission experiment. This may be a lab-type fixed energy soft X-ray source, producing say Mg Kα or Al Kα radiation (Chapter 3), or a synchrotron source which may be tuned through a range of X-ray wavelengths. The experiment is carried out in UHV, and the photoelectrons are conveniently collected using a hemispherical electron energy analyser (of the type discussed in Chapter 3). A typical experimental arrangement is shown in Figure 8.47 [49]. In order to collect the diffracted beams emerging in different directions from the sample, the analyser must be rotatable about two axes, in order to change both the exit angle relative to the surface normal, ϕ and the azimuthal angle ψ (Figure 8.48).

The intensities of the diffracted beams are measured as a function of a change in the diffraction conditions. This may be a change in the exit angle, $I(\phi)$, or a change in the azimuthal angle, $I(\psi)$, (as in the RHEED experiment, Section 2.3.3). This is known as 'angle resolved photoelectron diffraction' or APD. Plots of the azimuthal (ψ) dependence of the photoemission intensity display characteristic interference maxima and minima consistent with the symmetry of the surface. In practice, the location of an adsorbate relative to a substrate is determined by comparing measured XPD $I(\phi)$ and $I(\psi)$ scans to those calculated for different geometries. Use of a synchrotron source offers the opportunity to vary the incident radiation wavelength, through a change in its energy, $I(E)$. In this case, all angles are kept constant, as is the binding energy of the photoelectrons (i.e. only electrons coming from the same core levels are considered), while the energy of the incident radiation is varied. Most studies of this type have been carried out using photoelectrons emitted normal to the sample surface, and are usually termed NPD (normal photoelectron diffraction). Studies using a fixed emission angle which is not normal to the surface are termed OPD (off-normal photoelectron diffraction). Again, the results of NPD or OPD experiments are usually analysed by comparing them to results calculated for different possible surface structures.

One disadvantage common to all surface diffraction techniques is that analysis of the data involves an iterative procedure of obtaining the best possible fit between theoretical curves and experimental data. This is very expensive in

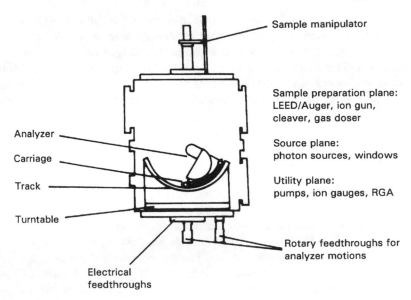

Figure 8.47. A typical experimental chamber for photoelectron diffraction, showing a rotatable hemispherical electron energy analyser mounted inside the vacuum chamber on a carriage allowing both azimuthal and in-plane rotations. Reproduced by kind permission from ref. [49]

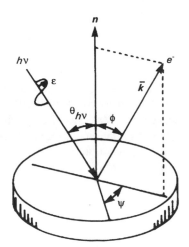

Figure 8.48. Schematic diagram of the experimental geometry used in photoelecton diffraction. ε is the polarisation vector of the incident radiation, $\theta_{h\nu}$ is the polar angle of the incident radiation, $\mathbf{k}$ is the photoelectron wavevector, and $\mathbf{n}$ is the surface normal. The polar angle relative to the surface normal, ϕ, of the emerging photoelectrons and their azimuthal angle, ψ, are varied in the experiment. Reproduced by kind permission from ref. [49]

computer time, and is rather subjective. One approach which has been taken is to adapt the Fourier transform techniques developed for treatment of EXAFS and SEXAFS data (Section 3.2.4), utilising the similarity which we have already noted between the SEXAFS and PD processes. (e.g. [50]). Peaks in the Fourier transform of NPD data are related to pathlength differences between the direct wave and scattered waves. This means that the analysis is only successful in cases where only a few pathlength differences contribute strongly to the scattering.

To date, the majority of PD studies have been of single crystal surfaces, or of overlayer adsorption on single crystal substrate combinations where the overlayer species is only present at the surface, allowing for the overlayer signal to be easily distinguished from substrate emission.

One important characteristic of the elastic scattering of relatively high energy electrons (typically 500–1000 eV or greater) is that it is primarily in the forward direction — i.e., the angle between the incident and scattered beams (the scattering angle) is close to zero, and most diffracted intensity is collected in a small cone around the direction of the incident beam. This can lead to series of pronounced maxima in plots of photoemission current vs polar exit angle, as overlayer atoms are brought into the correct geometry for forward scattering [51]. This means that provided the energy level from which photoemission originates is well-chosen, PD should be able to distinguish, for example, between say 'hollow' and 'on-top' sites for an adsorbate (Figure 8.49), and to determine the orientation of adsorbed molecules. This sensitivity will decrease with the separation between the surface and the overlayer, and in practice, one consequence of forward scattering is that XPD is rather insensitive to the position of an adsorbate if it sits well above the surface. At lower photon energies (typically 50–400eV), backscattering of the photoelectron wave is sufficiently large to produce significant photoelectron diffraction effects, and to allow the determination of the location of adsorbed atoms relative to the underlying substrate atoms.

An early example of the location of an adsorbate using PD azimuthal scans is shown in Figure 8.50 [52]. This used low energy synchrotron radiation to study the adsorption of Na on Ni (100). The figure shows the azimuthal rotation data for Na 2p photoemission obtained at a fixed emission angle of 30° and 80 eV photon energy. The scan reflects the symmetry of the Na adsorption site. The dotted curve shows theoretical intensities, assuming a Ni(001)c(2 × 2)-Na surface structure, with Na in hollow sites. After subtracting the minimum intensity from all points to enhance the anisotropy, the full curve is obtained. The dashed curve is the experimental data, which has been subjected to a four-fold symmetrisation. The agreement between theory and experiment is quite good, and even the secondary structure within the four main lobes is reproduced.

One of the first successful studies to apply EXAFS data analysis techniques to PD was the work of Shirley *et al.* on Ni(001)c(2 × 2)-S [53]. Here, two polar angles were used, 0° (NPD) and 45°, and the sample and analyser were also rotated in tandem, with respect to the light beam, to change the polarisation vector

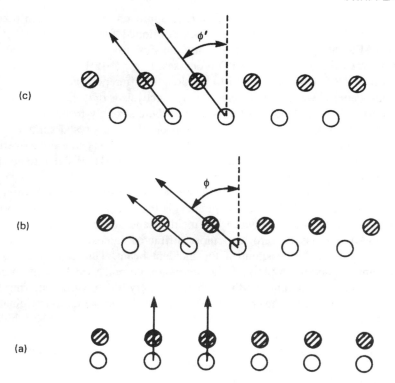

Figure 8.49. Scattering of relatively high energy electrons in PD is predominantly in the forward direction, in principle enabling us to distinguish 'hollow' from 'on-top' sites for an adsorbate. For the 'on-top' site, shown schematically in (a), we might expect strong scattering along the surface normal. For a 'hollow' site, strong scattering will be expected at an off-normal polar angle ϕ (b). If the adsorbate sits well above the surface, ϕ' is decreased, making it more difficult to distinguish between the sites (c)

of the light relative to the crystal. This is used to enhance the scattering from specific Ni atoms, making use of the fact that photoemitted intensity is peaked strongly in the direction of the electric vector of the light. Thus if the electric vector is polarised parallel to the surface normal, the photoemission current is directed into the sample, and strong scattering from substrate atoms occurs. If the vector is polarised parallel to the sample surface, strong intra-layer scattering from the within the topmost sample layer occurs. This effect is analogous to the polarisation dependencies observed in SEXAFS (Section 3.3.2). In this study, S 1s photoemission was measured, at kinetic energies up to around 500 eV Figure 8.51). Data processing followed the procedure for EXAFS (Sections 3.2.2 and 3.2.4); a background function representing the atomic contribution to the

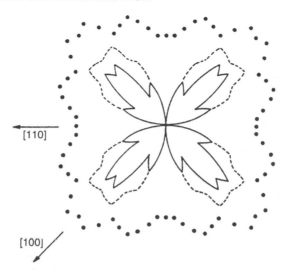

Figure 8.50. PD azimuthal scan for Ni(001)c(2 × 2)–Na at a fixed emission angle of 30° and photon energy of 80 eV, showing the comparison between theory (dotted and full lines), and experiment (dashed line, see text). Reproduced by kind permission of the American Institure of Physics from ref. [52]

photoemission current was subtracted from the experimental photoemission intensity, from which the normalised oscillatory part, $\chi(k)$ was obtained, and then Fourier transformed. The resulting transforms are shown in Figure 8.51. For the Ni/S surfaces, the first three peaks in the transformed data are related to scattering from the four nearest Ni neighbours around the S hollow site. The resulting Ni–S distance is 2.23 Å, in excellent agreement with SEXAFS data. Interpretation of the data for this system was then used to determine the geometry of an unknown system, Cu (001) p (2 × 2)–S. A similar four-fold hollow site was determined, with Cu–S distance of 2.28 Å.

In Section 2.2.4, we discussed the idea that all techniques based on the interference between coherent electron beams could lend themselves to holographic applications. PD is perhaps the area where the most dramatic progress has been made recently; there has been considerable interest in the idea that the angular distribution of photoelectrons emitted from a solid surface (caused by coherent interference between the directly emitted electron wavefunction and components elastically scattered by surrounding atoms) can be viewed as a photoelectron hologram. Photoelectron holography has recently been used to study a high temperature surface phase transition on Ge (111) [54]. Here, approximate direct imaging of atomic positions is achieved by a two-dimensional Fourier transform of large-scale data sets of intensity as a function of polar and azimuthal angles, obtained from X-ray photoelectron diffraction experiments. The combination of

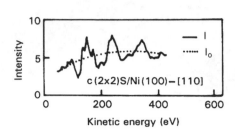
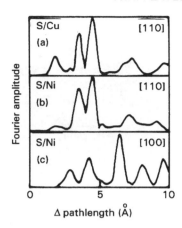

Figure 8.51. PD data for Ni (001) c (2 × 2)–S at the S K-edge, showing the application of EXAFS data analysis techniques. The left-hand panel shows the oscillations in the photoemission intensity, (with I_0 approximating to the atomic contribution to the photoemission current). The right-hand panel shows Fourier transforms of data recorded for different polarisations of the light relative to the sample, for the Ni system and for Cu (001) p (2 × 2)–S. Reproduced by kind permission of the American Institute of Physics from ref. [53]

PD and holography data indicate that by 1200 K the top 1–2 layers of Ge atoms are completely disordered.

Dramatic experiments in which the angular distribution of Auger electrons emitted from a Pt (111) surface has been studied [55], have indicated that it may be possible directly to image atoms at a surface in an even more direct way. In this experiment, the complete angular distribution of Auger electrons emitted from the surface on electron beam irradiation was measured. The resulting distributions plotted directly showed 'silhouettes' of the surface atoms, 'back-lit' by emission from atoms deeper in the solid. Very importantly, the angular distributions observed were those predicted theoretically if the atoms are regarded as spherical point emitters of uniform cross-section; i.e. no multiple-scattering or diffraction effects needed to be taken into account to explain the observed results. One suggestion for this simplicity of the data is that as Auger electrons, unlike photoelectrons are emitted at random times, this may prevent the formation of the plane waves required for efficient diffraction. The result is a technique of great simplicity, where the data may be interpreted very directly without complex Fourier analysis [55].

REFERENCES

Useful general texts are listed below; for more detailed information, the reader is referred to the articles refenced in the text.

GENERAL TEXTS

Andrew Zangwill, *Physics at Surfaces*, Cambridge University Press, Cambridge, (1988).
Giorgio Margaritondo, *Introduction to Synchrotron Radiation*, Oxford University Press, New York, (1988).
Martin Prutton, *Surface Physics*, Clarendon Press, Oxford (1983).
M.W. Roberts and C.S. McKee, *Chemistry of the Metal–Gas Interface*, Clarendon Press, Oxford, (1978).
Gabor A. Somorjai, *Chemistry in Two Dimensions: Surfaces*, Cornell University Press, London, (1981).
J.B. Pendry, *Low Energy Electron Diffraction*, John Wiley, New York, (1974).
I.K. Robinson and D.J. Tweet, *Surface X-ray Diffraction*, Rep. Prog. Phys., **55**, 599, (1992).

REFERENCES CITED IN TEXT

[1] M. Prutton, *Surface Physics*, Clarendon Press, Oxford, (1983).
[2] C.J. Davisson and L.H. Germer, *Phys. Rev. (2nd series)*, **30** (**6**), 705, (1927).
[3] E.J. Scheiber, L.H. Germer and C.D. Hartman, *Rev. Sci. Inst.*, **31**, 112, (1960).
[4] U. Scheithauer, G. Meyer and M. Henzler, *Surf. Sci.* **178**, 441, (1986).
[5] V.E. Henrich and P.A. Cox, *The surface science of metal oxides*, Cambridge University Press, Cambridge, (1994).
[6] J.B. Pendry *Low Energy Electron Diffraction*, John Wiley, New York, (1974).
[7] P.J. Rous and J.B. Pendry, *Surf. Sci.* **219**, 355 and 373, (1989).
[8] W. Oed, P.J. Rous and J.B. Pendry, *Surf. Sci.* **273**, 261, (1992).
[9] J.M. Powers, A. Wander, M.A. Van Hove and G.A. Somorjai, *Surf. Sci. Lett.* **260**, L7, (1992).
[10] J. Falta and M. Henzler, *Surf. Sci.*, **269/270**, 14, (1992).
[11] M. Sidoumou, T. Angot and J. Suzanne, *Surf. Sci.* **272**, 347, (1992).
[12] P. Zschack, J.B. Cohen and Y.W. Chung, *Surf. Sci.* **262**, 395, (1992).
[13] J.B. Hannon, E.W. Plummer, R.M. Wentzcovitch and Pui K. Lam, *Surf. Sci.*, **269/270**, 7, (1992).
[14] W.K. Ford, T. Guo, D.L. Lessor and C.B. Duke, *Phys. Rev. B*, **42**, 8952, (1990).
[15] P.L. de Andres, *Surf. Sci.*, **269/270**, 1, (1992)
     P. Hu and D.A. King, *Nature*, **360**, 655, (1992).
[16] J.B. Pendry in *X-ray Absorption Fine Structure*, (ed.) S.S. Hasnain, Ellis Horwood, Chichester, p. 3, (1991).
[17] J.H. Neave, B.A. Joyce, P.J. Dobson and N. Norton, *Appl. Phys.*, **A31**, 1, (1983).
[18] P.J. Dobson, N.G. Norton, J.H. Neave and B.A. Joyce, *Vacuum*, **33**, 593, (1983).
[19] H. Toyoshima, T. Shitara, J. Zhang, J.H. Neave and B.A. Joyce, *Surf. Sci.*, **264**, 10, (1992).
[20] Koichiro Saiki, Yuji Nakamura and Atsushi Koma, *Surf. Sci.*, **269/270**, 790, (1992).
[21] C.Y. Yang, J.M. Lee, M.A. Paesler and D.E. Sayers, *J. de Phys. Colloque*, **C8–47**, 387, (1986).
[22] J. Stöhr, in *Emission and Scattering Techniques, Proc. NATO ASI*, **73**, (ed.) P. Day, Dordrecht, D. Reidel, p. 215. (1981).
[23] Hisanobu Wakita, Seiichi Yamashita and Kazuo Taniguchi, in *X-ray absorption fine structure* (ed.) S.S. Hasnain, Ellis Horwood, Chichester p. 685, (1991).

[24] S.N. Gupta, K.K. Chaturvedi and Galav Shah, *X-ray absorption fine structure*, (ed.) S.S. Hasnain, Ellis Horwood, Chichester p. 688, (1991).

[25] G.M. Lamble, D.J. Holmes, D.A. King and D. Norman, *J. de. Phys. Colloque*, **C8-47**, 509, (1986).

[26] *X-ray absorption fine structure*, (ed.) S.S. Hasnain, Ellis Horwood, Chichester, (1991).

[27] G. Baker, A.J. Dent, G. Derbyshire, G.N. Greaves, C.R.A. Catlow, J.W. Couves and J.M. Thomas, in *X-ray absorption fine structure* (ed.) S.S. Hasnain, Ellis Horwood, New York, p. 738, (1991).

[28] G.N. Greaves, S. Pizzini, K.J. Roberts, N.T. Barrett and S. Kalbitter, in *X-ray absorption fine structure*, (ed.) S.S. Hasnain, Ellis Horwood, New York, p. 232, (1991).

[29] D.R. Warburton, G. Thornton, D. Norman, C.H. Richardson, R. McGrath and F. Sette, *Surf. Sci.*, **189/190**, 495, (1987).

[30] J. Haase, B. Hillert, L. Becker and M. Pedio, *Surf. Sci.*, **262**, 8, (1992).

[31] D. Norman, R.A. Tuck, H.B. Skinner, P.J. Wadsworth, T.M. Gardiner, I.W. Owen, C.H. Richardson and G. Thornton, *Phys. Rev. Lett.* **58**, 519, (1987).

[32] D. Norman, R.A. Tuck, H.B. Skinner, P.J. Wadsworth, T.M. Gardiner, I.W. Owen, C.H. Richardson and G. Thornton, *J. de Phys. Colloque*, **C8-47**, 529, (1986).

[33] J. Stöhr and R. Jaeger, *Phys. Rev. B*, **26**, 4111, (1982).

[34] D.R. Warburton, D. Purdie, C.A. Muryn, K. Prabhakaran, P.L. Wincott and G. Thornton, *Surf. Sci.* **269/270**, 305, (1992).

[35] C.A. Muryn, D. Purdie, P. Hardman, A.L. Johnson, N.S. Prakash, G.N. Raiker, G. Thornton and D.S. -L Law, *Faraday Discuss. Chem. Soc.*, **89**, 77, (1990).

[36] D. Purdie, C.A. Muryn, N.S. Prakash, K.G. Purcell, P.L. Wincott, G. Thornton and D.S. -L Law, *J Phys: Cond. Matt.*, **3**, 7751, (1991).

[37] E Vlieg, A Van't Ent, A.P. de Jongh, H. Neerings and J.F. Van der Veen, *Nuclear Instr. Meth. Phys. Res.*, **A262**, 522, (1987).

[38] A.R. Sandy, S.G.J. Mochrie, D.M. Zehner, O. Grübel, K.G. Huang and D. Gibbs, *Phys. Rev. Lett.*, **68**, 2192 (1992).

[39] H.A. van der Vegt, H.M. van Pinxteren, M. Lohmeier and E. Vlieg, *Phys. Rev. Lett.*, **68**, 3335, (1992).

[40] J. Bohr, R. Feidenhans'l, M. Nielsen, M. Toney, R.L. Johnson and I.K. Robinson, *Phys. Rev. Lett.*, **54**, 1275, (1985).

[41] R.G. van Silhout, M. Lohmeier, S. Zaima, J.F. van der Veen, P.B. Howes, C. Norris, J.M.C. Thornton and A.A. Williams, *Surf. Sci.*, **271**, 32, (1992).

[42] M. Kerkar, W.K. Walter, D.P. Woodruff, R.G. Jones, M.J. Ashwin and C. Morgon, *Surf. Sci.*, **268**, 36, (1992).

[43] D.P. Woodruff, D.L. Seymour, C.F. McConville, C.E. Riley, M.D. Crapper and N.P. Prince, *Phys. Rev. Lett.*, **58**, 1460, (1987).

[44] D.P. Woodruff, D.L. Seymour, C.F. McConville, C.E. Riley, M.D. Crapper and N.P. Prince, *Surf. Sci.*, **195**, 237, (1988).

[45] T. Yokoyama, Y. Takata, T. Ohta, M. Funabashi, Y. Kitajima and H. Kuroda, *Phys. Rev. B*, **42**, 7000, (1990).

[46] T. Jach and M.J. Bedzyk, *Phys. Rev. B*, **42**, 5399, (1990).

[47] J. Wang, M.J. Bedzyk, T.L. Penner and M. Caffrey, *Nature*, **354**, 377, (1991).

[48] A. Liebsch, *Phys. Rev. Lett.* **32**, 1203 (1974), *Phys. Rev. B*, **13**, 544, (1976).

[49] Y. Margoninski, *Contemp. Phys.*, **27**, 203, (1986).

[50] P.J. Orders and C.S. Fadley, *Phys. Rev. B*, **27**, 781, (1983).

[51] W.E. Engelhoff, *Phys. Rev. B*, **30**, 1052, (1984).

[52] D.P. Woodruff, D. Norman, B.W. Holland, N.V. Smith, H.H. Farrell and M.M. Traum, *Phys. Rev. Lett.* **41**, 1130, (1978).

[53] J.J. Barton, C. C, Bahr, Z. Hussain, S.W. Robey, L.E. Klebanoff and D.A. Shirley, *J. Vac. Sci. Technol.*, **A2**, 847, (1984).

[54] T.T. Tran, S. Thevuthasan, Y.J. Kim, G. S, Herman, D.J. Friedman and C.S. Fadley, *Phys. Rev. B*, **45**, 12106, (1992).

[55] D.G. Frank, N. Batina, T. Golden, F. Lu and A.T. Hubbard, *Science*, **247**, 182, (1990).

CHAPTER 9

Scanning Tunnelling Microscopy and Atomic Force Microscopy

GRAHAM J. LEGGETT

The University of Nottingham, UK

1 INTRODUCTION

The development of the scanning tunnelling microscope (STM) and its close relative the atomic force microscope (AFM) or, more generally, the scanning force microscope (SFM), has proved to be one of the more exciting developments in surface science over the past decade. Both techniques offer, in principle, the possibility for obtaining atomic-resolution images with, in many cases, a minimum of prior preparation of samples. The ease of construction and operation of scanning probe microscopes (SPMs) (a generic term covering both techniques, and a range of other, more recent, related techniques), has led to a proliferation of companies capable of providing effective instruments at relatively inexpensive prices (compared to electron microscopes capable of achieving comparable resolution). The 'observation' of individual atoms has become a commonplace event in many modern laboratories (for example, see Figure 9.1), and the potential exists for a revolution in our understanding of processes at surfaces. The consequence of this has been the application of SPM techniques across a broad spectrum of scientific research, from biological processes to solid state physics. This wide-ranging impact was undoubtedly responsible for the very rapid award of the Nobel Prize to the inventors of the STM, Binnig and Rohrer, in 1986, just five years after its invention.

The objectives of this chapter are three-old: firstly, to provide a basic understanding of the physical principles underlying STM and SFM; secondly, to provide a general introduction to the operation of scanning probe microscopes and other important experimental considerations; and thirdly, to give a flavour of the kinds of data which SPM techniques are capable of providing. Because of the vast literature which has arisen in the field, this overview is far from comprehensive in scope. Instead, we will examine a few of the fields in which

Surface Analysis — The Principal Techniques. Edited by John C. Vickerman
© 1997 John Wiley & Sons Ltd

G.J. LEGGETT

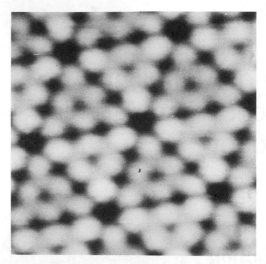

Figure 9.1. Atomic resolution STM image of the 7 × 7 reconstruction of a silicon(III) surface. Reproduced by kind permission of Dr. P.H. Beton, Department of Physics, The University of Nottingham

SPM techniques have had a particular impact, hoping to provide a starting point for more detailed reading on specific subjects.

2 BASIC PRINCIPLES OF OPERATION

The STM and the AFM are both stylus-type instruments, in which a sharp probe, scanned raster-fashion across the sample, is employed to detect changes in surface structure on the atomic scale. One of the most commonly employed analogies is with the operation of a record-player, in which a stylus moves up and down with the topography of the rotating record to generate a signal. In the case of the AFM, this simple analogy is not far from the truth: the interaction force between the probe and surface structural features is measured to reveal the surface topography in a fairly direct fashion. In the case of the STM, the analogy is less accurate: it is, in fact, the surface electron density that is measured. However, for surfaces which have relatively uniform electronic properties (and in the limits of dimensions >10 Å), the STM image effectively represents the surface topography. For electronically inhomogeneous surfaces, the electronic structure becomes important in determining the nature of the image. A lateral resolution of 0.1 Å and a vertical resolution of 0.01 Å are attainable with the STM. In principle, comparable lateral resolution is also possible for the AFM, although in practice, one generally finds that atomic resolution is more readily accessible with the STM — given a conducting sample, that is. The issue of atomic resolution in AFM is further complicated by controversy about the precise nature of the image formation mechanism; however, the sensitivity of the AFM in the vertical direction remains beyond question. In order to understand the origin of this astonishing degree of resolution, we need to pause to consider the operation of the instrument in more detail.

2.1 OPERATION OF THE STM

2.1.1 Quantum Tunnelling

The STM is an example of the practical exploitation of a strictly quantum mechanical phenomenon: quantum tunnelling. Tunnelling processes play an important role in a number of phenomena which occur at surfaces and interfaces. For example, in SIMS, incident ions may be neutralised prior to penetration of the solid surface by electron tunnelling, and, similarly, ejected secondary ions may be neutralised by tunnelling processes. Quantum mechanical tunnelling involves the penetration of a potential barrier by an electron wave function. The potential barrier may be a layer of insulating material (for example, an oxide layer in the case of a metallic electrode) or a vacuum gap (in the case of the STM). The basic physical process is best understood by consideration of the simple case of one-dimensional tunnelling (widely discussed in text books on quantum mechanics or solid state physics, for example see [1]). Consider the simple system (Figure 9.2(a)) in which an electron is incident upon an infinitely thick potential barrier of height V. The Schrödinger equation has two components. Where $x < 0$,

$$H = -(\hbar^2/2m)(\mathrm{d}^2/\mathrm{d}x^2) \tag{1}$$

Inside the barrier ($x > 0$),

$$H = -(\hbar^2/2m)(\mathrm{d}^2/\mathrm{d}x^2) + V. \tag{2}$$

The solutions of these equations are:

$$\psi = A\mathrm{e}^{ikx} + B\mathrm{e}^{-ikx}, \qquad k = (2mE/\hbar^2) \text{ inside the well, and} \tag{3}$$

$$\psi = C\mathrm{e}^{ik'x} + D\mathrm{e}^{-ik'x}, \qquad k = (2m(E-V)/\hbar^2)^{1/2} \text{ inside the barrier.} \tag{4}$$

The wave function inside the barrier has an imaginary part (which rises to infinity and is thus discounted) and a real part which decays exponentially with distance inside the barrier. This is a very important result, because where penetration is classically forbidden (for $E < V$), the quantum mechanical wavefunction is non-zero: the electron may tunnel into the potential barrier. There is, therefore, a finite probability that the electron will be found inside the barrier. A simple extension of this case involves the consideration of two potential wells close together (in other words, separated by a potential barrier of finite thickness). If the potential barrier is relatively narrow, there is a probability that an electron may penetrate it [1] and pass from one well to the other. Consider two nearby metallic

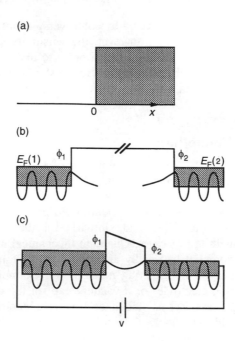

Figure 9.2. (a) An infinitely thick potential barrier. Potential $= V$ for $x > 0$ and 0 for $x < 0$. (b) Two potential wells which are spatially separated. (c) Two potential wells separated by a small distance, and with an applied potential difference

electrodes with a work function ϕ, separated by a large distance (Figure 9.2(b)). The effective overlap of the Fermi level wavefunctions is negligible, because of the exponential decay of the two separate wavefunctions. If these electrodes are brought close together, with some separation d (Figure 9.2(c)), then the overlap of the wavefunctions may be sufficiently great to facilitate quantum mechanical tunnelling and, under the influence of an applied potential difference, the passage of a measurable current. The magnitude of the tunnelling current I is a measure of the overlap of the two wavefunctions, and is given by

$$I \propto e^{(-2\kappa d)} \tag{5}$$

where κ is related to the local work function by:

$$\kappa = (2m\phi/\hbar^2)^{1/2}. \tag{6}$$

Unfortunately, the simple one-dimensional model is not really adequate for a full description of STM. The electronic structures of the tip and the surface are involved in a complex fashion which really demands a three-dimensional treatment. One of the most widely quoted treatments is due to Tersoff and Hamann

[2]. Here, the full general expression for the tunnelling current is:

$$I = (2\pi e/\hbar)e^2 V \sum_{\mu,\nu} |M\mu, \nu|2\delta(E_\nu - E_F)\delta(E_\mu - E_F) \tag{7}$$

where E_F is the Fermi energy, E_μ is the energy of the state ψ_μ in the absence of tunnelling, and $M_{\mu,\nu}$ is the tunnelling matrix element between states ψ_μ of the probe tip and ψ_ν of the sample, given by:

$$M_{\mu,\nu} = (\hbar^2/2m) \int dS(\psi_\mu^* \nabla \psi_\nu - \psi_\nu \nabla \psi_\mu^*) \tag{8}$$

Working on the assumption that the STM tip is spherical, one arrives at the following expression for the tunnelling current [2]:

$$I = 32\pi^3 \hbar^{-1} e^2 V \phi_o^2 D_t(E_F) R_t^2 \kappa^{-4} e^{2\kappa R_t} \sum |\psi_\nu(r_o)|2\delta(E_\nu - E_F) \tag{9}$$

in which ϕ_o is the work function and $D_t(E_F)$ is the density of states at the Fermi level per volume of the tip. Thus the tunnelling current is proportional to the local density of states at the Fermi level and at the centre of the STM tip. This means that the STM can provide a direct image of quantum mechanical electron states at the surface and, therefore, equation (9) provides a basis for the application of STM to atomic-scale surface spectroscopy (discussed below).

Equation (9) preserves the important result that the tunnel current depends exponentially upon the separation between the STM tip and the sample surface. This exponential dependence on the tip–sample separation, or the tunnelling gap, provides the basis for the astonishing resolution of the technique. To a rough approximation, the tunnelling current decreases by an order of magnitude for every increase of 1 Å in the tunnelling gap. For a sufficiently sharp tip, the bulk of the tunnelling current will flow through the very end of the tip (illustrated in Figure 9.3) and thus the effective diameter of the tip becomes very small (of the order of atomic dimensions [3]).

2.1.2 The Role of Tip Geometry

Ideally, we require a tip which has a single atom situated at its apex. This sounds like a very stringent constraint; however, tips sharp enough to generate high resolution images may be obtained fairly straightforwardly. Two methods are particularly popular. The first of these, mechanical preparation, is the simplest and commonly involves cutting a piece of fine wire (usually platinum–iridium or tungsten) at an oblique angle with a pair of sharp wire-cutters. The second method involves using a chemical agent (NaOH) to etch the wire; the lower portion of the wire drops off leaving a sharp tip. In addition to these methods, *in situ* methods of tip sharpening can be useful. Again, a number of methods exist. One technique involves applying a voltage pulse between the tip and the

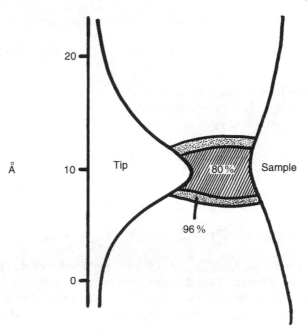

Figure 9.3. Schematic illustration of the current density distribution for tunnelling from a tip to a corrugated surface. After Binning and Rourer [3]

substrate [4], with the consequence that, according to one possible explanation [5], the electrical field in the gap draws atoms (for example, tungsten atoms on a tungsten tip) towards the apex of the tip. Often, this may result in a sharper tip profile, so that a damaged tip may be regenerated; however, the technique is also useful as a means of removing contaminants which may have become adsorbed to the tip during scanning of a contaminated or molecular sample. An alternative approach involves gently colliding the tip with a surface. Usually, this would result in terminal damage to the tip (known as 'crashing' the tip). However, it has been reported that gentle collisions between an STM tip and a silicon surface can result in the removal of a silicon cluster by the tip, leaving a small hole in the silicon surface. In a similar fashion, the wetting of tips by gold atoms has been reported [6] although these are of little value for imaging purposes.

Despite the ease with which tips capable of providing a very high level of resolution may be prepared, it nevertheless remains the case that tip structure plays an important role in determining the nature of the STM image. Variations in tip atomic structure may cause variations in the STM image. Equation (7), which gives an expression for the tunnelling current, involves the wavefunctions for both the tip and the substrate, and in reality the STM image represents a complex convolution of the structures of the tip and the substrate surface. As the required resolution is increased, the tip structure comes, increasingly, to determine the STM image. A rather simplistic illustration of the convolution of tip and surface structure is given in Figure 9.4. An atomically sharp tip scanning a region of the surface which contains an atomic protrusion would provide an accurate representation of the surface morphology (Figure 9.4(a)). However, if the tip is blunt, so that the surface feature has dimensions smaller than those the apex of the tip (Figure 9.4(b)), then the surface feature may effectively image the tip. Under such circumstances, it is not clear what the surface topography really is.

This rather simplistic example by no means represents the only kind of artefactual tip effect which may be observed, and, in fact, a number of different examples have been reported in the literature. Loenen et al. [7] reported the observation of ghost images of the STM tip in images of silicon surfaces. It was possible to determine the tip geometry from these ghost images, and in that particular case, the tip was found to be blunt over a length of 24 nm. Paradoxically, flat surface areas, steps and facets of the sample were imaged with atomic resolution, indicating that there were atomic protrusions from the top and edges of the tip. This example indicates the multitude of ways in which tip geometry can contribute to the formation of the STM image. In some cases the images can be so complex that interpretation is impossible, although most commonly the effects are more moderate and represent a potential interpretational pitfall.

(a)

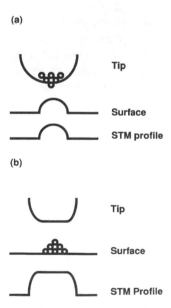

(b)

Figure 9.4. (a) Sharp tip imaging a small surface feature, yielding an STM image which accurately represents the surface topography. (b) Blunt tip imaging a sharp surface feature, with the consequence that an image of the tip (rather than the surface topography) is formed

Besides these rather gross effects on image generation, there are a variety of more subtle ways in which the STM image depends on the atomic structure of the tip. Tsukada *et al.* have discussed a first-principles theory of STM based upon the microscopic electronic state of both the tip and the sample [8]. Beginning with Bardeen's equation, as in the treatment above, and using the local density approximation, with a cluster model for the STM tip based upon 10–20 atoms, they calculated the electronic structure for a graphite surface and tungsten tips with geometries [111] and [110] using the LCAO method. They simulated images for a variety of different tips, and illustrated the kinds of variation which can occur in the STM image with small changes to the tip. Figure 9.5 illustrates this in the case of a [111] tungsten tip. Figure 9.5(a) shows the structure of graphite. There are two non-equivalent carbon atoms in the top-most layer. A second carbon atom lies directly underneath carbon atom A, in the next lowest plane of the solid, but the next carbon atom to lie directly beneath carbon atom B is in the second plane below the surface. In this respect, graphite provides an important example of the way in which the corrugation in the STM image reflects the local density of states (LDOS) at the Fermi level, and not the positions of atoms. The LDOS is lower at site A than it is at site B. Consequently, the honeycomb lattice is not seen and a hexagonal lattice is observed instead. For a typical W_{10} [111] tip such as that illustrated in Figure 9.5(b), the maximum of the tunnelling current contour map therefore lies over carbon B, with the minimum at the centre of the hexagonal ring of carbon atoms. However, if the atom at the apex of the tip is removed (Figure 9.5(c)), the image changes. Now the weakest current region is located at site B, and the maximum in the current contour map lies at carbon atom A. Tsukada *et al.* suggested the following explanation: in the second case, the tunnelling current reaches a maximum when the combined current contribution from the three equivalent topmost tungsten atoms is the largest, which will occur when the centre of the tip is positioned over site A. When the tip is centred over an A site, the three tungsten atoms nearest the surface are positioned over three B carbon atoms. When the tip is centred over site B, the three topmost tungsten atoms are located over the centres of three graphite hexagons, and the current is minimal.

Steps and defects in graphite, and grain boundaries, can perturb the electron density giving rise to the observation of superstructures. Large scale hexagonal superstructures have been observed with periods of 30, 4.2, 2.4 and 2.0 nm by Yang *et al.* [9], together with a strip-like periodicity with a period of 1 nm. These superperiodic features had an irregular appearance which the authors suggested was due to the presence of crystal defects. It was suggested that a Moiré effect due to rotational misorientation of two basal planes near the surface was responsible for the hexagonal superstructures. Such Moiré effects are not uncommon, and can produce hexagonal arrays of points of varying periods. The positions of the points may in fact match the expected positions of the points in an authentic atomic resolution image, although the corrugation amplitude may be much larger.

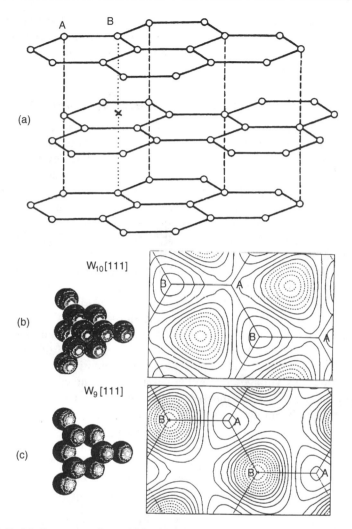

Figure 9.5. (a) Structure of graphite, showing characteristic layered structure. Note the two non-equivalent lattice sites, marked A and B. (b) Simulation of the image of graphite surface obtained with a $W_{10}[111]$ tip. (c) Simulation of the expected image after removal of the atom from the apex of the tip. Reproduced with kind permission from Ref 8 © 1991 Elsevier Science – NL. Sara Burgerhartstraat 25, 1055 kV Amsterdam, The Netherlands

In summary, giving consideration to the present theoretical understanding of image formation, and to the role of tip geometry, it is clear that an STM image, even at atomic resolution, is far from being a simple visual representation the spatial locations of atoms. What we see instead is a complex image, determined by a convolution of the electronic structures of the tip and the sample surface. However, when it is interpreted correctly, the STM image is capable of providing us with data on the structure and bonding of atoms and molecules at surfaces which are quite unique in nature.

2.1.3 Instrumentation and Basic Operating Parameters

Figure 9.6 illustrates the operation of an STM in a very schematic fashion. A wide variety of STM designs are in use, but certain features are common to all of them. The key components are the tip, some means of achieving very delicate movements in the x-, y- and z-directions and a computer system to control the whole operation. Fine control of the tip position (both laterally and vertically) is achieved by the use of a piezoelectric crystal, on which either the STM tip or the sample is mounted. The piezoelectric crystal moves in a well defined fashion in each of the three spatial directions as the applied potential difference in each direction is varied.

There are three basic modes of operation: constant current, constant voltage and spectroscopic modes. We will discuss spectroscopic operation in a separate section; for the meantime we concern ourselves with the two imaging modes. In both modes, the tip is brought close to the surface (a few Ångströms) and a potential difference is applied, with the consequence that a current begins to flow. This current is the tunnel current, and it is monitored as the tip begins to scan across the surface. In constant voltage mode, the potential between the tip and the surface (the bias voltage) is maintained at a constant value, and the image represents the variation in the measured tunnelling current with position (x_i, y_i). The more commonly employed mode is constant current mode, however. In this mode, the instantaneous tunnelling current is measured at each position (x_i, y_i) and the bias voltage is adjusted via a computer-controlled feedback loop such that the tunnelling current re-assumes some pre-set value. The STM image thus represents the variation in the z-voltage with coordinate (x_i, y_i). Adjustments to the bias voltage cause the piezoelectric crystal to move up and down, and if the

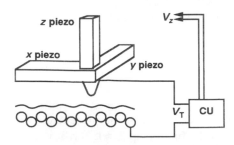

Figure 9.6. Operation of the STM. CU = control unit. V_T is the sample bias voltage. V_z is the voltage applied to the z piezo to maintain constant tunnelling current

displacement of the crystal is known for a given change δV in the bias voltage, then the image may be plotted as tip displacement (height), z_i against surface coordinate (x_i, y_i). Although the contours of this image are really formed by variations in electron density, in the limit of dimensions >10 Å, the image will provide a very good approximation to the surface topography.

There are two key experimental parameters: the tunnelling current and the tunnelling voltage. It is also sometimes helpful to think about the resistance of the tunnelling gap. The gap resistance clearly gives a measure of the distance between the tip and the surface. If the tip is close to the surface, the gap resistance will be relatively small; as the tip moves away from the surface, the gap resistance will increase. The distance of separation between the tip and the surface is in turn determined by the set-point for the tunnelling current: for large tunnelling current set-point values, the gap will be small; for low set-point values, the gap will be large. Typically, the tunnelling current lies in the range 10 pA–1 nA. At values much greater than 1 nA, there is an increasing likelihood that tip–surface interactions will become strong enough to change the surface morphology, although this is dependent upon the nature of the sample. For weakly bound adsorbates, the tip–surface interactions may become significant at much lower values, posing a number of problems which we will discuss in a little more detail below. Tip-induced damage to the surface can be quite severe, and thus sets upper limits on the magnitude of the tunnelling current which may be used. On the other hand, resolution is often better at higher tunnelling currents, and clearly a balance needs to be struck between the resolution of the recorded image on the one hand, and the likelihood of damage to the surface on the other. While tunnelling currents of the order of tens of pA may be employed for biological systems, currents of the order of several nA may be acceptable (and even necessary — see section 3.1.1) when imaging metallic surfaces.

Variations in the bias voltage also have a strong effect on the nature of the image recorded (we deal with specific examples in Sections 3.1.4, 3.1.6 and 3.2.1 below). At very high bias voltage values, alterations to the surface structure may be induced. It becomes possible to etch the sample surface, and if the tip is moved in a controlled fashion across the surface whilst a high bias voltage is maintained, it is possible to create nanometer scale surface structures. Even at low bias voltages, the electric field gradient in the tunnel gap is substantial; for a tip–surface separation of about 10^{-10} m, and a sample bias potential of 0.1 V, $E = 10^9$ V m^{-1}.

2.2 OPERATION OF THE AFM

We have already noted that at high tunnelling currents, the STM tip may interact physically with the surface in such a way that disruption of the surface structure occurs. In fact, there is always a finite interaction force between the tip and the sample surface, even at minimal tunnelling currents. Even at relatively low tunnelling currents, the interaction force is often of the same order of magnitude as interatomic forces or greater [10, 11]. Spence *et al.* [12] have studied the mechanical influence of an STM tip on a graphite surface during imaging. Using a microscopical arrangement very sensitive to surface strains, they observed the formation of strain fringes, extending some 200 nm from the tunnel junction, which moved with the tip as it scanned across the surface. Knowledge of the existence of these forces led Binnig, Quate and Gerber [13] to develop the AFM, in which the probe becomes a cantilever, placed parallel to the surface rather than normal to it. The cantilever of the AFM has a sharp, force-sensing tip at its end, and it is this that interacts with the surface. As the interaction force between the cantilever tip and the surface varies, deflections are produced in the cantilever. These deflections may be measured, and used to compile a topographic image of the surface. (The process is illustrated very schematically in Figure 9.7)

The first requirement for the construction of a force microscope is some suitably sensitive means by which the deflections of the cantilever may be measured. The first AFM utilised, effectively, an STM tip to monitor deflections using electron tunnelling. The most widely utilised approach in commercial instruments is to measure the deflection of a laser beam which is reflected from a mirror mounted on the back of the cantilever. Sarid and Elings [18] describe a number of arrangements in detail. The measured signal is then used to control movements of a piezoelectric crystal, on which the cantilever is mounted, via a feedback system — in just the same way that the movements of the STM tip are controlled. Again, by analogy with the STM, the microscope can operate in constant force mode (in which the piezo moves so that the tip–sample interaction force always equals some preset value) or in constant height mode. However, it

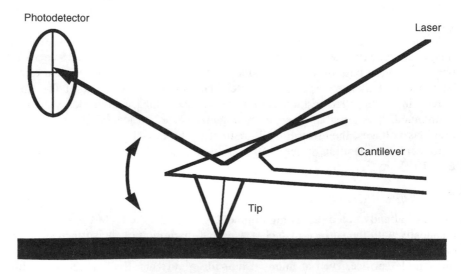

Figure 9.7. Schematic illustration of the operation of the AFM

is constant force mode that is the more widely used and which provides the most accurate data on sample topography.

The mechanical properties of the cantilever are important in controlling the performance of a force microscope. The cantilever deforms in response to the interaction forces between the probe tip and the surface. It is this deformation which determines the performance of the microscope. Of particular interest are the spring constant k and the resonant frequency ω_o. Generally, we require the microscope to be relatively insensitive to mechanical noise in its environment, and, generally speaking, the sensitivity of the cantilever to vibrations decreases with increasing resonant frequency. However, a small spring constant is required in order to facilitate the detection of small forces. Relatively stiff materials (silicon nitride, silicon or silicon oxide) are generally used for cantilever construction in order to ensure high resonant frequencies; in order to, at the same time, maintain a low spring constant, the cantilever must have a very small mass, so microfabrication techniques are used. Typical commercially produced cantilevers have lateral dimensions of the order of 100 μm with thicknesses of the order of 1 μm. Typical spring constants lie in the range 0.1–1 Nm^{-1}, while resonant frequencies are in the range 10–100 kHz.

Force microscopes have been designed which can monitor interactions due to a range of forces, including Van der Waals, electrostatic and magnetic forces. For example, the magnetic force microscope has a tip which possesses a magnetic moment and which therefore responds to the magnetic field of a magnetised sample [14]. The electrostatic force microscope senses isolated charges on surfaces; it is the electrostatic interaction between the charged tip and the sample which is measured. One application of this kind of instrument is in the measurement of charge distributions on the surfaces of polymers [15]. Scanning force microscopes (SFMs) generally measure forces in the range 10^{-9}–10^{-6} N, although the measurement of forces as low as 3×10^{-13} N has been reported [16].

There are two principal modes of interaction between the tip and the surface: attractive and repulsive. Figure 9.8 shows the variation with separation of the interatomic potential energy of two atoms. At atomic separations less than r_0, the force (which is equal to $-dV/dr$) is positive, and the interactions are repulsive. At separations greater than r_0, the force is negative and the interactions are attractive. In the SFM, r_0 corresponds to the transition from contact to non-contact mode (or mechanical point contact). The repulsive interactions are therefore short range,

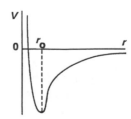

Figure 9.8. Potential energy diagram for the interaction of two atoms

whereas the attractive interactions are long range. SFMs have been constructed which may sense either the attractive or the repulsive forces. In repulsive mode, or contact mode, the AFM tip actually touches the sample, and scans across it in much the same way that the stylus of a record player moves across the surface of a record. A dc bias voltage is utilised to control the piezoelectric scanner. In non-contact mode, or ac imaging mode, the position of the cantilever is oscillated at a high frequency, close to the resonant frequency, with a large amplitude of oscillation. In this mode, the AFM tip senses long-range attractive forces, and the lateral force exerted during scanning (in the plane of the sample) is virtually zero. Non-contact mode thus offers better prospects for imaging weakly bound adsorbates or soft materials. By comparison, lateral forces may be quite significant in contact mode.

Contaminants at the surface can exert a considerable influence on the nature of the tip–surface interactions. For example, thin films of water at a surface can be the cause of attractive capillary forces (as large as 4×10^{-7} N on mica in air). Consequently, a number of groups have explored the possibility of controlling these effects by working under liquids. Operation in an aqueous environment can reduce some of the probe–sample interactions quite substantially. The Van der Waals forces are reduced quite significantly, and the surface charge which is resident on the probe tip is screened both by the attraction of counter-ions from solution, and by dielectric lowering of the interaction energy [17]. Clearly capillary forces can also be minimised by immersing the probe in a liquid medium. Against this, however, the presence of the liquid can create its own problems, through the adsorption of ions onto the sample surface, or through the creation of polarisation forces where the dielectric constant of the tip differs from that of the solvent [17]. Besides aqueous operation, the use of other liquid environments has been reported, including the use of organic liquids. In some circumstances, this may offer distinct advantages.

Measurements of the interaction force between the tip and the sample as the tip is first approached towards the sample and then withdrawn, known as force spectroscopy or force–distance measurement, are a valuable means by which the mechanical characteristics of a material may be probed. The rate of change in the interaction force F with decreasing nominal tip–sample separation s with the tip in contact with the surface gives a measure of the sample stiffness: materials with high moduli under compressive loading give rise to high values of dF/ds. By modulating the cantilever or sample position during scanning in contact mode, it is possible to map local variations in this dF/ds response. This mode of imaging, known as force modulation mode, has been used by Maivald *et al* to map variations in the stiffness of a carbon fibre/epoxy composite, with high contrast being observed over the fibre cross-sections (expected to exhibit the higher stiffness) [19]. Force–distance measurements do not only provide a means to probe mechanical properties; in some elegant recent work, force spectroscopy has been used to probe chemical interaction forces. Recently, a number of workers have attempted to control tip surface chemistries, in order to control or to measure specific tip–sample interactions. For example, Lee *et al.* have adsorbed biotinylated albumin to glass microspheres and measured the variation in force as these microspheres were approached towards, and withdrawn from, mica surfaces functionalised with streptavidin [20]. Biotin and streptavidin exhibit a very strong molecular recognition interaction. In a similar study, they attached DNA oligomers to a silica probe and measured the strength of the interaction between this probe and surface immobilised oligomers [21].

Measurement of lateral forces, in the plane of the sample surface, is straightforwardly possible using typical commercial SFM instruments which employ four-quadrant photodetectors. The light intensity reflected from the back of the cantilever is measured on each quadrant. In conventional constant force mode, the force normal to the sample surface is monitored by measuring the difference between the signals falling on the top and bottom halves of the photodetector. However, measurement of the difference between the signals falling on the left and right halves of the detector is usually possible simultaneously, yielding the force acting on the tip parallel to the sample surface. This lateral force provides a measure of the frictional interaction between the tip and the surface. Wilbur *et al.* used lateral force microscopy (LFM) to study patterned self-assembled monolayers containing regions functionalised with carboxylic acid and methyl terminated adsorbates [22]. The acid-functionalised regions yielded much greater contrast than the methyl-functionalised regions, relecting the much higher surface energy. Care must be exercised when interpreting LFM data, however, because the lateral force typically contains a component due to the sample topography as well as the frictional force [23]. This topographical component may be identified by comparing images recorded in the forwards and backwards directions [23, 24]. By subtracting the forwards and backwards images, the frictional force may be calculated [23, 25].

3   ATOMIC RESOLUTION AND SPECTROSCOPY: SURFACE CRYSTAL AND ELECTRONIC STRUCTURE

It was STM images of individual atoms at surfaces and, still more astonishing, images in which surface electronic structure was resolved, that first fired the imagination of the scientific community. These early studies revealed the enormous potential of scanning probe techniques, and provided the impetus for the exploration of other materials. Consequently, we begin our survey of the applications of STM/AFM with a brief overview of some of the important milestones in high resolution imaging of crystal surfaces. Studies of surface crystal and electronic structure still provide a fertile field of application for STM, so our interest here is not merely historical.

Generally speaking, a crystalline solid organises itself in such a way as to minimise its total energy. The resulting structure is usually accessible through X-ray diffraction analysis. When the crystal is cleaved, the atoms in the newly-formed surface rarely remain in their original positions. For a metal, smoothing of the surface electronic charge density leaves surface atoms out of electrostatic equilibrium with the newly asymmetrical screening distribution. Relaxation of the surface atoms occurs in order to re-establish equilibrium. Whereas metallic bonding is non-directional, the bonding in a semiconductor is highly directional. When a semiconductor crystal is cleaved, for example to give the Si(111) surface shown in Figure 9.9, the atoms in the surface layer are left with dangling bonds directed away from the surface. Considerable reconstruction can occur in order to facilitate the formation of fresh bonds. The reconstructed surface may be extremely complex, and the determination of its structure can be an extremely difficult task.

Several techniques do exist by which the surface crystal structure may be studied, chief amongst them being low energy electron diffraction, LEED. However, the possibility of being able to visualise directly the arrangement of atoms at a surface has considerable appeal, and it was in the determination of surface crystal structure that the first successes of the STM were achieved. The ability of the STM to provide atomic-scale topographic information, and atomic-scale electronic structure information (via spectroscopic operation) has enabled it to make a real and profound impact in this area.

3.1   STUDIES OF GOLD SURFACES

STM images which exhibited atomic rows, with atomic resolution perpendicular to the rows, were obtained first by Binnig *et al.* for the Au(110) surface [26].

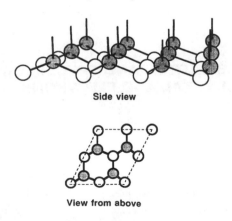

Side view

View from above

⬤ = atom with a dangling bond

Figure 9.9. Ideal surface of Si(111) seen from the side and above

The STM images showed parallel hills, running for several hundred Å in the [110] direction, and formed from a number of different facets, each of which exhibited a characteristic reconstruction Figure 9.10 shows an STM image of a region of an Au(110) surface together with the authors' interpretation of the data (in the inset). In a subsequent paper, Binnig *et al.* reported observations of the reconstruction of the Au(100) surface [27]. A large-scale STM image of a clean 1×5 reconstructed Au(100) surface revealed a flat topography with monatomic steps (see Figure 9.11). However, higher resolution images revealed inhomogeneities in the detailed structure of the five-periodicities, and the authors were able postulate a detailed structure based upon their STM data.

Although these studies had reported images of atomic rows, the individual atoms within the rows were not resolved. For a long while, this was thought to be impossible because of the intrinsic smoothness of electron density functions at the surfaces of metals. Au(111) was another close-packed surface to which this consideration was thought to apply. Although images had been obtained which exhibited extensive terraces separated by monolayer steps, no corrugation

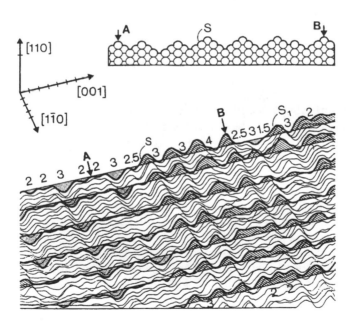

Figure 9.10. STM image of an Au(110) surface, exhibiting disorder. The straight lines help to visualise the terraced structure with monolayer steps (e.g. at S); below each line, the missing rows, and above each line, the remaining rows, are enhanced. The numbers in the top scan give distances between maxima in units of the bulk lattice spacing. The inset shows the proposed structural model for the observed corrugation between A and B. Reproduced with kind permission from Ref. 26. © 1983 Elsevier Science – NL

Figure 9.11. STM image of a clean 1 × 5 reconstructed Au(100) surface, showing monatomic steps. Reproduced with kind permission from Ref. 27. © 1984 Elsevier Science

had been observed along the terraces. However, important progress was made in 1987, when Hallmark *et al.* demonstrated that individual gold atoms could be resolved on an Au(111) surface with the STM [28]. Not only was this possible under UHV conditions, but under ambient conditions, too. This was principally because of the resistance of gold surfaces to oxidation in air. Most metals are rapidly oxidised on exposure to atmospheric oxygen, with the result that a passive oxide layer is formed. Besides the change in surface crystal structure which would be expected to accompany oxidation, the oxide layer formed on the surfaces of most metals under ambient conditions is sufficiently thick to preclude electron tunnelling.

Figure 9.12 shows a 25×25 Åregion of an Au(111) thin film in air, prepared by epitaxial evaporation of gold onto a heated mica substrate at 300°C. The atomic spacing in the image is 2.8 ± 0.3 Å, which compares well with the known gold interatomic spacing of 2.88 Å. Note that the image in Figure 9.12 is a tunnelling current image, rather than the usual z-voltage image. It was recorded with a dc level of 2 nA and a bias potential of + 50 mV, with the gap resistance estimated to be ca 10^7 Ω. Under these conditions, the tunnelling gap is relatively small. The authors attributed a good part of their success to the use of such a low gap resistance, compared to the larger values of around 10^9 Ω employed by other workers, and noted that increases in the gap resistance to around 2×10^8 Ω led to the non-observation of atomic corrugation [28].

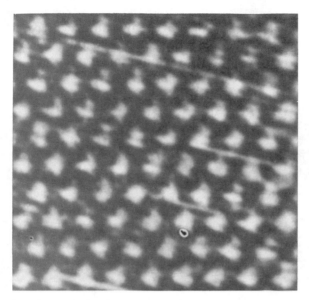

Figure 9.12. Atomic resolution image of an Au(111) surface. Reproduced with kind permission from Ref. 28. © 1987 American Institute of Physics

In fact, Binnig and Rohrer experienced considerable difficulties in their first studies of gold surfaces, and with hindsight, gold seems to have been a difficult choice to begin work on. The difficulties were not simply due to the smoothness of the surface electron density, either, but to the mobility of gold, which is so great that rough surfaces tend to smooth themselves out. Binnig *et al.* found that gold atoms were transferred to the STM tip, and the self-smoothing of these gold atoms led to a blunting of the tip with a concomitant loss of resolution. On occasions, they observed the resolution to jump unpredictably from high to low values, and attributed this to migrating adatoms locating themselves temporarily at the apex of the tip [29].

The effects of this gold mobility have been observed directly with the STM, using time-lapse imaging of the surface topography [30]. Indentations were created in an Au(111) surface by gently colliding an STM tip with it, creating a distinctive 'footprint'. This was achieved by applying a short-duration voltage pulse to the tip. At 30°C, steps were observed to move, and recessed regions were filled in. Figure 9.13 shows an initial image of the gold surface showing three steps which run vertically along the {112} direction, followed by fourteen images taken at successive intervals of eight minutes after colliding the tip with the surface.

The stability of gold in air clearly makes it an ideal material on which to examine molecular samples, provided it is possible to prepare surfaces which are atomically flat over large enough areas. A number of methods have been

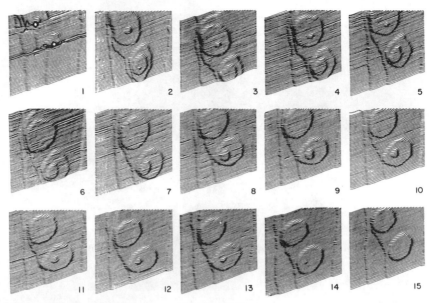

Figure 9.13. A series of time-lapse STM topographic images showing a 400 × 400 Å area of Au(111) after cleaning and annealing (1); creation of a 'footprint' with the STM tip (2); and thereafter at intervals of 8 minutes. During the two-hour period, the smaller craters in the footprint are filled in by diffusing gold atoms. Reproduced with kind permission from Ref. 30. © 1988 American Institute of Physics

investigated [31]. One of the most effective methods is the preparation of gold spheres by flame-annealing gold wire. Schneir *et al.* have described the formation of Au(111) facets on the surfaces of gold spheres formed by the melting of gold wire into an oxy-acetylene flame [32]. These facets are large enough to be observed quite straightforwardly using an optical microscope, and STM images confirmed them to be Au(111) surfaces which are atomically flat (or atomically flat with steps) over considerable distances (often extending as far as hundreds of nanometers). Although the facets do not cover the surface of the ball, they may be prepared routinely without any elaborate preparation procedures.

3.2 GRAPHITE SURFACES

The ability of the STM to obtain images in air (noted in Section 3.1.1 above) was demonstrated by Baro *et al.* [33] and subsequently exploited to good effect by Park and Quate [34] who obtained images of graphite surfaces with atomic resolution under ambient conditions. They studied the surface of highly oriented pyrolytic graphite (HOPG), supplied as a block which may easily be cleaved to yield a clean surface which is atomically smooth. The most popular method of preparation of HOPG surfaces is by removal of the top few layers of an HOPG block using adhesive tape. The clean surface which is exposed is stable in air for several days.

HOPG has become a highly popular substrate for STM studies, because of its relatively low cost and the ease of preparation of surfaces suitable for STM studies. However, the observation in recent years of a variety of graphite features which bear a close resemblance to helical molecules has led to a shift of interest towards other substrates (most notably, perhaps, the annealed gold substrates discussed in Section 3.1) in certain applications — especially biological applications (see Section 6).

Shortly afterwards, Sonnenfeld and Hansma showed that atomic resolution images could be obtained with the sample mounted in water [28]. Furthermore, it proved possible to operate the microscope in a saline solution. This latter capability indicated the potential role for the STM in studies of biological systems, where the possibility of performing high resolution microscopy in physiological-type buffer solutions became a reality for the first time. In order to facilitate studies under water, it was necessary to coat the bulk of the length of the STM tip with an insulating material, in order to minimise the area of the tip which could conduct current through the water and into the sample surface. This is a less demanding operation than might at first be expected, and many groups now operate the STM and AFM under water, with the consequence that *in situ* studies of interfacial electrochemical processes are now possible. A variety of tip-coating procedures have been reported, some of which are astonishingly simple (for example, coating the tip with wax).

3.3 THE SILICON (111) 7 × 7 RECONSTRUCTION

Shortly after their studies of the Au(110) reconstruction, Binnig *et al.* tackled a truly demanding problem which had been the subject of intense interest for some time: the 7×7 reconstruction of the Si(111) surface [36]. A number of models had been postulated for the structure of this surface, but none of these had been confirmed. In fact, the STM images did not fit any of the models exactly, but they matched one particular model, the adatom model, more closely than the others. It is a modified version of the adatom model which has ultimately come to be accepted, supported by data from other techniques, including transmission electron microscopy [37].

The STM images exhibited the expected rhombohedral 7×7 unit cell, bounded by lines of minima. These minima corresponded to empty adatom positions. Twelve maxima were observed inside each cell. These correspond to the positions of twelve adatoms, each bonded to three silicon atoms in the second layer of the reconstructed surface. This second layer is in fact formed from the topmost layer of the crystal during reconstruction. In the process of reconstruction, seven atoms are lost from the top layer (the reconstructed surface has only 42 atoms in its second layer). The unit cell of the reconstructed surface is customarily divided into two. In one half, the atoms in the second layer exhibit a stacking fault; the other half is unfaulted. It is along the row where these two halves of the unit cell are matched that the seven atoms are lost. Some of these atoms may occupy the adatom sites; alternatively, the adatoms may come from elsewhere in the crystal. In any case, the energetic gain is readily seen: 36 silicon atoms previously in the topmost layer of the solid are now incorporated into the second layer. Thus the number of dangling bonds in the surface is substantially reduced, even though the adatoms still possess dangling bonds, and this means an accompanying gain in energetic terms. The structure of the reconstructed surface is shown in Figure 9.14. It was the adatoms which were imaged by the STM; these are atoms which still possess dangling bonds in the reconstructed surface. The constant-current topograph was predominantly thought to be formed by tunnelling from the dangling bonds. An STM image of the Si(111)-(7 × 7) reconstruction is shown in Figure 9.15.

3.4 SCANNING TUNNELLING SPECTROSCOPY

Conventional surface-sensitive spectroscopic techniques (such as X-ray photoelectron spectroscopy (Chapter 3), Auger spectroscopy (Chapter 4) and vibrational spectroscopies (Chapter 7)) generate data which are averaged over an entire surface. However, many surface phenomena are strictly local in nature (for example, those associated with steps and defects). The images generated by the STM are determined by the electronic states at the surface and in the tip and, therefore, the STM can, in principle, map the electronic structure of a surface with atomic resolution. It is possible, in other words, to perform spectroscopic measurements on single atoms, facilitating detailed studies of local surface phenomena.

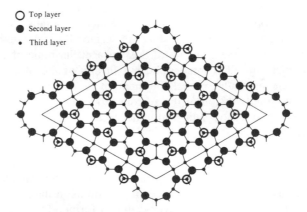

Figure 9.14. The structure of the Si (111)–(7 × 7) reconstruction. Reproduced with kind permission from A. Zangwill, "Physics at Surfaces", Cambridge University Press, Cambridge, 1988

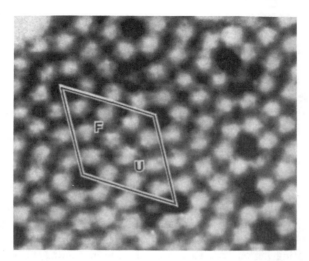

Figure 9.15. Topographic image of the Si (111)–(7 × 7) reconstruction, showing the unit cell. The faulted and unfaulted halves are marked F and U, respectively. Reprinted with kind permission from Ref. 39. © 1986 American Institute of Physics

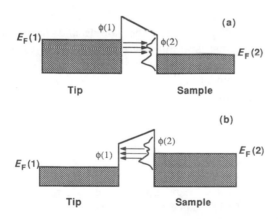

Figure 9.16. (a) Tip biased negative with respect to the sample, so that tunnelling is from the tip to the sample. (b) Polarity and direction of tunnelling reversed

The STM image is dependent upon the bias potential in a complex fashion. It is this dependence which forms the basis for the spectroscopic operation of the STM. To see this, consider the diagrams in Figure 9.16 for a hypothetical one-dimensional system. When the tip potential is negative with respect to the sample, electrons tunnel from occupied states of the tip to unoccupied states of the sample (Figure 9.16(a)). When the tip potential is positive with respect to the sample (negative sample bias), electrons tunnel from occupied states of the sample to unoccupied states of the tip (Figure 9.16(b)). Since states with the highest energy have the longest decay lengths into the vacuum, most of the tunnelling current arises from electrons lying near the Fermi level of the negatively biased electrode.

There are several spectroscopic modes of operation of the STM. The simplest of these is known as voltage-dependent STM imaging, which involves acquiring conventional STM images at different bias voltages, and which provides essentially qualitative information. Such a procedure can, however, prove very valuable. For example, consider the case of the Si(111)-(7 × 7) reconstruction. Images recorded at positive sample biases reveal 12 adatoms of equal height in each unit cell [36]. However, as we noted in Section 3.1.3, the unit cell of this reconstruction may be divided into two non-equivalent halves, one of which exhibits a stacking fault with the underlying atomic planes. Imaging of the surface at negative sample bias voltages results in the adatoms in the faulted half of the unit cell appearing higher than those in the unfaulted half [38]. Furthermore, in each half of the unit cell, the adatoms nearest the deep corner holes appear higher than the central adatoms.

The second spectroscopic mode of operation, the most common form of which is scanning tunnelling spectroscopy (STS), provides quantitative information through the use of a constant dc bias voltage with a superimposed high-frequency sinusoidal modulation voltage between the sample and the tip. The component of the tunnelling current that is in phase with the applied voltage modulation is measured, whilst maintaining a constant average tunnelling current. This enables the simultaneous measurement of dI/dV and the sample topography.

A third spectroscopic mode of operation involves the local measurement of tunnelling $I-V$ curves. These $I-V$ curves must be measured with atomic resolution at well defined locations, with a fixed tip–sample separation in order to correlate the surface topography with the local electronic structure. The first such measurements were made by Hamers *et al.* [39] who termed their technique current-imaging-tunnelling spectroscopy (CITS). Other variants have been described (for further discussion and a review of spectroscopic operation of the STM, see [40]). Using CITS, Hamers *et al.* were able to map out the electronic structure of the Si(111)-(7×7) unit cell with a lateral resolution of 3 Å [39, 40]. At voltages between −0.15 and −0.65 V, most of the current arises from dangling bond states of the twelve adatoms (see Figure 9.17(a)). More current comes from the adatoms in the faulted half of the unit cell (cf. the example of voltage-dependent imaging above) and, in each half-cell, more current comes from the adatoms adjacent to a corner hole than comes from the other adatoms. A differential current image recorded with bias voltages between −1.0 and −0.6 V (Figure 9.17(b)) revealed the positions of the six dangling bonds on atoms in the second layer (the Si(111)-(7 × 7) unit cell has a total of 18 dangling bonds).

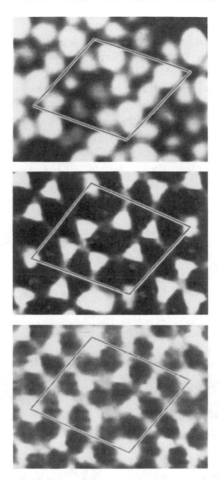

Figure 9.17. CITS images of occupied Si (111)–(7 × 7) surface states: (a) adatom state at −0.35 V; (b) dangling-bond state at −0.8 V; (c) backbond state at −1.7 V. Reproduced with kind permission from ref. 39. © 1986 American Institute of Physics

These dangling bonds exhibited a reflection symmetry which was attributed to the presence of the stacking fault.

A third occupied state was imaged as the differential current between −2.0 and −1.6 V (Figure 9.17(c)), revealing regions of higher current density, corresponding to Si–Si backbonds, and bright spots at the corner holes, where additional back bonds are exposed. These quite astonishing images exemplify the remarkable capabilities of the STM for the determination of surface electronic structure, and illustrate the power which the technique makes available to the surface scientist for the determination of surface structure.

3.5 CRYSTAL GROWTH

Much effort is expended by surface scientists in determining the mechanisms of film growth at surfaces. It is often of critical importance whether film formation proceeds in a layer-by-layer fashion, for example, or by island growth. A number of methodologies have been developed for the determination of growth mechanisms spectroscopically. However, the simplest approach would clearly be to attempt to visualise the growing film using STM. Knall and Pethica have taken just such an approach to the growth of germanium on Si(100) and Si(113) [41]. For Si(113), they examined the surface using STM at a range of coverages between 0.3 and 5 monolayers (see Figure 9.18). During the initial stages of growth, they observed the growth of islands with the same 3 × 2 reconstruction as the substrate in a layer-by-layer fashion. However, at coverages between 1 and

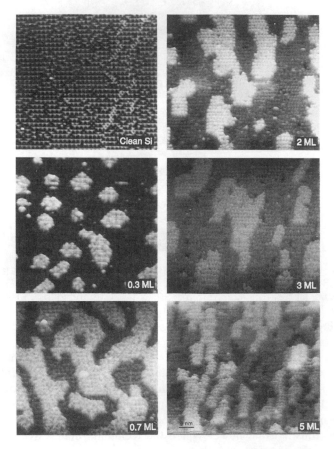

Figure 9.18. Growth of germanium on an Si(113). Reproduced with kind permission from ref. 41. © 1992 Elsevier Science – NL

2 monolayers, they observed the reconstruction of the surface to change, yielding an Si(113)2 × 2-Ge structure. The determination of the type of surface reconstruction was in this case facilitated by the ability of the STM to image either filled or empty surface states, depending upon the tunnelling parameters. The filled and empty state images revealed the locations of different bonds, resulting in a reversal of contrast. At coverages between two and five monolayers, the growth mechanism was still essentially two-dimensional, although some nucleation on top of incomplete layers was observed. Consequently, small holes were observed in the third layer which developed into valleys and ridges along the [332] direction, leading to an increase in the surface roughness. At coverages greater than six monolayers, the growth mechanism had become a three-dimensional island growth mechanism.

3.6 LITHOGRAPHY AND MICROMANIPULATION

We noted in Section 2.1.3 that the interactions between the STM tip and the sample surface can be quite substantial, and can cause alterations in the sample surface structure. These interactions may be utilised in order to modify the surface in a controlled fashion, giving rise to the possibility for nm-scale manipulation of surface structure. However, besides these mechanical interactions, there are a number of more subtle ways in which the STM may be used to modify a surface structure on the nanometer scale.

Mamin *et al* have described the formation of nanometer-sized gold features using the STM [42]. Gold tips were prepared by electrochemical etching in concentrated hydrochloric acid. Field evaporation from the tip was produced (at atmospheric pressure) by applying short voltage pulses (a few hundred nanoseconds or less), with the effect that small mounds could be created on an Au(111) surface. Arrays of mounds were deposited in ordered rows on the surface.

Lyo and Avouris have also used field-induced effects (generated by applying a voltage pulse of +3 V to the sample) to manipulate atoms on an Si(111)-(7 × 7) reconstructed surface [43]. They studied the formation of structures at varying field strengths and tip–sample separations. A lower threshold field strength was determined at which small mounds could be generated. At higher field strengths, the top atomic layer was removed from a small area of the sample. At small tip–sample separations, small mounds were created which were surrounded by moats. It proved possible to move these mounds. Application of an initial pulse of +3 V was required to form the mound (Figure 9.19(a)); a second subsequent pulse of +3 V resulted in the mound being picked up by the STM tip. The STM tip was then moved a short distance and the mound deposited by applying a pulse of −3 V (Figure 9.19(b)). Considerable fine control of the process was possible, to the extent that single atoms could be manipulated. In Figure 9.20(a), a region of the Si(111)-(7 × 7) surface is shown. A +1 V pulse was applied to the STM tip, removing one of the silicon atoms and creating the vacancy visible

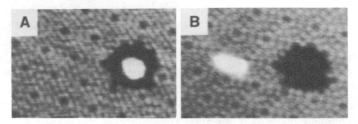

Figure 9.19. Creation and movement of a small mound on a silicon surface by pulsing the STM tip. Reproduced with kind permission from ref. 43. © 1991 American Association for the Advancement of Science

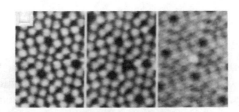

Figure 9.20. Removal (b) and replacement (c) of a single silicon atom on an Si(111) surface. Reproduced with kind permission from ref. 43. © 1991 American Association for the Advancement of Science

in Figure 9.20(b). Finally, the atom was redeposited on the surface at its original location (Figure 9.20(c)).

Eigler and Schweizer utilised a different technique to manipulate xenon atoms on a nickel (110) surface under UHV conditions, with the entire STM cooled to 4 K [44]. They reported that, at this temperature, the contamination rate of the surface was decreased, and the stability of the microscope increased, to such a degree that experiments could be performed on a single atom for days at a time.

The manipulative procedure involved sliding atoms across the surface of the nickel substrate using the STM tip. The tip was scanned across the surface until it was positioned directly above an atom. The tip–sample separation was then decreased by increasing the tunnel current, lowering the tip onto the atom. The atom was then dragged to its desired location and tip withdrawn, allowing the atom to be imaged in its new position. The process is illustrated in Figure 9.21. In Figure 9.21(a), we see a nickel (110) surface after dosing with xenon. In Figure 9.21(b–f), the atoms are rearranged to form the letters 'IBM'.

Zeppenfeld *et al* have shown that atomic manipulation using this sliding process is not restricted to the very weakly bound xenon atoms, but may be applied to CO, and even Pt atoms, on a Pt(111) surface [45]. Several structures were created by positioning individual CO molecules on the Pt surface, including the letters 'CO', a small hexagonal island, and a Molecular Man, built from 28 CO molecules and measuring 45 Å from head to foot (see Figure 9.22).

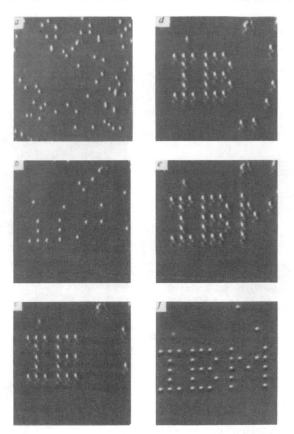

Figure 9.21. Clean nickel (110) surface dosed with xenon atoms (a) which are arranged using the STM tip to form the letters 'IBM'. Reproduced with kind permission from ref. 44. © 1990 Nature

On a slightly larger scale, the possibility of manipulating individual biological molecules is an attractive one. Jiang *et al.* have used an epifluorescence microscope to monitor the manipulation of individual DNA molecules [46]. They were able to observe directly the attachment of a DNA molecule to the STM tip following the application of a +1.5 V voltage pulse, its movement whilst attached to the STM tip, and its re-deposition on application of a −1.5 V pulse. The ability to perform such delicate operations opens the way, in principle, for a range of biological applications for the STM (for example, in the structural modification of cells or sub-cellular material).

A final intriguing application of the manipulative capabilities of the STM has been demonstrated by Eigler *et al.*: an atomic switch in which a xenon atom moves reversibly between stable positions on each of two contacts (formed by

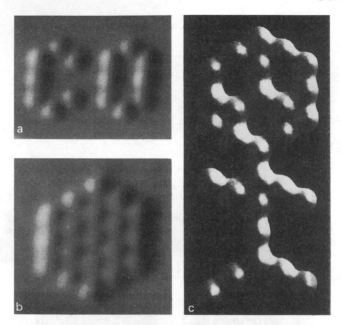

Figure 9.22. Features formed by sliding CO molecules across a Pt(111) surface. Reproduced with kind permission from ref. 45. © 1992 Elsevier Science

the STM tip and a nickel surface) [47]. Although far from constituting a workable electronic device, this example illustrates the potential which STM has to offer in the area of nanotechnology. As electronic devices become ever smaller, there is likely to be an increasing role for scanning probe techniques in both their fabrication and their characterisation.

3.7   ATOMIC RESOLUTION BY AFM

It was initially hoped that AFM would generate data with the same kind of high spatial resolution provided by STM. There were apparently some initial early successes, with contact mode images of HOPG being reported which appeared very similar to those obtained using STM. However, the origin of these images soon became shrouded in doubt. The popular image of a single atom at the apex of the tip in contact with the sample was rightly questioned: under these circumstances, with a load of 10 nN, the pressure exerted by the tip would be sufficient to significantly disrupt the structure of the sample surface (even supposing the tip material was strong enough to sustain the load). In reality, contact areas are thought to be a little larger — perhaps a few nm in diameter. If multi-atom contact was occurring, what might the mechanism of contrast formation be? Significantly, atomic defects were conspicuous by their absence from these early high resolution images. There was much speculation concerning their origin. For example, in the case of HOPG, it has been suggested that a graphite flake was dislodged and, sliding across the sample surface under the tip, gave rise to an interference effect which reflected the periodicity of the graphite surface structure. There was, for a while, doubt that true atomic resolution was possible using AFM; the doubt was dispersed in 1993 when Ohnesorge and Binnig reported atomically resolved images of inorganic crystals, inclusing atomic defects, using non-contact mode [48]. They employed a maximum attractive force of only -4.5×10^{-11} N. While the feasibility of performing such high resolution microscopy using the AFM was therefore demonstrated, the practical difficulties associated with working at such low forces mean that the levels of resolution obtained by AFM are typically less than those achievable with the STM.

4 MOLECULES AT SURFACES

There are many situations in which one might wish to visualise the interactions of molecules with surfaces, and many of these lie outside of the conventional boundaries of surface science. The STM and the AFM have made dramatic impacts on a number of fields in which it is molecular material which is of interest. In a large number of these applications, the STM and the AFM are able to provide structural data under circumstances in which other techniques would be rendered useless. Sections 4–6 therefore describe some of the progress which has been made in three important areas, in which it is the behaviour of complex materials at surfaces which is of interest. We begin by looking at organic liquids and liquid crystals.

4.1 LIQUID CRYSTALS AND ORGANIC LIQUIDS

Smith *et al.* have investigated the interaction of liquid crystals with graphite surfaces [49]. They examined alkylcyanobiphenyl molecules of the type 4'-n-alkyl-4-cyanobiphenyl, known by the abbreviated names mCB, where m is the number of carbons in the alkyl group. The liquid crystal molecule 8CB has the structure:

At first sight it might seem unlikely that it would be possible to obtain STM images of something so poorly conducting as a liquid crystalline film. In fact, however, a low intrinsic conductivity does not constitute an obstacle provided the sample is in the liquid state; the STM tip simply travels through the bulk of the liquid until it is sufficiently close to the surface for electron tunnelling to reach a measurable level. Practically, this effectively restricts the probe to the interfacial region, and it therefore means that the molecules which are imaged are those molecules closest to the substrate surface. The key constraint in the case of liquid crystals is that the samples are close to their smectic-liquid to solid-crystalline transition temperature. At higher temperatures, the films become completely liquid-like and no order is observable in the STM images. At lower temperatures, the films no longer wet the graphite.

With a tunnelling current of 100 pA and a bias voltage of 0.8 V, images of the liquid crystal/graphite interface revealed a complex interlocking arrangement of molecules (see Figure 9.23). Bright spots were observed which corresponded to the locations of the cyanobiphenyl headgroups. The alkyl chains were also observed, although they showed up less brightly, as patterns of points which corresponded to methylene groups. The cyano groups were observed as a small bright point at the opposite end of each molecule. Examination of high resolution images indicated that this bright spot was separated from the phenyl group to which it was joined by some distance, suggesting that it corresponded to the nitrogen atom of the cyano group in particular, and that the triply-bonded carbon atom of the cyano group therefore has a relatively low tunnelling conductance. The cyano groups pointed inwards towards each other, interdigitating slightly and increasing the packing density.

It was suggested that the two-dimensional lattice formed by the molecules was primarily the result of registry of the alkyl tails with the graphite lattice. By decreasing the gap resistance, and so driving the tip closer to the surface, it was possible to penetrate the film and to image the underlying substrate. The directions of the graphite lattice vectors were therefore determined, and it was found that the alkyl chain of each molecule aligned along a particular graphite lattice vector, whilst the cyano group aligned along a different lattice vector. At grain boundaries, the tails were often observed to rotate by 60°, matching the graphite symmetry.

Hara *et al.* have performed STM measurements on 8CB, but instead of using graphite as a substrate, they used molybdenum disulphide [50]. MoS_2 is electrically conducting, and clean surfaces may be prepared by cleaving an MoS_2 single crystal. In contrast to the bilayer structure which 8CB forms on graphite, it forms a periodic structure on MoS_2 in which cyanobiphenyl head groups and alkyl tails alternate in each row (see Figure 9.24). Each row has a width of around 21 Å, whereas the width of the bi-layer formed on graphite is around 38 Å. The adsorbed molecules are bent in opposite directions in successive rows, forming a packing arrangement which is in fact not favourable. A more favourable arrangement would place the aromatic groups in adjacent positions. The driving force for adsorption in this unfavourable arrangement is thought to be the strong interaction between the 8CB molecules and the substrate, and, in fact, the periodicities

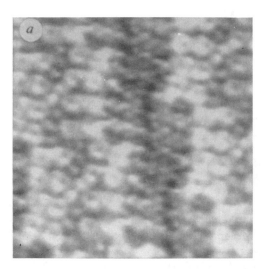

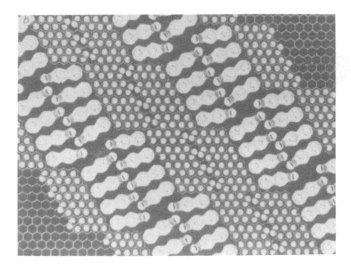

Figure 9.23. (a) 56 × 56 Å STM image of 8CB on graphite and (b) a model of the 8CB lattice showing the hydrogens of the alkyl chains registered with the hexagonal centres of the graphite lattice. Reproduced with kind permission from ref. 49. © 1990 Nature

observed in the liquid crystal film correlate closely with the lattice spacing of the MoS_2 substrate.

McGonigal *et al.* have studied liquid alkane layers at graphite surfaces, again operating the STM in a liquid environment [51]. They observed two-dimensional

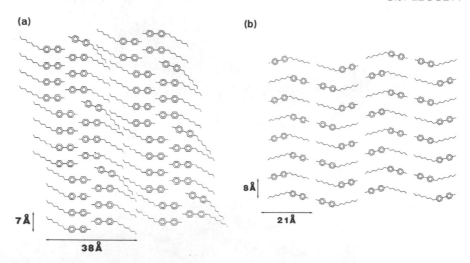

Figure 9.24. Models showing the packing of 8CB molecules on graphite (a) and MoS$_2$ (b). Reproduced with kind permission from ref. 50. © 1990 Nature

and highly ordered arrays of n-C$_{32}$H$_{66}$ molecules, arranged in rows with their long axes parallel to each other and perpendicular to the troughs between rows (Figure 9.25). Molecules in neighbouring rows were observed to be offset by a distance equal to half the separation between the molecules.

A local enhancement of the tunnelling current by the adsorbate was observed, with the consequence that the troughs between the rows of adsorbate molecules appeared lower than the rest of the surface. Importantly, the images did not directly show the positions of the methylene groups of the adsorbed molecules, but instead showed patterns of bright dots corresponding to type B graphite lattice positions (see discussion under 2.1.2 above) where the tunnelling current was enhanced by the presence of an adsorbate methylene group. No dots were observed corresponding to the methylene groups located over type A positions (see Figure 9.26). Thus the images were, in fact, dominated by features associated with the graphite substrate, suggesting that the adsorbate molecules were commensurate with the graphite lattice. If this was not the case, a more irregular image would have been expected.

Yeo *et al.* have studied n-decanol adsorbed on graphite [52] and gold [53]. STM images of n-decanol on graphite revealed individually resolved molecules in an ordered monolayer (Figure 9.27). The length of the molecule was measured to be 1.4 nm, consistent with n-decanol lying parallel to the substrate surface. The molecules were arranged into rows of pairs of molecules linked together by hydrogen bonds between OH head groups. The molecules were arranged at an angle of 60° with respect to the dark bands which separated the molecular rows.

Figure 9.25. Filtered STM image of n-$C_{32}H_{66}$ molecules adsorbed on graphite. Reproduced with kind permission from ref. 51. © 1990 American Institute of Physics

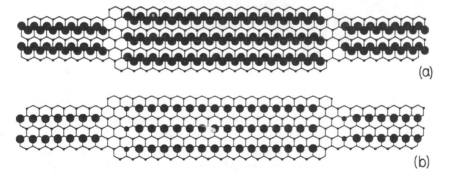

(a)

(b)

Figure 9.26. Schematic representation of the graphite lattice and the location of the adsorbate molecules. The small dots correspond to the B-type graphite sites, which are the ones usually imaged by the STM. In (a), the large dots correspond to methylene groups of the adsorbate. In (b), the large dots represent the positions of the dots observed in the STM images of the adsorbate (i.e. graphite sites from which tunnelling is enhanced by the presence of the adsorbate). Reproduced with kind permission from ref. 51. © 1990 American Institute of Physics

Figure 9.27. Filtered STM image of *n*-decanol molecules adsorbed on graphite. Image size: approximately 12 nm × 12 nm. Reproduced with kind permission from ref. 52. © 1991 American Vacuum Society

The voltage dependence of the images was investigated. At a bias voltage of 0.1 V and a tunnelling current of 0.66 nA, the graphite lattice could be seen, together with narrow bright bands. Increasing the bias voltage to 0.33 V led to the observation of the n-decanol molecules, and suggested that the bright bands observed at the lower bias voltage were associated with the hydroxyl groups. A further increase in the bias voltage, to 0.45 V, resulted in a marked change in the STM image, with the positions of the hydroxyl groups marked by dark holes, with diameters of 0.5 nm, equal to the width of the bright bands observed at a bias voltage of 0.1 V. These changes were explained in terms of the electronic structure of n-decanol. It was suggested that it is only an OH antibonding state which is accessible to the STM at a bias voltage of 0.1 V. At a bias voltage of 0.33 V, the electronic states of the n-alkane tails become accessible, and at 0.45 V, it is tunnelling to these states which predominates.

On Au(111), the behaviour was different [54]. The n-decanol molecules were observed to form a densely packed hexagonal array in which the hydroxyl head groups were adsorbed at inequivalent sites on the substrate surface. The nearest-neighbour spacing of approximately 0.5 nm, agreed well with the expected diameter of the hydrocarbon chain, suggesting that the molecules were standing on end with the hydroxyl groups located at the surface.

4.2 SELF-ASSEMBLED MONOLAYERS

Self-assembled monolayers, typically prepared by the chemisorption of either thiol molecules onto gold surfaces or silane molecules onto silica surfaces, have recently become the subject of intense interest in the surface chemistry community. Scanning probe techniques have been applied in studies of film morphology, ordering and interfacial interactions [54]. Sun and Crooks [55] studied monolayers of n-octadecylthiol ($C_{18}H_{37}SH$) on Au(111) using STM. They imaged films formed at a variety of time intervals ranging from 10^{-1} to 10^4 minutes. Copper was electrochemically deposited at defect sites, and the films were again imaged using STM. Islands of copper were formed at these defect sites, and the defect density then determined. The defect density varied with adsorption time, and resulted in a general picture of adsorption kinetics which was in agreement with models established by ellipsometry and contact angle measurements.

Defect structures are commonly observed in SPM images of self-assembled monolayers of alkylthiol molecules on gold surfaces. McCarley *et al.* observed defect structures by STM [56] and recorded an average pit depth of 2.5 Å irrespective of the length of the adsorbate molecule and the tunnelling parameters, indicating that the defect structures reflected imperfections in the gold substrate rather than regions from which the adsorbate was absent. Upon annealing the samples to 100°C, the pits were observed to be removed, without apparent damage to the monolayer. It was proposed that the Au-S bonds are in fact labile, with the consequence that Au atoms can diffuse on heating to fill the pits in the substrate surface.

Molecular resolution images of self-assembled monolayers (SAMs) have been obtained using AFM [57, 58], confirming the formation of the hexagonal lattice (as predicted by, for example, electron diffraction [59]), even when the substrate was not, in fact, crystalline [57]. Similar resolution proved difficult to achieve with the STM until the recent publication of images of $SH(CH_2)_{11}CH_3$ on gold recorded with a very high gap resistance (as high as 1 T Ω) [60]. Such high gap resistances were necessary to prevent deformation of the monolayer structure by the STM tip, which would clearly result in the loss of molecular resolution. It is important to note the contrast between the massive gap resistance (achieved using a tunnel current of 1.5 pA) required to image a monolayer of molecular adsorbates to high resolution, and the very small gap resistance (tunnel current around 10 nA) employed to image the clean gold surface to atomic resolution (see Section 3.1 above). These authors also achieved molecular resolution within defect structures, showing that the monolayer was intact there and lending support to McCarley's hypothesis [56] that the defect structures arise from the substrate structure.

Subsequent STM studies have revealed the structures of SAMs with great clarity. A c(4 × 2) superlattice model has been proposed for long chain alkylthiols adsorbed onto gold [61, 62], following the observation of two different molecular adsorption sites in STM images. These sites are thought to correspond to different

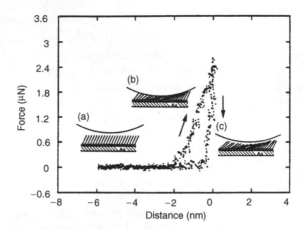

Figure 9.28. Interfacial force as a function of relative tip-sample separation. The schematic illustrations indicate the proposed film response during various stages of the cycle. Reproduced with kind permission from ref. 64. © 1992 American Institute of Physics

degrees of twist about the molecular axis, giving rise to a height difference of only 0.7 Å between the two different types of orientation [61]. Butanethiol, on the other hand, exhibits a p $\times \sqrt{3}$ structure [63]. Initially, the monolayer exhibits small ordered domains surrounded by a two-dimensional liquid phase; over the course of 127 hours, this disordered liquid phase gradually becomes ordered, adopting the p $\times \sqrt{3}$ structure [63].

The mechanical relaxation of n-hexadecanethiol ($C_{16}H_{33}SH$) has been studied using an interfacial force microscope (IFM) [64]. The IFM uses an rf bridge arrangement to measure the deflection of a tungsten tip mounted onto one end of a differential capacitor, balancing this deflection with an externally applied electrostatic force. Samples were imaged in constant repulsive force mode (at 2×10^{-7} N). Force profiles, showing the dependence of the interfacial force on tip–sample separation, were obtained (see Figure 9.28). These exhibited three distinct regions. Prior to tip contact, there was no evidence of the existence of significant attractive forces, due to the low-energy methyl-terminated surface of the self-assembled monolayer. After contact, the force profile shows first a gentle repulsive interaction (up to a force of 8×10^{-7} N) and then a much stronger repulsion as the tip–surface separation narrows. Finally, after withdrawing the tip, the interfacial force drops off to zero at distances smaller than the initial contact separation, with, again, no evidence of an attractive interaction.

4.3 OTHER ADSORBATE AND FILM STRUCTURES

Many other types of interactions between molecules and surfaces have been studied using scanning probe techniques. Some of these involve the interaction of gaseous molecules with well-defined surfaces. For example, Ohtani *et al.* have examined the co-adsorption of benzene and carbon monoxide onto Rh(111) surfaces [65]. They observed a well ordered array of ring-like features, which they associated with individual benzene molecules, but were not able to resolve the CO molecules. A number of groups have been interested in the structure of films of C_{60} at surfaces. STM has been used to image monolayer and multilayer structures of C_{60} grown on GaAs(110) at temperatures from 300 K to 470 K [66]. They found that C_{60} structures may be either commensurate or incommensurate with the substrate. At 300 K, multilayered structures were observed to have point defects, dislocations, domain boundaries and surface faceting; whereas multilayer structures formed at 470 K were highly ordered. Enhanced bonding was observed at step edges.

Howells *et al.* have studied C_{60} on Au(111) and have obtained high resolution images of single molecules [67]. Gold films were prepared by epitaxial deposition onto mica, and methyl isobutyl ketone (MIBK) was adsorbed onto the surface forming an ordered overlayer which was observed, using the STM, to be hexagonal with a nearest-neighbour spacing of around 0.41 nm. C_{60} was deposited onto the surface in an evaporation chamber. It was suggested that the MIBK layer fixed the fullerenes firmly in position, allowing individual molecules to be observed. The diameter of the adsorbed C_{60} molecules was found to be 1.1 nm. The molecules were observed to have a doughnut-shape, attributed to a strong electronic interaction with the gold substrate.

The ability of the AFM to probe surface and intermolecular forces down to the atomic scale, holds tremendous potential for deepening our understanding of intermolecular and surface interactions. This potential has begun to be exploited for a wide range of types of interaction. For example, Ducker *et al.* [68] have investigated forces between a silica-glass sphere and a large smooth silica surface in aqueous solution, as a function of salt concentration and pH. Leung and Goh [69] have observed the orientational ordering of polystyrene films, cast onto mica, by tip-surface interactions in the AFM. Slowly scanning the tip across the surface, whilst maintaining a constant interaction force, resulted in the gradual formation of oriented bundles of polymer molecules. They were also able to lift a polystyrene molecule upwards from the surface, together with polymer molecules with which it was entangled, forming features raised above the surrounding polymer film. Finally, Hamada and Kaneko [70] have examined the indentation of polymers using a point contact microscope. They were able to compare indentation depths with those measured using a conventional micro-Vickers hardness tester, finding very good agreement, and were able to examine the effects of surface distortions on frictional forces.

Langmuir–Blodgett films have attracted a great deal of interest. Smith *et al.* [71] reported images of fatty acid bi-layers forming a partially ordered film on graphite substrates, exhibiting a distribution of intermolecular distances averaging as 5.84 Å in one direction and 4.1 Å in another. Classical tunnelling theory cannot account for conduction through such thick films (the film thickness was estimated to be about 50 Å). Alternative explanations which have been proposed for the phenomenon suggest either that the tip penetrates into the film until the tip–substrate separation is sufficiently small for tunnelling to be possible, or that a revision of the theoretical models is required [60]. Meyer *et al.* [72] performed a similar study using an AFM. This time, the cadmium arachidate sample was prepared on an etched silicon surface. They were able to measure the intermolecular distances along the two lattice vectors and the packing density. Over distances of several thousand Ångströms, few defects were observed in the periodicity of the film; because the same images exhibited both the defect structures and the periodic arrangement of the molecules, it was concluded that the contact region between the tip and the sample must be as small as a few atoms. Garnaes *et al.* [73] have also imaged cadmium arachidate on silicon with the AFM. For two-to-five-layer films, they observed a rectangular lattice in which domain boundaries were observed. Surface buckling was also observed, with an amplitude which varied from domain to domain.

5 STUDIES OF POLYMER SURFACES

Scanning probe techniques are having a considerable impact within polymer science. Polymer samples pose particular problems for the STM, because on the whole they are electrically insulating. Consequently, the development of the AFM has been of considerable importance, facilitating the high resolution imaging of untreated polymeric materials and offering a quite unequalled analytical capability. However, we begin our discussion by considering a few areas of polymer science in which the STM has also been able to provide insights.

Electrode processes are interfacial processes and the performance of a conducting polymer is thus critically related to its surface structure. There is therefore growing interest in the use of STM and AFM to study conducting polymer materials. For example, Brown *et al.* [74] have examined the surface structure of osmium-containing metallopolymer films before and after electrochemical treatment in aqueous sulphuric acid electrolytes. Films of $Os(bipy)_2Cl_2$-loaded poly(4-vinylpyridine) were supported on polycrystalline graphite and were imaged in air. Before electrochemical cycling a fibrillar structure was observed, together with consistently-sized steps. After electrochemical cycling, a granular structure was observed, and the dimensions of the steps were observed to have changed.

It is in the study of electrode processes that the remarkable ability of the STM to operate in liquids sets it in a position unequalled by electron microscopies, providing the potential for examining electrode processes *in situ*. This capability enabled Yaniv *et al.* [75] to obtain images of polypyrrole-glucose oxidase electrodes in solution. The polymeric electrode film was deposited onto a polished glassy carbon disc by electropolymerisation and the effects of various parameters upon the morphology of the electrode surface were investigated. The electropoly-merisation potential was found to play an important role in controlling surface morphology. At low potentials, a film with a fibrous structure was observed. As the potential was increased, the fine structure was gradually lost and large holes were observed. Enzyme loading was found to have a strong influence on film structure too, and both effects were correlated with differences in electrode perfor-mance. However, deposition time had little effect on film structure and this was mirrored by a small effect on performance, too. Mantovani *et al.* [76] examined the surface morphology of emeraldine hydrochloride by STM. For electrochem-ically prepared material, they also observed surfaces which exhibited variation in ordering that depended upon the preparation conditions. Samples ranged from being fibrillar in appearance to being relatively featureless. Solution-cast films exhibited a generally fibrillar appearance. These data illustrate the ability of the STM to straightforwardly record data for polymeric materials in contact with their chemical environments; this ability is not simply restricted to electrode processes, but, through the AFM, is possible in quite diverse applications for non-conducting materials.

Stange *et al.* have studied the formation of polystyrene (PS) films during spin-coating onto silicon surfaces using STM and AFM [77]. They examined films cast from solutions of concentrations 0.0005 wt % through 0.075 wt % in toluene. Conductivity was retained in the resulting samples by carbon-coating, facilitating STM analysis. Similar experiments were performed with the AFM, but in this case carbon coating was unnecessary. At low PS concentrations, isolated molecules were observed. As the concentration of the casting solution was increased, molecules were observed to clump together in patterns on the surface, resulting in the formation of network pattern which, at a PS concentration of 0.025 wt %, resembled a Voronoi tesselation pattern. A Voronoi tessela-tion pattern is generated by forming discs at a uniform rate from a randomly distributed collection of points. Because no disc is allowed to impinge upon another, the discs eventually become polygons. Each edge of the polygons is equidistant from two of the original points; each vertex is equidistant from three. Upon heating above T_g, the network was observed to break up giving, ultimately, small aggregates of polymer molecules distributed across the surface. At higher PS concentrations, the polymer eventually formed a continuous film.

A very different, but quite distinct, type of behaviour is exhibited by polyethy-lene, during crystallisation from an entangled solution [78]. Polymer samples were deposited onto freshly cleaved mica and shadowed with Pt/C, thus providing good surface conductivity for STM analysis and suitable replicas for TEM imaging. In a comparative study, samples were imaged using STM and TEM. Branched structures were observed which, upon examination of higher magnifi-cation STM images, were found to be composed of globular features measuring some 20 × 100 nm. The lateral resolution was found to be comparable for both microscopies, but the STM yielded superior vertical resolution. It is this enhanced sensitivity in the z-direction which provides an impetus for the use of STM to study coated polymer film structures, which in other respects are accessible using electron miocroscopy. Where film formation takes place on the surface of thick, insulating substrates, or where analysis *in situ* without pre-treatment of the sample is required, the AFM comes into its own. For example, Collin *et al.* [79] have observed the formation of films of a poly(styrene-b-n-butyl methacrylate) di-block co-polymer cast from toluene solution onto silicon. They observed time evolution of the free surface with an AFM. Wang *et al.* [80] have reported remarkable AFM images of poly(butyl methacrylate) (PBMA) latex films formed on mica under varying experimental conditions, and have also investigated the effect of ageing. Thick latex films prepared by slow evaporation exhibited packing with a strong periodicity and few vacancies; order was found to persist, with some dislocations, over length scales of several micrometers. After annealing for six hours, vacancies were observed to appear in the film, thought to be due to the migration of sub-surface vacancies to the surface. Such vacancies were not observed in the surface of a freshly formed film. Besides the quality of the images which were obtained, this work exemplified the ability of the AFM to monitor the evolution of the surface structure of a single sample. It goes without saying that this kind of data is not readily accessible with other microscopical techniques.

Many of the considerations which apply to polymer film samples also hold true for polymer crystal samples. Here again, the earliest studies were of samples to which a metallic coating had been applied. Piner *et al.* [81] have presented STM images of gold- and chromium-coated polyethylene lamella. Their concern was to explore the feasibility of imaging coated polymer crystals using the STM. They experienced difficulties imaging large mounds of PE crystals, but on the edges of PE crystals, and for small outgrowths on top of lamellae, they obtained STM images successfully, yielding a variety of data regarding crystal structure. For example, they suggested that small outgrowths were formed by a screw dislocation mechanism; the chain-fold length of the larger lamellae was also determined.

The AFM again has the ability to image insulating samples with no pre-treatment requirement. Snetivy and Vancso [82] have obtained AFM images of PEO lamellae grown in toluene solutions, exhibiting screw-dislocations which showed a growth face at the diagonal. Typical lamellae had a square habit with a thickness of ca 12.5 nm and an edge length which was dependent upon the crystallisation time. They also observed the disintegration of these lamellae in air, leading to contraction of the polymer forming irregular, perforated patterns. These regions were found to be 5–8 nm thicker than the lamella thickness prior to disintegration. In this case, the AFM is providing high quality EM-type data; however, it is where information on the atomic scale is required that the AFM comes into its own. Although it is generally accepted that the ultimate resolution attainable with the AFM is practically (if not in principle) slightly poorer than that attainable with the STM, there are nevertheless examples which suggest that the magnitude of the gap is small. Magonov *et al.* [83] suggested that they were able to image the surfaces of lozenge-shaped crystals of poly(2-4-hexadienylene bis(p-fluorobenzene sulphonate)) with atomic resolution, providing data about surface structure in general agreement with crystallographic data. Magonov *et al.* [84] have also studied cold-extruded polyethylene with atomic resolution using the AFM, reporting a fibrillar structure composed of uniaxially oriented polymer molecules. Lin and Meier [85] have provided another clear example of the high resolution which is attainable with the AFM. They studied stacked arrays of lamellar polyethylene single crystals, prepared by drying a drop of a cooled solution of high density polyethylene in xylene onto a freshly cleaved mica substrate (Figure 9.29), and highly oriented, drawn high density polyethylene. The drawn material was imaged to high resolution, revealing individual polymer molecules and chain folds (Figure 9.30), and providing elegant proof for the existence of a chain-folded structure in drawn polyethylene. These data clearly illustrate the power of AFM for the determination of polymer structure on the molecular and the atomic scales, for even the most difficult samples.

In many respects, it is the ability of the AFM to probe local properties which holds the most promise for polymer surface characterisation, and it is in this area that future years are likely to see much development. Nisman *et al.* have provided an illustration of the sensitivity of the AFM to molecular structure in lateral force mode [86]. They imaged poly(oxymethylene) crystals using the

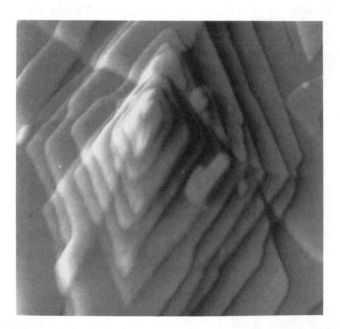

Figure 9.29. AFM image of polyethylene single crystals. Reproduced by kind permission of Dr. D.J. Meier

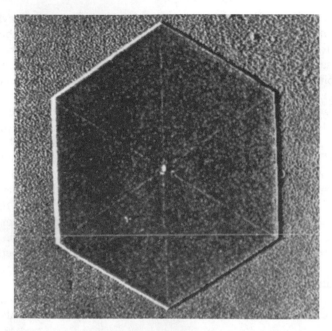

Figure 9.30. AFM image of drawn polyethylene, revealing individual polyethylene chains and chain folds. Reproduced by kind permission of Dr. D.J. Meier

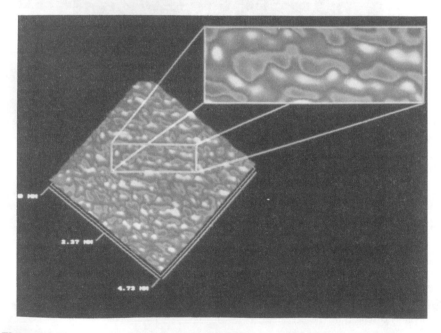

Figure 9.31. AFM image of a poly(oxymethylene) lamellar crystal, recorded in "deflection mode". Image size: 12.5 μm × 12.5 μm. Reproduced with kind permission from Ref. 86. © 1994 American Chemical Society

Figure 9.32. Images of a 1.5 µm × 1.5 µm region of the poly(oxymethylene) crystal recorded in height-mode kind (left) and lateral force mode (right). Reproduced with permission from Ref. 86. © 1994 American Chemical Society

conventional topographical (constant force mode), to reveal a very regular crystal morphology and a hexagonal habit (Figure 9.31). The crystal was observed to be divided into segments by lines of raised contrast joining each corner of a crystal to the centre. These followed the expected fold boundaries between segments of the crystal with different orientations. When the crystal was imaged using LFM, contrast differences were observed between adjacent fold domains (Figure 9.32). These were attributed to variations in the frictional force measured as the direction of orientation of the rows of chain folds changed with respect to the scan direction of the microscope. Lower friction would be expected along rows of chain folds than perpendicular to them.

6 BIOLOGICAL APPLICATIONS

We close this chapter by briefly examining some biological applications of scanning probe techniques. In the biological sphere, scanning probe techniques hold tremendous promise, and there is certain to be considerable development in this area in the next few years. However, some very difficult practical problems are also encountered, and much effort is currently being directed towards resolving them.

We have already noted, in Section 3.1.2, that the STM is capable of operating in a liquid environment. Not only is it possible to operate in water, but it is also possible to operate in salt solution. This is very important in a biological context. Although electron microscopy has been developed into a very powerful technique for the study of biomolecular structure, it is not capable of imaging molecules without some form of sample pre-treatment. Biological molecules are highly sensitive to environmental changes (for example, proteins are highly sensitive to changes in pH, and are easily denatured resulting in severe structural modifications), and many of the procedures employed to prepare samples for electron microscopy (for example, drying and freezing) are potentially the causes of structural modification in biomolecules.

The first STM images of DNA were obtained by Binnig and Rohrer [87]. Considerable interest was provoked, and early hopes were that STM might provide a probe for structure more than capable of competing with existing biophysical techniques. There were, indeed, even hopes that DNA sequencing might become possible using the STM. In 1989, Lindsay et al. [88] obtained the first images of DNA under water. The way had opened up for STM to provide the capability for visualising biomolecules and their interactions in physiological buffer solutions. The realisation of this possibility would constitute a major step forward in our ability to understand biological molecules and their interactions.

However, a re-evaluation of the situation was called for when it was discovered that features of graphite surfaces could mimic DNA (and other helical molecules, for that matter) [89]. Clemmer and Beebe presented images of freshly cleaved HOPG which bore an astonishing resemblance to helical molecules (Figure 9.33).

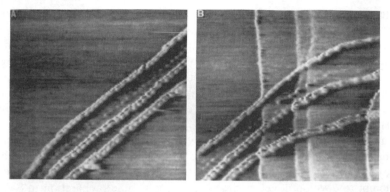

Figure 9.33. Graphite features which resemble helical molecules. Reproduced by kind permission from ref. 89. © 1991 American Association for the Advancement of Science.

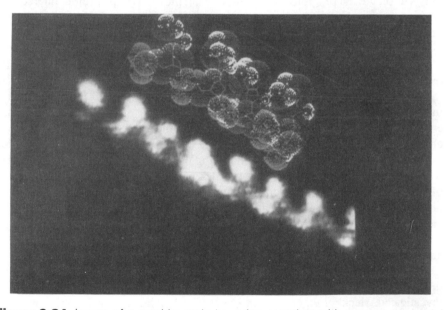

Figure 9.34. Image of a graphite grain boundary, together with a computer model of a DNA molecule. Reproduced with kind permission from ref. 90. © 1992 Elsevier Science – NL

Heckl and Binnig [90] have recorded images at graphite grain boundaries which also resemble DNA quite closely. Comparison with a computer-generated model (Figure 9.34) reveals just how close the resemblance is. Besides these problems with surface features which mimic biomolecules, it was also being recognised that the tip–sample interaction force during imaging was sufficiently large to move biomolecules. The consequence of this tip-induced movement is that the sample molecules are pushed outside of the scanned area of the surface, a phenomenon illustrated by Roberts *et al.* in the case of mucin proteins [91]. Moving the tip to a new spot simply reproduces the problem. Ultimately, it is often only possible to image molecules which are prevented from movement by some physical obstacle (for example, a step edge), but then it can become difficult to deconvolute the two structures.

A number of approaches have been developed to overcome these problems. The first has been a shift of interest away from graphite substrates to other substrates (for example, gold and mica). More importantly, a number of groups have begun to attempt to develop novel methodologies for sample preparation in STM. The most successful methodologies can be roughly divided into two: immobilisation by the application of a conducting coating, and immobilisation by coupling the sample molecule to a chemically functionalised surface. In either case, the objective is to secure the biomolecule so that it is not moved by tip–sample interactions during imaging, and to generate a uniform surface coverage. We consider some of these methodologies below.

Amrein *et al.* have obtained STM images of recA-DNA complexes coated with a conducting film, in this case Pt-Ir-C [92]. This represents the adaptation of a technique which is well established in electron microscopy: rotary shadowing. In fact, whereas shadowing at a large angle to the surface normal is most effective in transmission electron microscopy, the best results are obtained for normal evaporation in STM. The helical structure of the recA-DNA complexes was clearly resolved by Amrein *et al.* (see Figure 9.35). However, there is a resolution limitation inherent in the use of a conducting overlayer: the overlayer is granular, with typical grain sizes being of the order of a few nanometers. The potential of the STM for high resolution structure determination cannot be exploited. A justifiable criticism would therefore be that, if samples must be coated prior to analysis, then the STM offers little improvement over the existing microscopical techniques. What we require is some preparation method which will allow the naked molecule to be imaged.

A much better solution is to use some kind of coupling procedure to immobilise the biomolecule at the surface. A number of different approaches have been investigated. Heckl *et al.* created carboxyl functionalities at an HOPG surface by oxidation with chromium trioxide [93]. Treatment with thionyl chloride yielded an acid chloride which was coupled to 1,10-decanediol. Oxidation of the free hydroxyl function with potassium permanganate yielded an acid function which could be coupled to a lipid, which was then imaged using STM. Lyubchenko *et al.* also used chromium trioxide oxidation and thionyl chloride treatment of HOPG as a starting point [94], introducing cystamine as the nucleophile; subsequent treatment with mercaptoethanol to cleave the adduct yielded a surface-bound thiol function. This thiol group was then coupled to mercurated DNA.

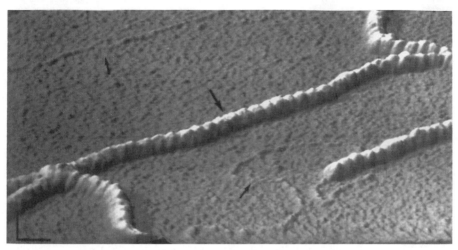

Figure 9.35. STM image of freeze-dried recA-DNA complexes, coated with a PT-Ir-C film. Reproduced with kind permission from ref. 92. © American Association for the Advancement of Science

A slightly different approach has been applied by Bottomley *et al.* [95], who adsorbed N,N-dimethyl-2-mercaptoethylamine onto gold substrates. Subsequent treatment with plasmid DNA in the pH range 5.0–9.5 allowed the exploitation of electrostatic interactions between the DNA molecules and the charged nitrogen head group of the adsorbate to bind the DNA to gold surfaces. Luttrull *et al.* also used covalent coupling procedures to bind molecules to gold surfaces [96]. They prepared porphyrin-based molecules with pendant isocyano groups which could bind to gold.

Self-assembled monolayers provide an effective means by which ordered monolayers of molecules with specific reactive tail-groups may be prepared. Leggett *et al.* prepared monolayers of carboxylic-acid-terminated thiol molecules on gold surfaces, and attached protein molecules using a coupling procedure which involved the use of a water-soluble carbodiimide reagent [97]. Island structures formed from immobilised protein molecules were observed, and the veracity of this interpretation of the data confirmed by observing a differential contrast change, following reversal of the polarity of the sample bias voltage, for substrate and protein-covered regions of the surface.

The very fact that STM images may be obtained of uncoated biological molecules poses some very interesting problems. A protein may be of the order of a few nanometers thick, and yet one may still be able to record a tunnelling current. Tunnelling directly through the biomolecule in the conventional sense is ruled out, because the tunnelling gap would simply be too great (remember that the tunnelling current is reduced by an order of magnitude for every 1 Å increase in the tip–surface separation). A number of competing models have been proposed, and the issue remains the subject of a great deal of contention. One explanation, due to Lindsay [17], is that deformation of the sample molecule by the STM tip alters its electronic structure, with the consequence that states are created at the Fermi level of the substrate. Under these conditions, resonant tunnelling can occur and an enhancement of the tunnel current will be observed. Although it has clear applications in explaining data obtained under UHV conditions, such a process would result in a substantial disruption of the molecular structure, with a concomitant reduction in the usefulness of the data obtained.

Alternatively, it has been suggested that water, present on the surface of the biomolecule, provides the means for conduction. This latter hypothesis is supported by evidence provided by Guckenberger *et al.* [98], in studies of two-dimensional protein crystals. They found that image acquisition was difficult if the relative humidity (RH) inside the STM chamber was less than about 33%, implying that hydration of a biological sample was necessary for the operation of the STM in air. Significantly, they also reported that very low tunnel currents (typically no larger than a few pA, giving an extremely large gap resistance) were necessary. A rationalisation for these observations was provided by Yuan *et al.* [99] who proposed an essentially electrochemical image formation mechanism in which ions were transported through the film of water covering the protein molecule. A water bridge linked the STM tip to the sample.

This model was tested further by Leggett *et al.* who examined the effects of dehydration of the STM chamber on image contrast. Protein molecules were immobilised by covalent attachment to thiol monolayers on gold substrates [97, 100], and imaged under ambient conditions (RH about 33%). A desiccant was then inserted into the STM chamber, and a series of images was recorded as the relative humidity fell, over several hours to 5% [100]. The image contrast gradually changed until at 5% RH, a reversal of contrast had occurred. The protein structures, which under ambient conditions had appeared to be raised some 6 nm above the substrate, now appeared as troughs. The STM measured a sharp drop in conductance as the tip traversed the dehydrated protein molecules, as compared to a rise under ambient conditions. This observation lends strong support to the hypothesis that water provides the conducting path in STM. However, it also provides a salutory reminder that seeing is not necessarily believing in STM: image contrast does not necessarily bear a simple relationship to surface topography, and without an adequate theoretical basis for image interpretation, STM data for biological systems must be treated with extreme caution.

Some of the most spectacular successes to date have been achieved with the AFM. Here, the requirement for sample conductivity is removed, and it becomes possible to study quite large biological objects. Again, too, the interest is not simply in the obtaining of images, but the obtaining of images in biological environments and, importantly, in real time. An impressive example was provided by Drake *et al.* who used the AFM to image the thrombin-induced polymerisation of fibrin in a physiological buffer in real time [101]. Although the resolution was poor, individual molecules could be observed forming chains as the reaction proceeded. Haberle *et al.* examined the behaviour of monkey kidney cells infected with pox viruses, under normal cell growth conditions, using the AFM [102]. They observed an initial and short-lived softening of the cell membrane immediately after infection. Subsequently, protrusions were observed to grow out

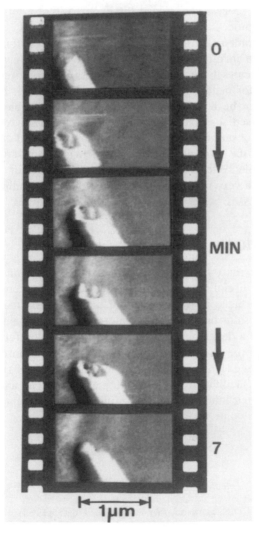

Figure 9.36. Exocytic process seen about 19 hours after infection of a monkey kidney cell with pox virus. Reproduced with kind permission from ref. 102. © 1992 Elsevier Science – NL

of the cell membrane and then disappear again. Ultimately, after about 20 hours, a different kind of event was observed, which was interpreted as the exocytosis (birth) of progeny viruses. This latter process was observed over a period of seven minutes, to yield the astonishing images seen in Figure 9.36. An unusually long finger-like structure is visible, at the end of which a protrusion appears and then disappears again after about three minutes. The size of the protrusion, the time-scale of the event, and comparison with data from electron microscopy suggested that this process was indeed the exocytosis of a virus.

Although these processes occur on a relatively large scale, the potential exists for studying phenomena on a much smaller scale. Recently, there have been great strides in the imaging of DNA by AFM, with the result that images of helical macromolecules have been recorded with a resolution at least as good as that attainable for metal-shadowed molecules with STM. If sample preparation procedures can be developed to such an extent that biomolecules can be routinely immobilised at surfaces and imaged with the STM/AFM, there remains the possibility for observing their responses to physical and chemical stimuli in real time. Even if the ultimate resolution which is attainable is poorer than is the case for metallic systems, this should nevertheless result in scanning probe techniques having a very important impact on our understanding of a wide range of biological processes.

7 CONCLUSIONS

Scanning probe microscopies have begun to have a significant impact in a broad range of scientific disciplines where the structures of surfaces, the adsorption of molecules and interfacial interactions are of interest. Early naivite regarding the potential difficulties surrounding image interpretation and sample preparation is being displaced by a developing understanding of both the fundamental theoretical principles and the nature of the experimental operation of scanning probe microscopes. Although great strides in our understanding of surface phenomena have already been facilitated by the development of STM and AFM, the techniques are still in their infancy, and the coming years promise a wealth of exciting development yet to come.

REFERENCES

[1]   P.W. Atkins: *Molecular Quantum Mechanics*, Oxford University Press, Oxford, pp 41 (1983).

[2]   J. Tersoff and D.R. Hamann, *Phys. Rev. B*, **31** 805, (1985).

[3]   G. Binnig and H. Rohrer, *IBM J. Res. Develop.* **30** 355 (1986).

[4]   J. Winterlin, J. Wiechers, H. Burne, T. Gritsch, H. Hofer and R.J. Behm, R. *J., Phys. Rev. Lett.* **62** 59, (1989).

[5]   C.J. Chen, *J. Vac. Sci. Technol.* **A9** 44, (1991).

[6] U. Landman and W.D. Luedtke, *Appl. Surf. Sci.* **60/61**, 1 (1992).
[7] E.J. van Loenen, D. Dijkkamp, A.J. Hoeven, J.M. Lenssinck and J. Dieleman, *Appl. Phys. Lett.* **56** 1755 (1990).
[8] M. Tsukada, K. Kobayashi, I. Nobuyuki and H. Kageshima, *Surf. Sci. Rep.* **8** 265, (1991).
[9] X. Yang, Ch. Bromm, U. Geyer and G. von Minnigerode, *Ann. Physik.* **1**, 3 (1992).
[10] J.M. Coombs and J.B. Pethiza, *IBM J. Res. Dev.* **30**, 455, (1986).
[11] H.J. Mamin, E. Ganz, D.W. Abraham, R.E. Thompson and J. Clarke *Phys. Rev.* **B34** 9015, (1986).
[12] J.C.H. Spence, W. Lo and M. Kuwabara *Ultramicroscopy* **33** 69, (1990).
[13] G. Binnig, C.F. Quate and C. Gerber, *Phys. Rev. Lett.* **56** 930, (1986).
[14] H. Mamin, D. Rugar, J. Stern, R. Fontana and P. Kasiraj, *Appl. Phys. Lett.* **55** 318, (1989).
[15] B. Terris, J. Stern, D. Rugar and H. Mamin, *Appl. Phys. Lett.* **63** 2669, (1989) R.M. Tromp, R.J. Hamers and J.E. Demuth, *Phys. Rev.* **B34**, 1388, (1986).
[16] Y. Martin, C.C. Williams and K. Wickramsinghe, *J. Appl. Phys.* **61**, 4723 (1987).
[17] S.M. Lindsay in D.A. Bonnell (ed.) *Scanning Tunnelling Microscopy: Theory, Techniques and Applications*, VCH, New York, 335–408, (1993).
[18] D. Sarid and V. Elings, *J. Vac. Sci. Technol.* **B9** 431, (1991).
[19] P. Maivald, H.J. Butt, S.A.C. Gould, C.B. Prater, B. Drake, J.A. Gurley, V.B. Elings and P.K. Hansma, *Nanotechnology* **2** 103, (1991).
[20] G. U. Lee, D. A. Kidwell and R. J. Colton, *Langmuir* **10** 354 (1994).
[21] G. U. Lee, L. A. Chrisey and R. J. Colton, *Science* **266** 771 (1994).
[22] J. L. Wilbur, H. A. Biebuyck, J. C. MacDonald and G. M. Whitesides, *Langmuir* **11**, 825 (1995).
[23] S. Grafstrom, M. Neizert, T. Hagen, J. Ackerman, R. Neumann, O. Probst and M. Wortge, *Nanotechnology* **4** 143 (1993).
[24] J. S. G. Ling and G. J. Leggett, *Polymer*, in press.
[25] G. Haugstad, W.L. Gladfelter, E.B. Weberg, R.T. Weberg and R.R. Jones, *Langmuir* **11** 3473, (1995).
[26] G. Binnig, H. Rohrer, Ch. Gerber and E. Weibel, *Surf. Sci.* **131** L379, (1983).
[27] G. Binnig, H. Rohrer, Ch. Gerber and E. Stoll, *Surf. Sci.* **144** 321, (1984).
[28] V.M. Hallmark, S. Chiang, J.F. Rabolt, J.D. Swalen and R.J. Wilson, *Phys. Rev. Lett.* **59** 2879, (1987).
[29] G. Binnig and H. Rohrer, *Rev. Mod. Phys* **59** 615, (1987).
[30] R.C. Jaklevic and L. Elie, *Phys. Rev. Lett.* **60** 120, (1988).
[31] C.R. Clemmer and T.P. Beebe, Jnr, *Scan. Microsc.* **6** 319, (1992).
[32] J. Schneir, R. Sonnenfeld, O. Marti, P.K. Hansma, J.E. Demuth and R.J. Hamers, *J. Appl. Phys.* **63** 717, (1988).
[33] A. M. Baro, R. Miranda, J. Alaman, N. Garcia, G. Binnig, H. Rohrer, Ch. Gerber and J. L. Carrascosa, *Nature* **315** 253 (1985).
[34] S. Park and C.F. Quate, C. F., *Appl. Phys. Lett.* **48** 112 (1986).
[35] R. Sonnenfeld and P. K. Hansma, *Science* **232** 211, (1986).
[36] G. Binnig, H. Rohrer, Ch. Gerber and E. Weibel, *Phys. Rev. Lett.* **50**, 120 (1983).
[37] K. Takayanagi, Y. Tanishiro, M. Takahashi and S. Takahashi, *J. Vac. Sci. Technol.* **A3**, 1502 (1985).
[38] R.M. Tromp, R.J. Hamers and J.E. Demuth, *Phys. Rev.* **B34**, 1388 (1986).
[39] R.J. Hamers, R.M. Tromp and J.E. Demuth, *Phys. Rev. Lett.* **56**, 1972 (1986).
[40] R.J. Hamers, *Ann. Rev. Phys. Chem.* **40**, 531 (1989).
[41] J. Knall and J.B. Pethica, *Surf. Sci.* **265**, 156 (1992).
[42] H.J. Mamin, P.H. Gueturer and D. Rugar, *Phys. Rev. Lett.* **65**, 2418 (1990).

[43]　I.-W. Lyo and P. Avouris, *Science* **253**, 173 (1991).

[44]　D.M. Eigler and E.K. Schweizer, *Nature* **344**, 524 (1990).

[45]　P. Zeppenfeld, C.P. Lutz and D.M. Eigler, *Ultramicroscopy* **42–44**, 128 (1992).

[46]　Y. Jiang, C. Juang, D. Keller, C. Bustumante and D. Beach, *Nanotechnology*, submitted.

[47]　D.M. Eigler, C.R. Lutz and W.E. Rudge, *Nature*, **352**, 600 (1991).

[48]　F. Ohnesorge and G. Binnig, *Science* **260**, 1451, (1993).

[49]　D.P.E. Smith, J.K.H. Horber, G. Binnig and H. Nejoh, *Nature* **344**, 641 (1990).

[50]　M. Hara, Y. Iwakabe, K. Tochigi, H. Sasabe, A.F. Garito and A. Yamada, *Nature*, **344**, 228 (1990).

[51]　G.C. McGonigal, R.H. Bernhardt and D.J. Thomson, *Appl. Phys. Lett.* **57**, 28 (1990).

[52]　G.C. McGonigal, R.H. Bernhardt, Y.H. Yeo and D.J. Thompson, *J. Vac. Sci. Technol.* **B9**, 1107, (1991)

[53]　Y. H. Yeo, G.C. McGonigal, K. Yackoboski, C.X. Guo and D.J. Thomson, *J. Phys. Chem.* **96**, 6110 (1992).

[54]　A.J. Bard and F.-R. Fan, *Faraday Discuss.* **94**, 1 (1992).

[55]　L. Sun and R.M. Crooks, *J. Electrochem. Soc.* **138**, L23 (1991).

[56]　R.L. McCarley, D.J. Dunaway and R.J. Willicut, *Langmuir* **9**, 2775 (1993).

[57]　H.-J. Butt, K. Seifert and E. Bamberg, *J. Phys. Chem.* **97**, 7316 (1993).

[58]　J. Pan, N. Tao and S.M. Lindsay, *Langmuir*, **9**, 1556 (1993).

[59]　L. Strong and G.M. Whitesides, *Langmuir* **4**, 546 (1988).

[60]　C. Schonenberger, J.A.M. Sondag-Huethorst, J. Jorritsma and L.G.J. Fokkink, *Langmuir*, **10**, 611 (1994).

[61]　E. Delmarche, B. Michel, Ch. Gerber, D. Anselmetti, H.-J. Guntherodt, H. Wolf and H. Ringsdorf, *Langmuir*, 2869 (1994).

[62]　G.E. Poirier and M.J. Tarlov, *Langmuir*, **10**, 2853 (1994).

[63]　G.E. Poirier, M.J. Tarlov and H.E. Rushmeier, *Langmuir*, **10**, 3383 (1994).

[64]　S.A. Joyce, R.C. Thomas, J.E. Houston, T.A. Michalske and R.M. Crooks, *Phys. Rev. Lett.*, **68**, 2790 (1992).

[65]　H. Ohtani, R.J. Wilson, S. Chiang and C.M. Mate, *Phys. Rev. Lett.* **60**, 2398 (1988).

[66]　Y.Z. Li, M. Chander, J.C. Patrin, J.H. Weaver, L.P.F. Chibante and R.E. Smalley, *Science*, **253**, 429 (1991).

[67]　S. Howells, T. Chen, M. Gallagher, D. Sarid, D.L. Lichtenberger, L. Wright, C.D. Ray, D.R. Huffman and L.D. Lamb, *Surf. Sci.* **274**, 141 (1992).

[68]　W.A. Ducker, T.J. Senden and R.M. Pashley, *Langmuir* **8**, 1831 (1992).

[69]　O.M. Leung and M.C. Goh, *Science*, **255**, 64 (1992).

[70]　E. Hamada and R. Kaneko, *Ultramicroscopy*, **42–44**, 184 (1992).

[71]　D.P.E. Smith, A. Bryant, C.F. Quate, J.P. Rabe, C. Gerber and J.D. Swalen *Proc. Natl. Acad. Sci. USA*, **84**, 969 (1987).

[72]　E. Meyer, L. Howald, R.M. Overney, H. Heinzelman, J. Frommer, H.-J. Guntherodt, T. Wagner, H. Schier and S. Roth, *Nature*, **349**, 398 (1991).

[73]　J. Garnaes, D.K. Schwartz, R. Viswanathan and J.A.N. Zasadzinski, *Nature* **357**, 54, (1992).

[74]　N.M.D. Brown, H.X. You, R.J. Forster and J.G. Vos, *J. Mater. Chem.* **1**, 517 (1991).

[75]　D.R. Yaniv, L. McCormick, J. Wang and N. Naser, *J. Electroanal. Chem.* **314**, 353 (1991).

[76]　J.G. Mantovani, R.J. Warmack, B.K. Annis, A.G. MacDiarmid and E. Scherr, *J. Appl. Polym. Sci* **40**, 1693 (1990).

[77] T.G. Stange, R. Mathew, D.F. Evans and W.A. Hendrickson, *Langmuir*, **8**, 920 (1992).
[78] D.H. Reneker, J. Schneir, B. Howell and H. Harary, *Polym. Comm.* **31**, 167 (1990).
[79] B. Collin, D. Chatenay, G. Coulon, D. Aussere and Y. Gallot, *Macromolecules*, **25**, 1621 (1992).
[80] Y. Wang, D. Juhue, M.A. Winnik, O. Leung and M.C. Goh, *Langmuir*, **8**, 760 (1992).
[81] R. Piner, R. Reifenberger, C. Martin, E.L. Thomas and R.P. Aparian, *J. Polym. Sci. C: Polym. Lett.* **28**, 399 (1990).
[82] D. Snetivy and G.J. Vancso, *Polymer*, **33**, 432 (1992).
[83] S.N. Magonov, G. Bar, H.-J. Cantow, H.D. Bauer, I. Muller and M. Schwoerer, *Polym. Bull.*, **26**, 223 (1991).
[84] S.N. Magonov, K. Qvarnstrom, V. Elings and H.-J. Cantow, *Polym. Bull.*, **25**, 689 (1991).
[85] F. Lin and D.J. Meier, *Langmuir*, **10**, 1660 (1994); F. Lin, N.V. Gvozdic and D.J. Meier, Topometrix Applications, Note no. 2–1092–001.
[86] R. Nisman, P. Smith and G.J. Vancso, *Langmuir*, **10**, 1667 (1994).
[87] G. Binnig and H. Rohrer in Janka, J. and Pantoflicek, J. (eds): *Trends in Physics, Europ. Phys. Soc., The Hague* pp 38 (1984).
[88] S.M. Lindsay, T. Thundat, L. Nagahara, U. Knipping and R.L. Rill, *Science*, **244** 1063, (1989).
[89] C.R. Clemmer and T.P. Beebe, *Jnr, Science*, **251**, 640 (1991).
[90] W.M. Heckl and G. Binnig, *Ultramicroscopy*, **42–44**, 1073 (1992).
[91] C.J. Roberts, M. Sekowski, M.C. Davies, D.E. Jackson, M.R. Price and S.J.B. Tendler, *Biochem. J.*, **283**, 181 (1992).
[92] M. Amrein, A. Stasiak, H. Gross, E. Stoll and G. Travaglini, *G., Science*, **240**, 514 (1988).
[93] W.M. Heckl, K.M.R. Kallury, M. Thompson, Ch Gerber, H.J. Horber and G. Binnig *Langmuir*, **5**, 1433 (1989).
[94] L.A. Bottomley, J.N. Haseltine, D.P. Allison, R.J. Warmack, T. Thundat, R.A. Sachlebenm, G.M. Brown, R.P. Woychik, K.B. Jacobson and T.L. Ferrell, *J. Vac. Sci. Technol.* **A10**, 591 (1992).
[95] Y.L. Lyubchenko, S.M. Lindsay, J.A. DeRose and T. Thundat, *J. Vac. Sci. Technol.* **B9**, 1288 (1991).
[96] D.K. Luttrull, J. Graham, J.A. DeRose, D. Gust, T.A. Moore and S.M. Lindsay, *Langmuir*, **8**, 765 (1992).
[97] G.J. Leggett, C.J. Roberts, P.M. Williams, M.C. Davies, D .E. Jackson and S.J.B. Tendler, *Langmuir*, **9**, 2356 (1993).
[98] R. Guckenberger, W. Wiegrabe, A. Hillebrand, T. Hartmenn, Z. Wang and W. Baumeister, *Ulramicrosc.* **31**, 327 (1989).
[99] J.-Y. Juan, Z. Shao and C. Gao, *Phys. Rev. Lett.* **67**, 863 (1991).
[100] G.J. Leggett, M.C. Davies, D.E. Jackson, C.J. Roberts, S.J.B. Tendler and P.M. Williams, *J. Phys. Chem.* **97**, 8852 (1993).
[101] B. Drake, C.B. Prater, A.L. Weishorn, S.A.C. Gould and T.R. Albrecht, *Science* **243**, 1586 (1989).
[102] W. Habarle, J.K.H. Horber, F. Ohnesorge, D.P.E. Smith and G. Binnig, *Ultramicroscopy* **42–44**, 1161 (1992).
[103] S. Xu and M.F. Arnsdorf, *J. Microsc.* **173**, 199 (1994).
[104] N.A. Burnham and R.J. Colton in D.A. Bonnell (ed.) *Scanning Tunneling Microscopy: Theory, Techniques and Applications*, VCH, New York, 191–249 (1993).

QUESTIONS

1. The magnitude of the tunnel current in the STM is exponentially dependent upon the tip–sample separation d, and is given approximately by the equation

$$I \propto e^{(-2\kappa d)}$$

(see Section 2.1.1). The high spatial resolution of the STM is a consequence of this exponential dependence. The total measured tunnelling current is the sum of all the tunnelling interactions between tip and sample; however, the tunnelling interactions between the tip apex and the sample have much higher probabilities than those between other regions of the tip and the surface, and so contribute a greater proportion of the total measured current.

   Explore this dependence in the following way. Plot the following tip profiles:

(a) $y = 2 + x^2$
(b) $y = 7 - (25 - x^2)^{1/2}$
(c) $y = 2 + x^2/4$.

Now assume that the tunnel current is the sum of a number of elemental tunnelling currents δI between regions dx of the tip surface and the sample. Assume that the current density is uniform across the surface, and arbitrarily assigned the value 1. Assume that the elemental tunnelling current between any point x and the tip is proportional to e^y where y is the separation between tip and sample. Now plot, for each tip, the elemental tunnelling currents. Which tip will provide the highest resolution?

2. The STM tips above represent very simplistic models; real tips rarely exhibit such idealised geometries, and atomic asperities are quite possible. AFM tips are typically much broader, however, and may have curved profiles which approximate much better to a spherical shape. The radius of curvature of the AFM tip may be substantial, and the consequence of this is that small features may have an apparent size which is somewhat larger than their real size. A number of authors have derived expressions which it is hoped can quantify the extent of broadening of small features in the AFM image. Many of these treatments are based upon a simple geometrical analysis of the tip–sample interaction. One example is the following expression [91] which, for features with a circular cross-section smaller than the tip radius of curvature R_t, relates the real radius r of a feature to its radius in the AFM image R (see Figure 9.37):

$$r = R^2/(4R_t)$$

(a) Calculate the measured radius of a DNA molecule (real radius 2 nm) when imaged by tips of radius of curvature 45 nm and 100 nm. Evaluate the % broadening for each tip.

(b) For a tip of radius of curvature 45 nm, calculate the measured radius of features with diameters 20 and 5 nm. Evaluate the % broadening for each feature.

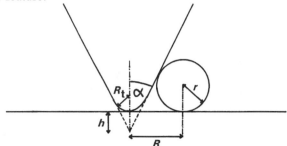

3. From question 2 it will be clear that far from being a vanishing quantity, the area of contact between the FM and a flat sample is substantial on the atomic scale. For some flat materials, 'atomic' resolution images have been recorded, showing arrays of points with atomic dimensions which appear very similar to atomic resolution STM images. The precise origin of these points is the subject of debate, and the reader is referred to detailed discussions elsewhere for further information [92]. Even if a sufficiently sharp tip could be produced, there would be theoretical difficulties in attempting to image individual atoms. Suppose that a minimal load of 1 nN is applied to the FM tip, with a contact area of radius 10 pm. Estimate the mean pressure applied. Compare this figure with the yield strength for silicon nitride, 550 MPa. What is the likelihood of single atom contacts giving rise to the AFM image?

Index